Principles of Environmental Chemistry

Principles of Environmental Chemistry

Roy M Harrison
School of Geography, Earth and Environmental Sciences,
University of Birmingham, Birmingham, UK

RSCPublishing

ISBN-13: 978-0-85404-371-2

A catalogue record for this book is available from the British Library

Published by The Royal Society of Chemistry,
Thomas Graham House, Science Park, Milton Road,
Cambridge CB4 0WF, UK

Registered Charity Number 207890

For further information see our web site at www.rsc.org

Typeset by Macmillan India Ltd, Bangalore, India
Printed by Henry Ling Ltd, Dorchester, Dorset, UK

Preface

While this book is in its first edition, it nonetheless has a lengthy pedigree, which derives from a book entitled *Understanding Our Environment: An Introduction to Environmental Chemistry and Pollution*, which ran to three editions, the last of which was published in 1999. *Understanding Our Environment* has proved very popular as a student textbook, but changes in the way that the subject is taught had necessitated its splitting into two separate books.

When *Understanding Our Environment* was first published, neither environmental chemistry nor pollution was taught in many universities, and most of those courses which existed were relatively rudimentary. In many cases, no clear distinction was drawn between environmental chemistry and pollution and the two were taught largely hand in hand. Nowadays, the subjects are taught in far more institutions and in a far more sophisticated way. There is consequently a need to reflect these changes in what would have been the fourth edition of *Understanding Our Environment*, and after discussion with contributors to the third edition and with the Royal Society of Chemistry, it was decided to divide the former book into two and create new books under the titles respectively of *An Introduction to Pollution Science* and *Principles of Environmental Chemistry*. Because of the authoritative status of the authors of *Understanding Our Environment* and highly positive feedback which we had received on the book, it was decided to retain the existing chapters where possible while updating the new structure to enhance them through the inclusion of further chapters.

This division of the earlier book into two new titles is designed to accommodate the needs of what are now two rather separate markets. *An Introduction to Pollution Science* is designed for courses within degrees in environmental sciences, environmental studies and related areas including taught postgraduate courses, which are not embedded in a specific physical science or life science discipline such as chemistry, physics or biology. The level of basic scientific knowledge assumed of the reader is therefore only that of the generalist and the book should be accessible to a very wide readership including those outside of the academic world wishing to acquire a broadly based knowledge of pollution phenomena. The second title, *Principles of Environmental Chemistry* assumes a significant knowledge of chemistry and is aimed far more at courses on environmental chemistry which are embedded within chemistry degree courses. The book will therefore be suitable for students taking second or third year option courses in environmental chemistry or those taking specialised Masters' courses, having studied the chemical sciences at first-degree level.

In this volume I have been fortunate to retain the services of a number of authors from *Understanding Our Environment*. The approach has been to update chapters from that book where possible, although some of the new authors have decided to take a completely different approach. The book initially deals with the atmosphere, freshwaters, the oceans and the solid earth as separate compartments. There are certain common crosscutting features such as non-ideal solution chemistry, and where possible these are dealt with in detail where they first occur, with suitable cross-referencing when they re-appear at later points. Chemicals in the environment do not respect compartmental boundaries, and indeed many important phenomena occur as a result of transfers between compartments. The book therefore contains subsequent chapters on environmental organic chemistry, which emphasises the complex behaviour of persistent organic pollutants, and on biogeochemical cycling of pollutants, including major processes affecting both organic and inorganic chemical species.

I am grateful to the authors for making available their great depth and breadth of experience to the production of this book and for tolerating my many editorial quibbles. I believe that their contributions have created a book of widespread appeal, which will find many eager readers both on taught courses and in professional practice.

Roy M. Harrison
Birmingham, UK

Contents

CHAPTER 1

Introduction

ROY M. HARRISON

Division of Environmental Health and Risk Management, School of Geography, Earth and Environmental Sciences, University of Birmingham, Edgbaston, B15 2TT, Birmingham, UK

1.1 THE ENVIRONMENTAL SCIENCES

It may surprise the student of today to learn that 'the environment' has not always been topical and indeed that environmental issues have become a matter of widespread public concern only over the past 20 years or so. Nonetheless, basic environmental science has existed as a facet of human scientific endeavour since the earliest days of scientific investigation. In the physical sciences, disciplines such as geology, geophysics, meteorology, oceanography, and hydrology, and in the life sciences, ecology, have a long and proud scientific tradition. These fundamental environmental sciences underpin our understanding of the natural world and its current-day counterpart perturbed by human activity, in which we all live.

The environmental physical sciences have traditionally been concerned with individual environmental compartments. Thus, geology is centred primarily on the solid earth, meteorology on the atmosphere, oceanography upon the salt-water basins, and hydrology upon the behaviour of freshwaters. In general (but not exclusively) it has been the *physical* behaviour of these media which has been traditionally perceived as important. Accordingly, dynamic meteorology is concerned primarily with the physical processes responsible for atmospheric motion, and climatology with temporal and spatial patterns in physical properties of the atmosphere (temperature, rainfall, *etc.*). It is only more recently that *chemical* behaviour has been perceived as being important in many of these areas. Thus, while atmospheric chemical processes are at least as important as physical processes in many environmental problems such as stratospheric ozone depletion, the lack of chemical knowledge has been

1

extremely acute as atmospheric chemistry (beyond major component ratios) only became a matter of serious scientific study in the 1950s.

There are two major reasons why environmental chemistry has flourished as a discipline only rather recently. Firstly, it was not previously perceived as important. If environmental chemical composition is relatively invariant in time, as it was believed to be, there is little obvious relevance to continuing research. Once, however, it is perceived that composition is changing (*e.g.* CO_2 in the atmosphere; ^{137}Cs in the Irish Sea) and that such changes may have consequences for humankind, the relevance becomes obvious. The idea that using an aerosol spray in your home might damage the stratosphere, although obvious to us today, would stretch the credulity of someone unaccustomed to the concept. Secondly, the rate of advance has in many instances been limited by the available technology. Thus, for example, it was only in the 1960s that sensitive reliable instrumentation became widely available for measurement of trace concentrations of metals in the environment. This led to a massive expansion in research in this field and a substantial *downward* revision of agreed typical concentration levels due to improved methodology in analysis. It was only as a result of James Lovelock's invention of the electron capture detector that CFCs were recognised as minor atmospheric constituents and it became possible to monitor increases in their concentrations (see Table 1). The table exemplifies the sensitivity of analysis required since concentrations are at the ppt level (1 ppt is one part in 10^{12} by volume in the atmosphere) as well as the substantial increasing trends in atmospheric halocarbon concentrations, as measured up to 1990. The implementation of the Montreal Protocol, which requires controls on production of CFCs and some other halocarbons, has led to a slowing and even a reversal of annual concentration trends since 1992 (see Table 1).

Table 1 *Atmospheric halocarbon concentrations and trends[a]*

Halocarbon	Concentration (ppt)		Annual change (ppt)		Lifetime (years)
	Pre-industrial	2000	To 1990	1999–2000	
CCl_3F (CFC-11)	0	261	+9.5	−1.1	50
CCl_2F_2 (CFC-12)	0	543	+16.5	+2.3	102
$CClF_3$ (CFC-113)	0	3.5			400
$C_2Cl_2F_4$ (CFC-113)	0	82	+4–5	−0.35	85
$C_2Cl_2F_4$ (CFC-114)	0	16.5			300
C_2ClF_5 (CFC-115)	0	8.1		+0.16	1700
CCl_4	0	96.1	+2.0	−0.94	42
CH_3CCl_3	0	45.4	+6.0	−8.7	4.9

[a] Data from: World Meteorological Organization, *Scientific Assessment of Ozone Depletion: 2002*, WHO, Geneva, 2002.

1.2 ENVIRONMENTAL CHEMICAL PROCESSES

The chemical reactions affecting trace gases in the atmosphere generally have quite significant activation energies and thus occur on a timescale of minutes, days, weeks, or years. Consequently, the change to such chemicals is determined by the rates of their reactions and atmospheric chemistry is intimately concerned with the study of reactions kinetics. On the other hand, some processes in aquatic systems have very low activation energies and reactions occur extremely rapidly. In such circumstances, provided there is good mixing, the chemical state of matter may be determined far more by the thermodynamic properties of the system than by the rates of chemical processes and therefore chemical kinetics.

The environment contains many trace substances at a wide range of concentrations and under different temperature and pressure conditions. At very high temperatures such as can occur at depth in the solid earth, thermodynamics may also prove important in determining, for example, the release of trace gases from volcanic magma. Thus, the study of environmental chemistry requires a basic knowledge of both chemical thermodynamics and chemical kinetics and an appreciation of why one or other is important under particular circumstances. As a broad generalisation it may be seen that much of the chapter on atmospheric chemistry is dependent on knowledge of reaction rates and underpinned by chemical kinetics, whereas the chapters on freshwater and ocean chemistry and the aqueous aspects of the soils are very much concerned with equilibrium processes and hence chemical thermodynamics. It should not however be assumed that these generalisations are universally true. For example, the breakdown of persistent organic pollutants in the aquatic environment is determined largely by chemical kinetics, although the partitioning of such substances between different environmental media (air, water, soil) is determined primarily by their thermodynamic properties and to a lesser degree by their rates of transfer.

1.3 ENVIRONMENTAL CHEMICALS

This book is not concerned explicitly with chemicals as pollutants. This is a topic covered by a companion volume on *Pollution Science*. This book, however, is nonetheless highly relevant to the understanding of chemical pollution phenomena. The major areas of coverage are as follows:

(i) *The chemistry of freshwaters.* Freshwaters comprise three different major components. The first is the water itself, which inevitably contains dissolved substances, both inorganic and organic. Its properties are to a very significant degree determined by the inorganic solutes, and particularly those which determine its hardness and alkalinity. The second component is suspended sediment, also referred to as suspended solids. These are particles, which are sufficiently small to remain suspended with the water column for significant periods of time where they provide a surface onto which dissolved substances may deposit or from which material may dissolve. The third major component of the system is the bottom sediment. This is an accumulation of particles and associated pore water, which has deposited out of the water column onto the bed of the stream, river, or lake. The size of the sediment grains is determined by the speed and turbulence of the water above. A fast-flowing river will retain small particles in suspension and only large particles (sand or gravel) will remain on the bottom. In relatively stagnant lake water, however, very small particles can sediment out and join the bottom sediment. In waters of this kind, sediment accumulates over time and therefore the surface sediments in contact with the water column contain recently deposited material while the sediment at greater depths contains material deposited tens or hundreds of years previously. In the absence of significant mixing by burrowing organisms, the depth profile of some chemicals within a lake bottom sediment can provide a very valuable historical record of inputs of that substance to the lake. Ingenious ways have been devised for determining the age of specific bands of sediment. While the waters at the surface of a lake are normally in contact with the atmosphere and therefore well aerated, water at depth and the pore water within the bottom sediment may have a very poor oxygen supply and therefore become oxygen-depleted and are then referred to as anoxic or anaerobic. This can affect the behaviour of redox-active chemicals such as transition elements, and therefore the redox properties of freshwaters and their sediments are an important consideration.

(ii) *Salt waters.* The waters of seas and oceans differ substantially from freshwaters by virtue of their very high content of dissolved inorganic material and their very great depth at some points on the globe. These facets confer properties, which although overlapping with those of freshwaters, can be quite distinct. Some inorganic components will behave quite differently in a very high salinity environment than in a low ionic strength freshwater. Historically, therefore, the properties of seawater have traditionally been studied separately from those of freshwaters and are presented separately, although the important overlaps such as in the area of carbonate equilibria are highlighted.

(iii) *The chemistry of soils and rocks*. There are very significant overlaps with freshwater chemistry but the main differences arise from the very large quantities of solid matter providing very large surfaces and often restricting access of oxygen so that conditions readily become anoxic. However, many of the basic issues such as carbonate equilibria and redox properties overlap very strongly with the field of freshwater chemistry. Soils can, however, vary very greatly according to their location and the physical and chemical processes which have affected them during and since their formation.

(iv) *Environmental organic chemistry*. Much of the traditional study of the aquatic and soil environment has been concerned with its inorganic constituents. Increasingly, however, it is recognised that organic matter plays a very important role both in terms of the contribution of natural organic substances to the properties of waters and soils, but also that specific organic compounds, many of them deriving from human activity, show properties in the environment which are not easily understood from traditional approaches and therefore these have become a rather distinct area of study.

(v) *Atmospheric chemistry*. The atmosphere contains both gas phase and particulate material. The study of both is important and the two interact very substantially. However, as outlined previously, chemical processes in the atmosphere tend to be very strongly influenced by kinetic factors, and to a large extent are concerned with rather small molecules, which play only a minor part in the chemistry of the aquatic environment or solid earth. Inevitably, there are important processes at the interface between the atmosphere and the land surface or oceans, but these are dealt with more substantially in the companion volume on *Pollution Science*.

1.4 UNITS OF CONCENTRATION

1.4.1 Atmospheric Chemistry

Concentrations of trace gases and particles in the atmosphere can be expressed as mass per unit volume, typically $\mu g\ m^{-3}$. The difficulty with this unit is that it is not independent of temperature and pressure. Thus, as an airmass becomes warmer or colder, or changes in pressure, so its volume will change, but the mass of the trace gas will not. Therefore, air containing $1\ \mu g\ m^{-3}$ of sulfur dioxide in air at $0°C$ will contain less than $1\ \mu g\ m^{-3}$ of sulfur dioxide in air if heated to $25°C$. For gases (but not particles), this difficulty is overcome by expressing the concentration of the trace gas as a volume mixing ratio. Thus, 1 cm^3 of pure sulfur dioxide dispersed in 1 m^3 of polluted air would be described as a concentration of 1 ppm. Reference to the gas laws tells us that not only is this one part per 10^6 by volume, it is also one molecule in 10^6 molecules and one mole in 10^6 moles, as well as a partial pressure of 10^{-6} atm. Additionally, if the temperature and pressure of the airmass change, this affects the trace gas in the same way as the air in which it is contained and the volume-mixing ratio does not change. Thus, ozone in the stratosphere is present in air at considerably higher mixing ratios than in the lower atmosphere (troposphere), but if the concentrations are expressed in µg m^{-3} they are little different because of the much lower density of air at stratospheric attitudes. Chemical kineticists often express atmospheric concentrations in molecules per cubic centimetre (molec cm^{-3}), which has the same problem as the mass per unit volume units.

Worked Example

The concentration of nitrogen dioxide in polluted air is 85 ppb. Express this concentration in units of µg m^{-3} and molec cm^{-3} if the air temperature is 20°C and the pressure 1005 mb (1.005×10^5 Pa). Relative molecular mass of NO$_2$ is 46; Avogadro number is 6.022×10^{23}.

The concentration of NO$_2$ is 85 µL m^{-3}. At 20°C and 1005 mb,

$$85 \,\mu L \; NO_2 \; \text{weigh} \; 46 \times \frac{85 \times 10^{-6}}{22.41} \times \frac{273}{293} \times \frac{1005}{1013}$$

$$= 161 \times 10^{-6} \, g$$

NO$_2$ concentration $= 161$ µg m^{-3}

This is equivalent to 161 pg cm^{-3}, and

$$161 \, pg \; NO_2 \; \text{contain} \; 6.022 \times 10^{23} \times \frac{161 \times 10^{-12}}{46}$$

$$= 2.1 \times 10^{12} \, \text{molecules}$$

and NO$_2$ concentration $= 2.1 \times 10^{12}$ molec cm^{-3}.

1.4.2 Soils and Waters

Concentrations of pollutants in soils are most usually expressed in mass per unit mass, for example, milligrams of lead per kilogram of soil. Similarly, concentrations in vegetation are also expressed in mg kg^{-1} or µg kg^{-1}. In the case of vegetation and soils, it is important to distinguish between wet and dry weight concentrations, in other words, whether the kilogram of vegetation or soil is determined before or after drying. Since the moisture content of vegetation can easily exceed 50%, the data can be very sensitive to this correction.

In aquatic systems, concentrations can also be expressed as mass per unit mass and in the oceans some trace constituents are present at concentrations of ng kg^{-1} or µg kg^{-1}. More often, however, sample sizes are measured by volume and concentrations expressed as ng L^{-1} or µg L^{-1}. In the case of freshwaters, especially, concentrations expressed as mass per litre will be almost identical to those expressed as mass per kilogram. As a kind of shorthand, however, water chemists sometimes refer to concentrations as if they were ratios by weight, thus, mg L^{-1} are expressed as ppm, µg L^{-1} as ppb and ng L^{-1} as ppt. This is unfortunate as it leads to confusion with the same units used in atmospheric chemistry with a quite different meaning.

1.5 THE ENVIRONMENT AS A WHOLE

A facet of the chemically centred study of the environment is a greater integration of the treatment of environmental media. Traditional boundaries between atmosphere and waters, for example, are not a deterrent to the transfer of chemicals (in either direction), and indeed many important and interesting processes occur at these phase boundaries.

In this book, the treatment first follows traditional compartments (Chapters 2, 3, 4, and 5) although some exchanges with other compartments are considered. Fundamental aspects of the science of atmosphere, waters, and soils are described, together with current environmental questions, exemplified by case studies. Subsequently, the organic chemistry of the environment is considered in Chapter 6, and quantitative aspects of transfer across phase boundaries are described in Chapter 7, where examples are given of biogeochemical cycles.

REFERENCES

1. For readers requiring knowledge of basic chemical principles R.M. Harrison and S.J. de Mora, *Introductory Chemistry for the Environmental Sciences*, 2nd edn, Cambridge University Press, Cambridge, 1996.
2. For more detailed information upon pollution phenomena *Pollution: Causes, Effects and Control*, 4th edn, R.M. Harrison (ed), RSC, Cambridge, 2001 or R.M. Harrison (ed), *Introduction to Pollution Science*, RSC, Cambridge, 2006.

CHAPTER 2

Chemistry of the Atmosphere

PAUL S. MONKS

Department of Chemistry, University of Leicester, LE1 7RH, Leicester, UK

2.1 INTRODUCTION

The thin gaseous envelope that surrounds our planet is integral to the maintenance of life on earth. The composition of the atmosphere is predominately determined by biological processes acting in concert with physical and chemical change. Though the concentrations of the major atmospheric constituents oxygen and nitrogen remain the same, the concentration of trace species, which are key to many atmospheric processes are changing. It is becoming apparent that man's activities are beginning to change the composition of the atmosphere over a range of scales, leading to, for example, increased acid deposition, local and regional ozone episodes, stratospheric ozone loss and potentially climate change. In this chapter, we will look at the fundamental chemistry of the atmosphere derived from observations and their rationalisation.

In order to understand the chemistry of the atmosphere we need to be able to map the different regions of the atmosphere. The atmosphere can be conveniently classified into a number of different regions which are distinguished by different characteristics of the dynamical motions of the air (see Figure 1). The lowest region, from the earth's surface to the tropopause at a height of 10–15 km, is termed the troposphere. The troposphere is the region of the active weather systems which determine the climate at the surface of the earth. The part of the troposphere at the earth's surface, the planetary boundary layer, is that which is influenced on a daily basis by the underlying surface.

Above the troposphere lies the stratosphere, a quiescent region of the atmosphere where vertical transport of material is slow and radiative transfer of energy dominates. In this region lies the ozone layer which

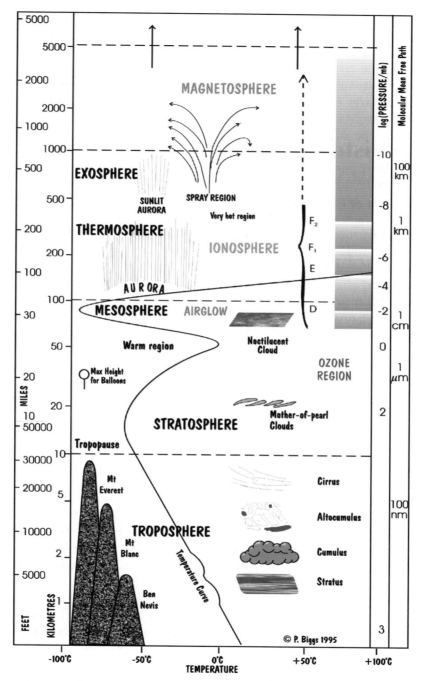

Figure 1 *Vertical structure of the atmosphere. The vertical profile of temperature can be used to define the different atmospheric layers*

has an important property of absorbing ultraviolet (UV) radiation from the sun, which would otherwise be harmful to life on earth. The stratopause at approximately 50 km altitude marks the boundary between the stratosphere and the mesosphere, which extends upwards to the mesopause at approximately 90 km altitude. The mesosphere is a region of large temperature extremes and strong turbulent motion in the atmosphere over large spatial scales.[1]

Above the mesopause is a region characterised by a rapid rise in temperature, known as the thermosphere.[2] In the thermosphere, the atmospheric gases, N_2 and O_2, are dissociated to a significant extent into atoms so the mean molecular mass of the atmospheric species falls. The pressure is low and thermal energies are significantly departed from the Boltzmann equilibrium. Above 160 km gravitational separation of the constituents becomes significant and atomic hydrogen atoms, the lightest neutral species, moves to the top of the atmosphere. The other characteristic of the atmosphere from mesosphere upwards is that above 60 km, ionisation is important. This region is called the ionosphere. It is subdivided into three regimes, the D, E and F region, characterised by the types of dominant photo ionisation.[3]

With respect to atmospheric chemistry, though there is a great deal of interesting chemistry taking place higher up in the atmosphere,[1–3] we shall focus in the main on the chemistry of the troposphere and stratosphere.

2.2 SOURCES OF TRACE GASES IN THE ATMOSPHERE

As previously described, the troposphere is the lowest region of the atmosphere extending from the earth's surface to the tropopause at 10–18 km. About 90% of the total atmospheric mass resides in the troposphere and the greater part of the trace gas burden is found there. The troposphere is well mixed and its bulk composition is 78% N_2, 21% O_2, 1% Ar and 0.036% CO_2 with varying amounts of water vapour depending on temperature and altitude. The majority of the trace species found in the atmosphere are emitted into the troposphere from the surface and are subject to a complex series of chemical and physical transformations. Trace species emitted directly into the atmosphere are termed to have *primary* sources, *e.g.* trace gases such as SO_2, NO and CO. Those trace species formed as a product of chemical and/or physical transformation of primary pollutants in the atmosphere, *e.g.* ozone, are referred to as having *secondary* sources or being *secondary* species.

Emissions into the atmosphere are often broken down into broad categories of anthropogenic or "man-made sources" and biogenic or natural sources with some gases also having geogenic sources. Table 1 lists a selection of the trace gases and their major sources.[4] For the individual emission of a primary pollutant there are a number of factors that need to be taken into account in order to estimate the emission strength, these include the range and type of sources and the spatio- and temporal-distribution of the sources. Often these factors are compiled into the so-called emission inventories that combine the rate of emission of various sources with the number and type of each source and the time over which the emissions occur. Figure 2 shows the UK emission inventory for a range of primary pollutants ascribed to different source categories (see caption of Figure 2). It is clear from the data in Figure 2 that, for example, SO_2 has strong sources from public power generation whereas ammonia has strong sources from agriculture. Figure 3 shows the (2002) 1×1 km emission inventories for SO_2 and NO_2 for the UK. In essence, the data presented in Figure 2 has been apportioned spatially according to magnitude of each source category (*e.g.* road transport, combustion in energy production and transformation, solvent use). For example, in Figure 3a, the major road routes are clearly visible, showing NO_2 has a major automotive source (*cf.* Figure 2). It is possible to scale the budgets of many trace gases to a global scale.

It is worth noting that there are a number of sources that do not occur within the boundary layer (the decoupled lowest layer of the troposphere, see Figure 1), such as lightning production of nitrogen oxides and a range of pollutants emitted from the combustion-taking place in aircraft engines. The non-surface sources often have a different chemical impact owing to their direct injection into the *free* troposphere (the part of the troposphere that overlays the boundary layer).

In summary, there are a range of trace species present in the atmosphere with a myriad of sources varying both spatially and temporally.[5] It is the chemistry of the atmosphere that acts to transform the primary pollutants into simpler chemical species.

2.3 INITIATION OF PHOTOCHEMISTRY BY LIGHT

Photodissociation of atmospheric molecules by solar radiation plays a fundamental role in the chemistry of the atmosphere. The photodissociation of trace species such as ozone and formaldehyde contributes to their removal from the atmosphere, but probably the most important role played by these photoprocesses is the generation of highly reactive atoms and radicals. Photodissociation of trace species and the subsequent reaction of the photoproducts with other molecules is the prime initiator and driver for the bulk of atmospheric chemistry.

Table 1 *Natural and anthropogenic sources of a selection of trace gases*

Compound	Natural sources	Anthropogenic sources
Carbon-containing compounds		
Carbon dioxide (CO_2)	Respiration; oxidation of natural CO; destruction of forests	Combustion of oil, gas, coal and wood; limestone burning
Methane (CH_4)	Enteric fermentation in wild animals; emissions from swamps, bogs, *etc.*, natural wet land areas; oceans	Enteric fermentation in domesticated ruminants; emissions from paddy fields; natural gas leakage; sewerage gas; colliery gas; combustion sources
Carbon monoxide (CO)	Forest fires; atmospheric oxidation of natural hydrocarbons and methane	Incomplete combustion of fossil fuels and wood, in particular motor vehicles, oxidation of hydrocarbons; industrial processes; blast furnaces
Light paraffins, C_2–C_6	Aerobic biological source	Natural gas leakage; motor vehicle evaporative emissions; refinery emissions
Olefins, C_2–C_6	Photochemical degradation of dissolved oceanic organic material	Motor vehicle exhaust; diesel engine exhaust
Aromatic hydrocarbons	Insignificant	Motor vehicle exhaust; evaporative emissions; paints, gasoline, solvents
Terpenes ($C_{10}H_{16}$)	Trees (broadleaf and coniferous); plants	
CFCs and HFCs	None	Refrigerants; blowing agents; propellants
Nitrogen-containing trace gases		
Nitric oxide (NO)	Forest fires; anaerobic processes in soil; electric storms	Combustion of oil, gas and coal
Nitrogen dioxide (NO_2)	Forest fires; electric storms	Combustion of oil, gas and coal; atmospheric transformation of NO
Nitrous oxide (N_2O)	Emissions from denitrifying bacteria in soil; oceans	Combustion of oil and coal
Ammonia (NH_3)	Aerobic biological source in soil. Breakdown of amino acids in organic waste material	Coal and fuel oil combustion; waste treatment
Sulfur-containing trace gases		
Dimethyl sulfide (DMS)	Phytoplankton	Landfill gas
Sulfur dioxide (SO_2)	Oxidation of H_2S; volcanic activity	Combustion of oil and coal; roasting sulfide ores

(Continued)

Table 1   (*Continued*)

Compound	Natural sources	Anthropogenic sources
Other minor trace gases		
Hydrogen	Oceans, soils; methane oxidation, isoprene and terpenes *via* HCHO	Motor vehicle exhaust; oxidation of methane *via* formaldehyde (HCHO)
Ozone	In the stratosphere; natural $NO-NO_2$ conversion	Man-made $NO-NO_2$ conversion; supersonic aircraft
Water (H_2O)	Evaporation from oceans	Insignificant

Source: From ref. 4.

Sources of emission in the UK

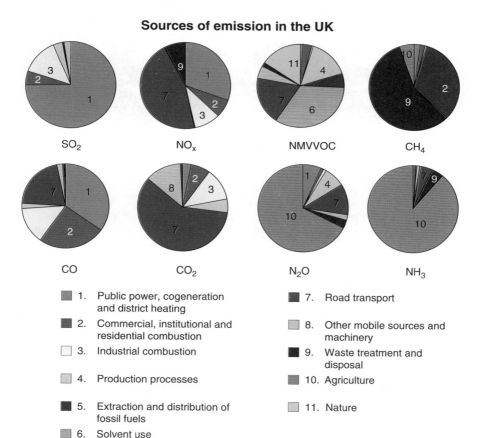

1. Public power, cogeneration and district heating
2. Commercial, institutional and residential combustion
3. Industrial combustion
4. Production processes
5. Extraction and distribution of fossil fuels
6. Solvent use
7. Road transport
8. Other mobile sources and machinery
9. Waste treatment and disposal
10. Agriculture
11. Nature

Figure 2   *UK emission statistics by UNECE source category (1) Combustion in Energy production and transformation; (2) Combustion in commercial, institutional, residential and agriculture; (3) Combustion in industry; (4) Production processes; (5) Extraction and distribution of fossil fuels; (6) Solvent use; (7) Road transport; (8) Other transport and mobile machinery; (9) Waste treatment and disposal; (10) Agriculture, forestry and land use change; (11) Nature*

Figure 3   *UK emission maps (2002) on a 1 × 1 km grid for (a) NO₂ and (b) SO₂ and in kg (data from UK NAIE, http://www.naei.org.uk/)*

The light source for photochemistry in the atmosphere is the sun. At the top of the atmosphere there is *ca.* 1370 W m^{-2} of energy over a wide spectral range, from X-rays through the visible to longer wavelength. By the time the incident light reaches the troposphere much of the more energetic, shorter wavelength light has been absorbed by molecules such as oxygen, ozone and water vapour or scattered higher in the atmosphere. Typically, in the surface layers, light of wavelengths longer than 290 nm are available (see Figure 4). In the troposphere, the wavelength at which the intensity of light drops to zero is termed the *atmospheric cut-off.* For the troposphere, this wavelength is determined by the overhead stratospheric ozone column (absorbs *ca.* $\lambda \leq 310$ nm) and the aerosol loading. In the mid- to upper-stratosphere, the amount of O_3 absorption in the "window" region at 200 nm between the O_3 and O_2 absorptions controls the availability of short-wavelength radiation that can photodissociate molecules that are stable in the troposphere. In the stratosphere (at 50 km), there is typically no radiation of wavelength shorter than 183 nm.

The light capable of causing photochemical reactions is termed the actinic flux, $F_\lambda(\lambda)$ (cm^{-2} s^{-1} nm^{-1}), which is also known as the scalar

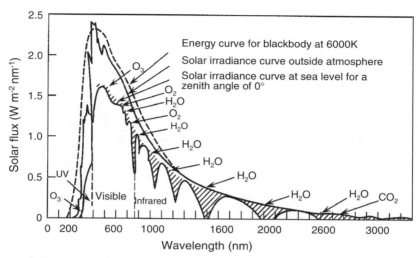

Figure 4 *Solar flux outside the atmosphere and at sea level, respectively. The emission of a blackbody at 6000 K is included for comparison. The species responsible for light absorption in the various regions (O_2, H_2O, etc.) are also shown (after ref. 60)*

intensity or spherical radiant flux, *viz*

$$F_\lambda(\lambda) = \int L_\lambda(\lambda, \vartheta, \varphi)\, d\omega \qquad (2.1)$$

where $L_\lambda(\lambda)$ ($cm^{-2}\,s^{-1}\,sr^{-1}\,nm^{-1}$) denotes the spectral photon radiance, ω is the solid angle and (ϑ, φ) are the polar and azimuthal angles of incidence of the radiation interacting with the molecule of interest. In essence, all angles of incident light must be considered when measuring or calculating the actinic flux (i.e. $\vartheta = 0$–$180°$, $\omega = 0$–$360°$). Photolysis rates are often expressed as a first-order loss process, *e.g.* in the photolysis of NO_2

$$NO_2 + h\nu \rightarrow NO + O(^3P) \qquad (2.2)$$

i.e.

$$-\frac{d[NO_2]}{dt} = j_{2.2}[NO_2] \qquad (2.3)$$

where the photolysis frequency, j, can be expressed as

$$j = \int_{\lambda_{min}}^{\lambda_{max}} \sigma(\lambda, T)\phi(\lambda, T)F_\lambda(\lambda)\, d\lambda \qquad (2.4)$$

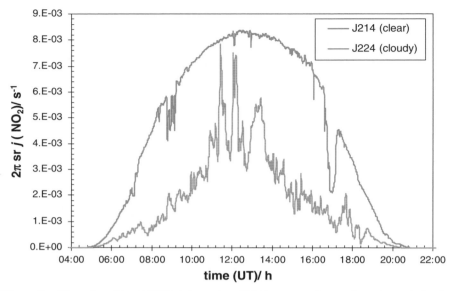

Figure 5   *Diurnal profile of $j(NO_2)$ (reaction (2.2)) measured on a clear sky and cloudy day*[61]

where σ the absorption cross-section (cm^2), ϕ the quantum yield of the photoproducts and T the temperature. Figure 5 shows a typical measured photolysis rate for NO_2 (reaction (2.2)) in the atmosphere. The photolysis rate reaches a maximum at solar noon concomitant with the maximum in solar radiation.

Example for the calculation of photolysis rates
From the following data calculate the photolysis rate of O_3 into $O(^1D)$ at $T = 298$ K.

Wavelength (nm)	$F(\lambda)$ (photon cm^{-2} s^{-1})	σ (cm^2 molec^{-1})	ϕ
295–300	2.66×10^{13}	59.95×10^{-20}	1.0
300–305	4.20×10^{14}	26.50×10^{-20}	1.0
305–310	1.04×10^{15}	15.65×10^{-20}	0.83
310–315	1.77×10^{15}	7.95×10^{-20}	0.26
315–320	1.89×10^{15}	4.15×10^{-20}	0.21

The data used represent a gross approximation, owing to the large changes in all the quantities in the region of the atmospheric cut-off, but are illustrative of the method and controlling quantities for ozone photolysis.

The most commonly used form of Equation (2.4) becomes

$$j(s^{-1}) = \sum_{\lambda_{min}}^{\lambda_{max}} \sigma_{av}(\lambda)\phi_{av}(\lambda)F_{av}(\lambda)$$

The av denotes that average of wavelength intervals. Therefore, $j = 3.14 \times 10^{-4}$ s^{-1}.

2.4 TROPOSPHERIC CHEMISTRY

In general, tropospheric chemistry is analogous to a low-temperature combustion system, the overall reaction is given by

$$CH_4 + 2O_2 \rightarrow CO_2 + 2H_2O \tag{2.5}$$

Unlike combustion, this is not a thermally initiated process but a process initiated and propagated by photochemistry. The chemistry that takes place in the troposphere, and in particular, the photochemistry is intrinsically linked to the chemistry of ozone. Tropospheric ozone acts as initiator, reactant and product in much of the oxidation chemistry that takes place in the troposphere and stratospheric ozone determines the amount of short wavelength radiation available to initiate photochemistry. Figure 6 shows a typical ozone profile through the atmosphere illustrating a number of interesting points. First, 90% of atmospheric ozone can be found in the stratosphere (see Section 2.10); on average about 10% can be found in the troposphere. Second, the troposphere in the simplest sense consists of two regions. The lowest kilometre or so contains the planetary boundary layer[6] and inversion layers which can act as pre-concentrators for atmospheric emissions from the surface and hinder exchange to the so-called *free* troposphere, the larger part by volume, that sits above the boundary layer.

For a long time, transport from the stratosphere to the troposphere was thought to be the dominant source of ozone in the troposphere.[7,8] Early in the 1970s, it was first suggested[9,10] that tropospheric ozone originated mainly from production within the troposphere by photochemical oxidation of CO and hydrocarbons catalysed by HO$_x$ and NO$_x$. These sources are balanced by *in-situ* photochemical destruction of ozone and by dry deposition at the earth's surface. Many studies, both experimental- and model-based have set about determining the

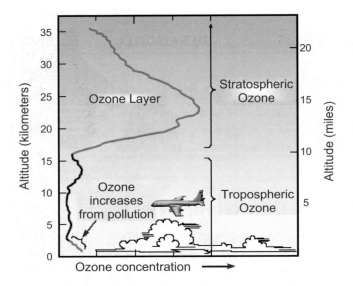

Figure 6 *A typical atmospheric ozone profile through the atmosphere.[58] The concentration is expressed as a volume mixing ratio*

contribution of both chemistry and transport to the tropospheric ozone budget on many different spatial and temporal scales.

There is growing evidence that the composition of the troposphere is changing.[11] For example, analysis of historical ozone records has indicated that tropospheric ozone levels in both hemispheres have increased by a factor of 3–4 over the last century. Methane concentrations have effectively doubled over the past 150 years and N_2O levels have risen by 15% since pre-industrial times.[12] Measurements of halocarbons have shown that this group of chemically and radiatively important gases to be increasing in concentration until relatively recently.[12]

One of the difficulties about discussing tropospheric chemistry in general terms is that by the very nature of the troposphere being the lowest layer of the atmosphere it has complex multi-phase interactions with the earth's surface, which can vary considerably between expanses of ocean to deserts (see Figure 7). The fate of any chemical species (C_i) in the atmosphere can be represented as a continuity or mass balance equation such as

$$\frac{\mathrm{d}C_i}{\mathrm{d}t} = \frac{\mathrm{d}uC_i}{\mathrm{d}x} - \frac{\mathrm{d}vC_i}{\mathrm{d}y} - \frac{\mathrm{d}wC_i}{\mathrm{d}z} + K_z\frac{\mathrm{d}C_i}{\mathrm{d}z} + P_i - L_i + S_i$$
$$+ \left(\frac{\mathrm{d}C_i}{\mathrm{d}t}\right)_{\text{clouds}} \tag{2.6}$$

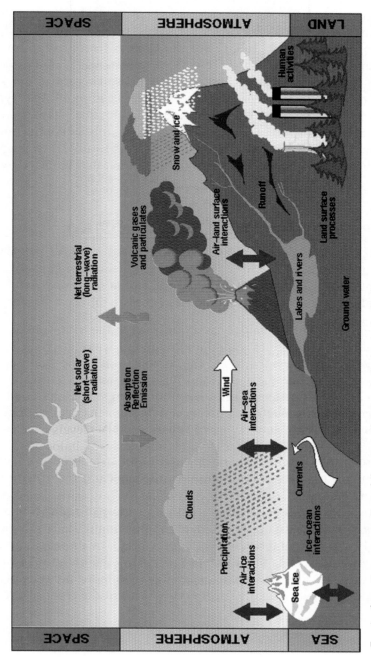

Figure 7 *A schematic representation of the atmosphere's role in the earth system*

where t is the time, u, v and w are the components of the wind vector in x, y and z accounting for the horizontal and vertical large-scale transport. Small-scale turbulence can be accounted for using K_z, the turbulent diffusion coefficient, P_i and L_i are the chemical production and loss terms and S_i are the sources owing to emissions. Cloud processes (vertical transport, washout and aqueous phase chemistry) are represented in a cloud processing term. The application of this type of equation is the basis of chemical modelling. In this chapter we will concentrate mainly on the chemical terms in this equation and the processes that control them, but inherently, as the study of tropospheric photochemistry is driven by observations, these must be placed within the framework of Equation (2.6).

2.5 TROPOSPHERIC OXIDATION CHEMISTRY

Though atmospheric composition is dominated by both oxygen and nitrogen, it is not the amount of oxygen that defines the capacity of the troposphere to oxidise a trace gas. The "*oxidising capacity*" of the troposphere is a somewhat nebulous term probably best described by Thompson.[13]

> The total atmospheric burden of O_3, OH and H_2O_2 determines the "*oxidising capacity*" of the atmosphere. As a result of the multiple interactions among the three oxidants and the multiphase activity of H_2O_2, there is no single expression that defines the earth's oxidising capacity. Some researchers take the term to mean the total global OH, although even this parameter is not defined unambiguously.

Figure 8 gives a schematic representation of tropospheric chemistry, representing the links between emissions, chemical transformation and sinks for a range of trace gases.

Atmospheric photochemistry produces a variety of radicals that exert a substantial influence on the ultimate composition of the atmosphere.[14] Probably the most important of these in terms of its reactivity is the hydroxyl radical, OH. The formation of OH is the initiator of radical-chain oxidation. Photolysis of ozone by UV light in the presence of water vapour is the main source of hydroxyl radicals in the troposphere, *viz*

$$O_3 + hv(\lambda < 340 \text{ nm}) \rightarrow O(^1D) + O_2(^1\Delta_g) \qquad (2.7)$$

$$O(^1D) + H_2O \rightarrow OH + OH \qquad (2.8)$$

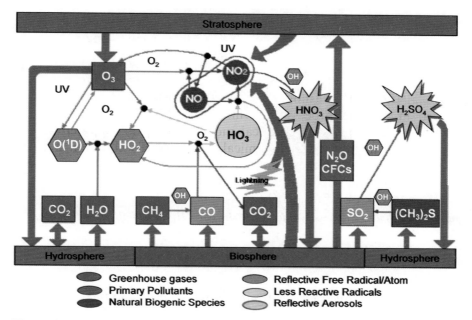

Figure 8 *A simplified scheme of tropospheric chemistry. The figure illustrates the interconnections in the chemistry, as well as the role of sources, chemical transformation and sinks*

The fate of the bulk of the $O(^1D)$ atoms produced *via* reaction (2.7) is collisional quenching back to ground-state oxygen atoms, *viz*

$$O(^1D) + N_2 \rightarrow O(^3P) + N_2 \qquad (2.9)$$

$$O(^1D) + O_2 \rightarrow O(^3P) + O_2 \qquad (2.10)$$

The fraction of $O(^1D)$ atoms that form OH is dependent on pressure and the concentration of H_2O; typically in the marine boundary layer (MBL) about 10% of the $O(^1D)$ generate OH. Reactions (2.7 and 2.8) are the primary source of OH in the troposphere, but there are a number of other reactions and photolysis routes capable of forming OH directly or indirectly. As these compounds are often products of OH radical initiated oxidation they are often termed secondary sources of OH and include the photolysis of HONO, HCHO, H_2O_2 and acetone and the reaction of $O(^1D)$ with methane (see Figure 9). Table 2 illustrates the average contribution of various formation routes with altitude in a standard atmosphere.

Two important features of OH chemistry make it critical to the chemistry of the troposphere. The first is its inherent reactivity; the second is its

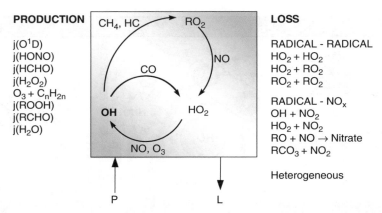

Figure 9 *The sources, interconversions and sinks for HO_x (and ROx) in the tropo-sphere*[14]

Table 2 *Calculated fractional contribution of various photolysis rates to radical production with altitude*

Altitude	$j(O(^1D)) +$ H_2O	$j(O(^1D)) +$ CH_4	$j(Acetone)$	$j(H_2O_2)$	$j(HCHO)$
Ground	0.68	0.0	Neg.	0.15	0.17
Mid-troposphere	0.52	Neg.	0.03	0.20	0.25
Upper-troposphere	0.35	0.02	0.1	0.25	0.28
Lower stratosphere	0.40	0.1	0.25	0.1	0.15

Note: (Neg: Negligible).

relatively high concentration given its high reactivity. The hydroxyl radical is ubiquitous throughout the troposphere owing to the widespread nature of ozone and water. In relatively unpolluted regimes (low NO_x) the main fate for the hydroxyl radical is reaction with either carbon monoxide or methane to produce peroxy radicals such as HO_2 and CH_3O_2, *viz*

$$OH + CO \rightarrow H + CO_2 \qquad (2.11)$$

$$H + O_2 + M \rightarrow HO_2 + M \qquad (2.12)$$

and

$$OH + CH_4 \rightarrow CH_3 + H_2O \qquad (2.13)$$

$$CH_3 + O_2 + M \rightarrow CH_3O_2 + M \qquad (2.14)$$

In low-NO_x conditions, HO_2 can react with ozone leading to further destruction of ozone in a chain sequence involving production of hydroxyl radicals.

$$HO_2 + O_3 \rightarrow OH + 2O_2 \qquad (2.15)$$

$$OH + O_3 \rightarrow HO_2 + O_2 \qquad (2.16)$$

Alternatively, it can recombine to form hydrogen peroxide (H_2O_2)

$$HO_2 + HO_2 \rightarrow H_2O_2 + O_2 \qquad (2.17)$$

or react with organic peroxy radicals such as CH_3O_2 to form organic hydroperoxides,

$$CH_3O_2 + HO_2 \rightarrow CH_3O_2H + O_2 \qquad (2.18)$$

The formation of peroxides is effectively a chain termination reaction, as under most conditions these peroxides can act as effective sinks for HO_x. In more polluted conditions (high-NO_x), peroxy radicals catalyse the oxidation of NO to NO_2

$$HO_2 + NO \rightarrow OH + NO_2 \qquad (2.19)$$

leading to the production of ozone from the subsequent photolysis of nitrogen dioxide and reaction of the photoproducts, *i.e.*

$$NO_2 + hv(\lambda < 420 \text{ nm}) \rightarrow NO + O(^3P) \qquad (2.2)$$

$$O + O_2 + M \rightarrow O_3 + M \qquad (2.20)$$

Hydroxyl radicals produced in reaction (2.19) can go on to form more peroxy radicals (*e.g. via* reactions (2.11) and (2.13)). Similarly to HO_2, CH_3O_2 can also oxidise NO to NO_2.

$$CH_3O_2 + NO \rightarrow CH_3O + NO_2 \qquad (2.21)$$

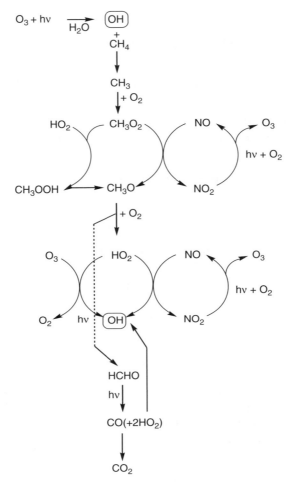

Figure 10 *Simplified mechanism for the photochemical oxidation of CH$_4$ in the troposphere[62]*

The resulting methoxy radical reacts rapidly with O$_2$ to form formaldehyde and HO$_2$.

$$CH_3O + O_2 \rightarrow HCHO + HO_2 \qquad (2.22)$$

The oxidation of methane is summarised schematically in Figure 10. The OH radical may have another fate, dependant on the concentration of NO$_2$, it can react with NO$_2$ to form nitric acid.

$$OH + NO_2 + M \rightarrow HNO_3 + M \qquad (2.23)$$

The formation of HNO$_3$ represents an effective loss mechanism for both HO$_x$ and NO$_x$.

Example lifetime calculation

For the reaction, $CH_4 + OH \rightarrow CH_3 + H_2O$ that is the key loss process for CH_4 (a greenhouse gas), the rate coefficient for the reaction at atmospheric temperatures is given by $k = 8.4 \times 10^{-15}$ cm^3 molecule^{-1} s^{-1}. Given the mean atmospheric concentration is $[OH] = 5 \times 10^5$ molecule cm^{-3}, what is the atmospheric lifetime of CH_4?

Answer: For a reaction of the type $A + B \rightarrow P$ (*i.e.* second-order) the atmospheric lifetime is given by

$$\tau_{OH}^{CH_4} = \frac{1}{k[OH]} = \frac{1}{(6.3 \times 10^{-15})(5 \times 10^5)} = 3.1 \times 10^8 \, s = 9.9 \, years$$

The type of calculations are useful as they give an indication of the likely chemical lifetime (*i.e.* the amount of time it will take before a molecule is reacted away) of a molecule in the atmosphere. Clearly, this form of calculation does not take into account any other chemical loss routes other than reaction with OH or any other physical process that may remove a molecule.

Temperature dependence of a reaction

The lifetime of a compound in the atmosphere will depend on how fast it reacts with main atmospheric oxidants. Many reaction rates vary with temperature, therefore in the atmosphere the lifetime will vary with altitude. For the reaction between OH and CH_4, the temperature dependence of the reaction is given by $k = 1.85 \times 10^{-12}\exp(-1690/T)$. How does the reaction rate vary between 0.1 and 10 km and therefore effect the lifetime? For the lifetime calculation see previous description.

Answer:

Region	Altitude (km)	T (°C)	k (cm^3 molecule^{-1} s^{-1})	τ_{OH} (years)
Boundary layer	0.1	25	6.3×10^{-15}	9.9
Lower troposphere	1	9	4.6×10^{-15}	13.8
Middle troposphere	6	−24	2.0×10^{-15}	30.3
Lower stratosphere	10	−56	7.6×10^{-16}	82.7

2.5.1 Nitrogen Oxides and the Photostationary State

From the preceding discussion, it can be seen that the chemistry of nitrogen oxides are an integral part of tropospheric oxidation and photochemical processes. Nitrogen oxides are released into the troposphere from a variety of biogenic and anthropogenic sources including fossil fuel combustion, biomass burning, microbial activity in soils and lightning discharges (see Figure 2). About 30% of the global budget of NO_x, *i.e.* ($NO + NO_2$) comes from fossil fuel combustion with almost 86% of the NO_x emitted in one form or the other into the planetary boundary layer from surface processes.[5] Typical NO/NO_2 ratios in surface air are 0.2–0.5 during the day tending to zero at night. Over the timescales of hours to days NO_x is converted to nitric acid (reaction (2.23)) and nitrates, which are subsequently removed by rain and dry deposition.

The photolysis of NO_2 to NO and the subsequent regeneration of NO_2 *via* reaction of NO with ozone is sufficiently fast, in the moderately polluted environment, for these species to be in dynamic equilibrium, *viz*

$$NO_2 + hv \rightarrow NO + O(^3P) \tag{2.2}$$

$$O(^3P) + O_2 + M \rightarrow O_3 + M \tag{2.20}$$

$$O_3 + NO \rightarrow NO_2 + O_2 \tag{2.24}$$

Therefore, at suitable concentrations, ambient NO, NO_2 and O_3 can be said to be in a photochemical steady-state or photostationary state (PSS),[15] provided that they are isolated from local sources of NO_x and that sunlight intensity is relatively constant, therefore

$$[O_3] = \frac{j_{2.2}[NO_2]}{k_{2.24}[NO]} \tag{2.25}$$

The reactions (2.2, 2.20 and 2.24) constitute a cycle with no net chemistry. The PSS expression is sometimes expressed as a ratio, *viz*

$$\phi = \frac{j_{2.2}[NO_2]}{k_{2.24}[NO][O_3]} \tag{2.26}$$

If ozone is the sole oxidant for NO to NO_2 then $\phi = 1$. This situation often pertains in urban areas where NO_x levels are high and other potential oxidants of NO to NO_2 such as peroxy radicals are suppressed. In the presence of peroxy radicals, Equation (2.25) has to be modified as the NO/NO_2 partitioning is shifted to favour NO_2, *viz*

$$\frac{[NO_2]}{[NO]} = (k_{2.24}[O_3] + k_{2.19}[HO_2] + k_{2.21}[RO_2])/j_{2.2} \qquad (2.27)$$

Though the radical concentrations are typically *ca.* 1000 times smaller than the $[O_3]$, the rate of the radical oxidation of NO to NO_2 is *ca.* 500 times larger than the corresponding oxidation by reaction with O_3. We shall return to the significance of the peroxy radical catalysed oxidation of NO to NO_2 when considering photochemical ozone production and destruction. From the preceding discussion, it can be seen that the behaviour of NO and NO_2 are strongly coupled through both photolytic and chemical equilibria. Because of their rapid interconversion they are often referred to as NO_x. NO_x, *i.e.* (NO + NO_2) is also sometimes referred to as "active nitrogen".

Photostationary state
At what NO_2/NO ratio will the PSS ratio be equal to 1 for midday conditions ($j_{2.2} = 1 \times 10^{-3}$ s^{-1}) given that $k_{2.24} = 1.7 \times 10^{-14}$ cm^3 molecule^{-1} s^{-1} and that $O_3 = 30$ ppbv.
Using the Equation (2.26)

$$\phi = \frac{j_{2.2}[NO_2]}{k_{2.24}[NO][O_3]}$$

Converting O_3 from ppbv to molecule cm^{-3}, as 1 ppbv = 2.46×10^{10} molecule cm^{-3} (at 25°C and 1 atm) 30 ppbv = 7.38×10^{11} molecule cm^{-3}. Given that $\phi = 1$ then

$$\frac{[NO_2]}{[NO]} = \frac{k_{2.24}[O_3]}{j_{2.2}}$$

$[NO_2]/[NO] = 12.55$

The extent of the influence of NO_x in any given atmospheric situation depends on its sources, reservoir species and sinks. Therefore, an important atmospheric quantity is the lifetime of NO_x. If nitric acid formation is considered to be the main loss process for NO_x (*i.e.* NO_2), then the lifetime of NO_x (τ_{NO_x}) can be expressed as the time constant for reaction (2.23), the NO_2 to HNO_3 conversion.

$$\tau_{NO_x} = \frac{1}{k_{2.23}[OH]} \left(1 + \frac{[NO]}{[NO_2]}\right) \qquad (2.28)$$

Therefore, using this simplification, the lifetime of NO_x is dependent on the [OH] and [NO]/[NO_2] ratio. Calculating τ_{NO_x} under typical upper tropospheric conditions gives lifetimes in the order of 4–7 days and lifetimes in the order of days in the lower free troposphere. In the boundary layer, the situation is more complex as there are other NO_x loss and transformation processes other than those considered in Equation (2.29), which can make τ_{NO_x} as short as 1 h. Integrally linked to the lifetime of NO_x and therefore the role of nitrogen oxides in the troposphere is its relation to odd nitrogen reservoir species, *i.e.* NO_y. The sum of total reactive nitrogen or total odd nitrogen is often referred to as NO_y and can be defined as $NO_y = NO_x + NO_3 + 2N_2O_5 + HNO_3 + HNO_4 + HONO + PAN +$ nitrate aerosol + alkyl nitrate, where PAN is peroxyacetlynitrate (see Section 2.5.4). NO_y can also be thought of as NO_x plus all the compounds that are products of the atmospheric oxidation of NO_x. NO_y is not a conserved quantity in the atmosphere owing to the potential for some of its constituents (*e.g.* HNO_3) to be efficiently removed by deposition processes. Mixing of air masses may also lead to dilution of NO_y. The concept of NO_y is useful in considering the budget of odd nitrogen and evaluating the partitioning of NO_x and its reservoirs in the troposphere.[16]

In summary, the concentration of NO_x in the troposphere

- determines the catalytic efficiency of ozone production;
- determines the partitioning of OH and HO_2;
- determines the amount of HNO_3 and nitrates produced; and
- determines the magnitude and sign of net photochemical production or destruction of ozone (see Section 2.5.2).

2.5.2 Production and Destruction of Ozone

From the preceding discussion of atmospheric photochemistry and NO_x chemistry, it can be seen that the fate of the peroxy radicals can have a marked effect on the ability of the atmosphere either to produce or to destroy ozone. Photolysis of NO_2 and the subsequent reaction of the photoproducts with O_2 (reactions (2.4) and (2.20)) are the only known way of producing ozone in the troposphere. In the presence of NO_x the following cycle for the production of ozone can take place:

$$NO_2 + hv \rightarrow O(^3P) + NO \tag{2.2}$$

$$O(^3P) + O_2 + M \rightarrow O_3 + M \tag{2.20}$$

$$OH + CO \rightarrow H + CO_2 \tag{2.11}$$

$$H + O_2 + M \rightarrow HO_2 + M \qquad (2.12)$$

$$\underline{HO_2 + NO \rightarrow OH + NO_2} \qquad (2.19)$$

$$\text{NET: } CO + 2O_2 + h\nu \rightarrow CO_2 + O_3 \qquad (2.29)$$

Similar chain reactions can be written for reactions involving RO_2. In contrast, when relatively little NO_x is present, as in the remote atmosphere, the following cycle can dominate over ozone production leading to the catalytic destruction of ozone, *viz*

$$HO_2 + O_3 \rightarrow OH + 2O_2 \qquad (2.15)$$

$$OH + CO \rightarrow H + CO_2 \qquad (2.11)$$

$$\underline{H + O_2 + M \rightarrow HO_2 + M} \qquad (2.12)$$

$$\text{NET: } CO + O_3 \rightarrow CO_2 + O_2 \qquad (2.30)$$

Clearly, there is a balance between photochemical ozone production and ozone loss dependent on the concentrations of HO_x and NO_x. Figure 11 shows the dependence of the production of ozone on NO_x taken from a numerical model. There are distinct regions in terms of $N(O_3)$ *vs.* $[NO_x]$

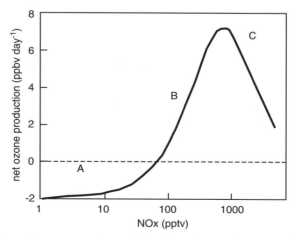

Figure 11 *Schematic representation of the dependence of the net ozone ($N(O_3)$) production (or destruction) on the concentration of NO_x*

on Figure 11. For example, in region A the loss of ozone ($L(O_3)$) is greater than the production of ozone ($P(O_3)$), hence the net product of this process, *i.e.*

$$N(O_3) = P(O_3) - L(O_3) \tag{2.31}$$

leads to a net ozone loss. The photochemical loss of ozone can be represented as

$$L(O_3) = (f \cdot j_{2.7}(O^1D) \cdot [O_3]) + k_{2.15}[HO_2] + k_{2.16}[OH] \tag{2.32}$$

where f is the fraction of $O(^1D)$ atoms that react with water vapour (reaction (2.8)) rather than are deactivated to $O(^3P)$ (reactions (2.9) and (2.10)). Evaluation of Equation (2.32) is effectively a lower limit for the ozone loss rate as it neglects any other potential chemical loss processes for ozone such as cloud chemistry,[17] NO_3 chemistry (see Section 2.6) or halogen chemistry (see Section 2.9). The balance point, *i.e.* where $N(O_3)$ = 0 is often referred to, somewhat misleadingly, as the compensation point and occurs at a critical concentration of NO_x. Above the compensation point $P(O_3) > L(O_3)$ and therefore $N(O_3)$ is positive and the system is forming ozone. The *in-situ* formation rate for ozone is approximately given by the rate at which the peroxy radicals (HO_2 and RO_2) oxidise NO to NO_2. This is followed by the rapid photolysis of NO_2 (reaction (2.4)) to yield the oxygen atom required to produce O_3.

$$P(O_3) = [NO] \cdot (k_{2.19}[HO_2] + \Sigma k_i[RO_2]_i) \tag{2.33}$$

It is also worth noting that $P(O_3)$ can also be expressed in terms of the concentrations of NO_x, $j_{2.2}(NO_2)$, O_3 and temperature by substitution of Equation (2.27) into Equation (2.33) to give

$$P(O_3) = j_{2.2}[NO_2] - k_{2.24}[NO][O_3] \tag{2.34}$$

At some concentration of NO_x the system reaches a maximum production rate for ozone at $dP(O_3)/d(NO_x) = 0$ and even though $P(O_3)$ is still significantly larger than $L(O_3)$ the net production rate begins to fall off with increasing NO_x. Until this maximum is reached the system is said to be NO_x limited with respect to the production of ozone. The turn-over, *i.e.* $dP(O_3)/d(NO_x) = 0$ is caused by the increased competition for NO_x by the reaction

$$OH + NO_2 + M \rightarrow HNO_3 + M \tag{2.23}$$

In reality, the situation is somewhat more complicated owing to the presence at high concentrations of NO_x of increased levels of non-methane hydrocarbons (NMHCs), especially in places such as the urban atmosphere. The oxidation of NMHCs in common with much of tropospheric oxidation chemistry, is initiated by reaction with OH, leading to the rapid sequence of chain reactions.

$$OH + RH \rightarrow R + H_2O \tag{2.35}$$

$$R + O_2 + M \rightarrow RO_2 + M \tag{2.36}$$

$$RO_2 + NO \rightarrow RO + NO_2 \tag{2.37}$$

$$RO \rightarrow \text{carbonyl products} + HO_2 \tag{2.38}$$

$$HO_2 + NO \rightarrow OH + NO_2 \tag{2.19}$$

This cycle is similar to the preceding one for the oxidation of CO, in that it is catalytic with respect to OH, R, RO and RO_2, with HO_2 acting as the chain propagating radical. The mechanism of reaction (2.38) is strongly dependent on the structure of RO.

Photochemical production of ozone

Calculate the photochemical production rate of ozone at midday, given that $j_{2.2} = 1 \times 10^{-3}$ s^{-1}, $O_3 = 30$ ppbv and $k_{2.24} = 1.7 \times 10^{-14}$ cm^3 molecule^{-1} s^{-1} for a 2:1 ratio of NO_2/NO.

Using the equation

$$P(O_3) = j_{2.2}[NO_2] - k_{2.24}[NO][O_3]$$

As the rate constant is in the units of cm^3 molecule^{-1} s^{-1}, all the concentrations should be converted into molecule cm^{-3} (as per PSS example). The answer should be converted into ppb h^{-1} by dividing the answer in cm^3 molecule^{-1} s^{-1} into ppbv and turning the seconds into hours. For $NO_2 = 2$ ppbv, $NO = 1$ ppbv, $P(O_3) = 7.1$ ppbv h^{-1}. The equation scales as per the absolute values of NO_2/NO used.

With the involvement of volatile organic compounds (VOCs) in the oxidation chemistry, Figure 11 represents a slice through an *n*-dimensional surface where there should be a third axis to represent the concentration of VOCs. The peak initial concentrations of ozone generated from various initial concentrations of NO_x and VOCs are usually represented as an "O_3 isopleth diagram", an example of which is shown in Figure 12.[18] In an isopleth diagram, initial mixture compositions giving rise to the same peak O_3 concentration are connected by the appropriate isopleth. An isopleth plot shows that ozone production is a highly non-linear process in relation to NO_x and VOC, but picks out many of the features already highlighted in Figure 11, *i.e.* when NO_x is "low" the rate of ozone formation increases with increasing NO_x in a near-linear fashion. On the isopleth, the local maximum in the ozone formation rate with respect to NO_x is the same feature as the turn over in $N(O_3)$ in Figure 11. The ridgeline along the local maximum separates two different regimes, the so-called NO_x-*sensitive* regime, *i.e.* $N(O_3) \propto$ (NO_x) and the VOC-*sensitive* (or NO_x *saturated* regime), *i.e.* $N(O_3) \propto$ (VOC) and increases with increasing NO_x. The relationship between NO_x, VOCs and ozone embodied in the isopleth diagram indicates one of the problems in the development of air quality policy with respect to ozone. Reductions in VOC are only effective in reducing ozone under VOC-*sensitive* chemistry (high NO_x) and reductions in NO_x will only be effective if NO_x-*sensitive* chemistry predominates and may actually increase ozone in VOC-*sensitive* regions. In general, as an air mass moves away from emission sources, *e.g.* in an urban region, the chemistry tends to move from VOC-*sensitive* to NO_x-*sensitive* chemistry.

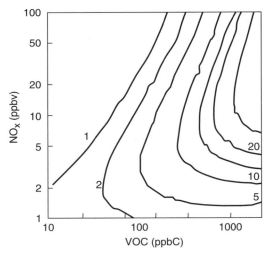

Figure 12   *Isopleths giving net rate of ozone production (ppb h⁻¹) as a function of VOC (ppbC) and NO_x (ppbv) for mean summer daytime meteorology and clear skies[18]*

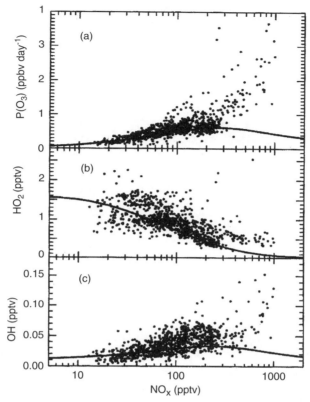

Figure 13 *Observed ozone production rates ($P(O_3)$) and concentrations of HO_2 and OH during SONEX (8–12 km altitude, 40–60°N latitude) plotted a function of the NO_x concentration (NO_x = observed NO + modelled NO_2). The observed rates and concentrations are averaged over 24 h, using diel factors obtained from a locally constrained box model. The lines on the three panels correspond to model-calculated values for median upper tropospheric conditions as encountered during SONEX[19]*

Case Study I – *A practical example of the dependence of ozone production on NO_x* – An example of the experimental determination of these relationships is shown in Figure 13, a comparison of observed ozone production rates ($P(O_3)$) and concentrations of HO_2 and OH from the NASA SONEX mission[19] plotted as a function of NO_x. The data were taken from a suite of aircraft measurements between 8 and 12 km altitude at latitudes between 40 and 60°N. The model data suggest that $P(O_3)$ becomes independent of NO_x above 70 pptv and the turn-over point into a NO_x-saturated regime occurs at about 300 pptv. The bulk of the experimental observation below $[NO_x] < 300$

pptv observe the $P(O_3)$ dependency predicted by the model, but above $[NO_x] \sim 300$ pptv $P(O_3)$, computed from the measured HO_2 and NO, continues to increase with NO_x, suggesting a NO_x-limited regime.

Case Study II – *Photochemical control of ozone in the remote marine boundary layer* (MBL) – An elegant piece of experimental evidence for the photochemical destruction of ozone comes from studies in the remote MBL over the southern ocean at Cape Grim, Tasmania $(41°S)$.[20] In the MBL, the photochemical processes are coupled to physical processes that affect the observed ozone concentrations, namely deposition to the available surfaces and entrainment from the free troposphere. The sum of these processes can be represented in the form of an ozone continuity equation (a simplified version of Equation 2.6), *viz*

$$\frac{d[O_3]}{dt} = C + \frac{E_v([O_3]_{ft} - [O_3])}{H} + \frac{v_d[O_3]}{H} \qquad (2.39)$$

where C is a term representative of the photochemistry (the net result of production, $P(O_3)$, minus destruction, $L(O_3)$ Equation (2.32)), E_v the entrainment velocity (a measure of the rate of ozone transport into the boundary layer), $[O_3]_{ft}$ the concentration of free-tropospheric ozone, v_d the dry deposition velocity (a measure of the physical loss of ozone to a surface) and H the height of the boundary layer. In general, the MBL is particularly suitable for making photochemical measurements owing to its stable and chemically simple nature. Figure 14 shows the average diurnal cycle of ozone and total peroxide (mainly H_2O_2) in clean oceanic air as measured at Cape Grim during January 1992. During the sunlit hours an ozone loss of about 1.6 ppbv occurs between mid-morning and late afternoon. This loss of ozone is followed by an overnight replenishment to a similar starting point. In contrast, the peroxide concentration decreases overnight from 900 to 600 pptv and then increases from 600 to 900 pptv between midmorning and late afternoon. It is worth noting that the magnitude of this anti-correlation of ozone and peroxide is dependent on season. The daytime anti-correlation between O_3 and peroxide can be interpreted as experimental evidence for the photochemical destruction of ozone, as the ozone is destroyed *via* reactions (2.7, 2.15 and 2.16) while simultaneously peroxide is formed from chemistry involving the odd-hydrogen radicals OH and HO_2 (reactions (2.11, 2.12 and 2.17)).

The night-time replenishment of ozone is caused by entrainment of ozone from the *free* troposphere into the boundary layer. The overnight loss of peroxide is due to deposition over the sea surface (and heterogeneous loss to the aerosol surface), as peroxide has a significant physical loss rate, in contrast to ozone which does not. Therefore, the daytime anti-correlation of ozone and peroxide is indicative of the net photochemical destruction of ozone.

2.5.3 Role of Hydrocarbons

The discussion up to this point has focused in the main on the role of CO and CH_4 as the fuels for atmospheric oxidation. It is clear that there are many more carbon compounds in the atmosphere than just these two.[21] One of the roles of atmospheric photochemistry is to "cleanse" the troposphere of a wide-range of these compounds. Table 3 illustrates the global turnover of a range of trace gases including hydrocarbons and illustrates, for a number of trace gases, the primary role played by OH in their removal. NMHC have a range of both biogenic and anthropogenic sources.[5,21] Carbon monoxide chemistry is not independent from NMHC chemistry as 40–60% of surface CO levels over the continents, slightly less over the oceans, and 30–60% of CO levels in the free troposphere, are estimated to come from NMHC oxidation.[22]

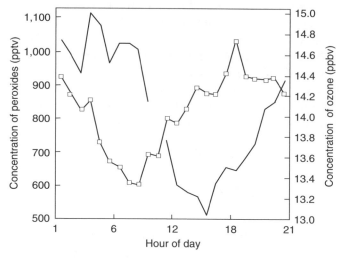

Figure 14 *Average diurnal cycles for peroxide (open squares) and ozone (filled squares) in baseline air at Cape Grim, Tasmania (41°S) for January 1992*[20]

Table 3   *Global turnover of tropospheric gases and fraction removed by reaction*
           with OH

Trace gas	Global emission rate($Tg\ yr^{-1}$)	Removal by OH^a (%)
CO	2800	85
CH_4	530	90
C_2H_6	20	90
Isoprene	570	90
Terpenes	140	50
NO_2	150	50
SO_2	300	30
$(CH_3)_2S$	30	90
$CFCl_3$	0.3	0

^{a} Assuming mean global [OH] = 1×10^6 molecule cm^{-3}.
Source: After ref. 59.

A major component of the reactive hydrocarbon loading are the biogenic hydrocarbons.[21] As previously indicated, the hydrocarbon oxidation chemistry is integral to the production of ozone. Globally, the contribution of NMHC to net photochemical production of ozone is estimated to be about 40%.[23]

There are a number of inorganic molecules such as NO_2 and SO_2 (See Table 3) which are also lost *via* reaction with OH. A number of halocarbons also exist that posses insubstantial tropospheric sinks and have importance in the chemistry of the stratosphere (see Section 2.10).

The ultimate products of the oxidation of any hydrocarbon are carbon dioxide and water vapour, but there are many relatively stable partially oxidised organic species such as aldehydes, ketones and carbon monoxide that are produced as intermediate products during this process, with ozone produced as a by-product of the oxidation process. Figure 10 shows a schematic representation of the free radical catalysed oxidation of methane, which is analogous to that of a hydrocarbon. As previously discussed, the oxidation is initiated by reaction of the hydrocarbon with OH and follows a mechanism in with the alkoxy and peroxy radicals are chain propagators and OH is effectively catalytic, *viz*

$$OH + RH \rightarrow R + H_2O \tag{2.35}$$

$$R + O_2 + M \rightarrow RO_2 + M \tag{2.36}$$

$$RO_2 + NO \rightarrow RO + NO_2 \tag{2.37}$$

$$RO \rightarrow carbonyl\ products + HO_2 \tag{2.38}$$

$$HO_2 + NO \rightarrow OH + NO_2 \qquad (2.19)$$

As both reactions (2.37) and (2.19) lead to the oxidation of NO to NO_2 the subsequent photolysis leads to the formation of ozone (see reactions (2.2) and (2.20)). The individual reaction mechanism depends on the identity of the organic compounds and the level of complexity of the mechanism. Although OH is the main tropospheric oxidation initiator, reaction with NO_3, O_3, $O(^3P)$ or photolysis may be an important loss route for some NMHCs or the partially oxygenated products produced as intermediates in the oxidation (see reaction (2.38)).

In summary, the rate of oxidation of VOCs and therefore by inference the production of ozone is governed by the concentration of the catalytic HO_x radicals. There are a large variety of VOCs with a range of reactivities; therefore this remains a complex area.

2.5.4 Urban Chemistry

In some respects, the story of atmospheric chemistry and particularly ozone photochemistry begins with urban chemistry and photochemical smog. The term smog arises from a combination of the words smoke and fog. In the 1940s, it became apparent that cities like Los Angeles (LA) were severely afflicted with a noxious haze.[24] Though, at the time it was thought to be a relatively local phenomenon, with the understanding of its chemistry came the development of a photochemical theory for the whole of the troposphere. The LA smog is often termed photochemical smog and is quite different in origin to the London smogs of the 19th and 20th centuries, which have their origins in abnormally high concentrations of smoke particles and sulfur dioxide. The London smogs have been alleviated with the effective application of legislation that has reduced the burning of coal in the London area. The major features of photochemical smog are high levels of oxidant concentration in particular ozone and peroxidic compounds, produced by photochemical reactions. The principal effects of smog are eye and bronchial irritation as well as plant and material damage. The basic reaction scheme for the formation of photochemical smog is

$$VOC + h\nu \rightarrow VOC + R \qquad (2.40)$$

$$R + NO \rightarrow NO_2 \qquad (2.41)$$

$$NO_2 + h\nu(\lambda < 420 \text{ mm}) \rightarrow NO + O_3 \qquad (2.2)$$

Table 4 *Percentage of NMHC classes measured in the morning at various locations*

	Urban Los Angeles	Urban Boston	Rural Alabama
Alkanes	42	36	9
Alkenes	7	10	43[a]
Aromatics	19	30	2
Other	33	24	46[b]

Source: From ref. 31.
[a] Large contribution from biogenic alkenes.
[b] Mainly oxygen containing.

$$NO + O_3 \rightarrow NO_2 + O_2 \tag{2.24}$$

$$R + R \rightarrow R \tag{2.42}$$

$$R + NO_2 \rightarrow NO_y \tag{2.43}$$

There is a large range of available VOCs in the urban atmosphere, driven by the range of anthropogenic and biogenic sources.[5,21,31] Table 4 illustrates the different loadings of the major classes of NMHC in urban and rural locations. These NMHC loadings must be coupled to measures of reactivity and the degradation mechanisms of the NMHC to give a representative picture of urban photochemistry. The oxidation of the VOCs drives, *via* the formation of peroxy radicals, the oxidation of NO to NO_2, where under the sunlit conditions the NO_2 can be dissociated to form ozone (reaction (2.2)). Pre-existing ozone can also drive the NO to NO_2 conversion (reaction (2.24)).

The basic chemistry responsible for urban photochemistry is essentially the same as that takes place in the unpolluted atmosphere (*see* Section 2.5). It is the range and concentrations of NMHC fuels and the concentrations of NO_x coupled to the addition of some photochemical accelerants that can lead to the excesses of urban chemistry. For example, in the Los Angeles basin it is estimated that 3333 ton day^{-1} of organic compounds are emitted as well as 890 ton day^{-1} of NO_x. In addition, to the reactions forming OH in the background troposphere, *i.e. via* the reaction of $O(^1D)$ with H_2O, *viz*

$$O_3 + h\nu(\lambda < 340 \text{ nm}) \rightarrow O_2 + O(^1D) \tag{2.7}$$

$$O(^1D) + H_2O \rightarrow 2OH \tag{2.8}$$

under urban conditions OH may be formed from secondary sources such as

$$HONO + h\nu(\lambda < 400 \text{ nm}) \rightarrow OH + NO \qquad (2.44)$$

where the HONO can be emitted in small quantities from automobiles or formed from a number of heterogeneous pathways,[25] as well as gas-phase routes. OH produced from HONO has been shown to be the dominant OH source in the morning under some urban conditions, where the HONO has built up to significant levels overnight. Another key urban source of OH can come from the photolysis of the aldehydes and ketones produced in the NMHC oxidation chemistry, in particular formaldehyde, *viz.*

$$HCHO + h\nu(\lambda < 334 \text{ nm}) \rightarrow H + HCO \qquad (2.45)$$

$$H + O_2 + M \rightarrow HO_2 + M \qquad (2.12)$$

$$\underline{HCO + O_2 \rightarrow HO_2 + CO} \qquad (2.46)$$

Net

$$HCHO + 2O_2 + h\nu \rightarrow 2HO_2 + CO \qquad (2.47)$$

Smog chamber experiments have shown that the addition of aldehyde significantly increases the formation rates of ozone and the conversion rates of NO and NO_2 under simulated urban conditions.[3]

A marked by-product of oxidation in the urban atmosphere, often associated, but not exclusive to, urban air pollution is PAN. PAN is formed by

$$OH + CH_3CHO \rightarrow CH_3CO + H_2O \qquad (2.48)$$

$$CH_3CO + O_2 \rightarrow CH_2CO{\cdot}O_2 \qquad (2.49)$$

addition of NO_2 to the peroxyacetyl radical ($RCO{\cdot}O_2$) leading to the formation of PAN.

$$CH_3CO{\cdot}O_2 + NO_2 + M \rightleftharpoons CH_3CO{\cdot}O_2{\cdot}NO_2 + M \qquad (2.50)$$

PAN is often used as an unambiguous marker for tropospheric chemistry. The lifetime of PAN in the troposphere is very much dependant on the temperature dependence of the equilibrium in reaction (2.50), the lifetime varying from 30 min at $T = 298$ K to 8 h at $T = 273$ K. At mid-troposphere temperature and pressures PAN has thermal decomposition lifetimes in the order of 46 days. It is also worth noting that peroxyacyl radical like peroxy radicals can oxidise NO to NO_2

$$CH_3CO\cdot O_2 + NO \rightarrow CH_3CO\cdot O + NO_2 \qquad (2.51)$$

PAN can be an important component of NO_y in the troposphere (see Section 2.5.1) and has the potential to act as a temporary reservoir for NO_x and in particular it has the potential to transport NO_x from polluted regions into the background/remote atmosphere.[26]

There are other side-products of urban air pollution that are giving increasing cause for concern. These include the production of both aerosols (both organic and inorganic) and particles.

The effects of photochemical smog/urban air pollution remain on the political agenda owing to their potential impact on human health and the economy. In summary, urban photochemistry is not substantially different from tropospheric photochemistry. It is the range and concentrations of the VOCs involved in oxidation coupled to the concentration of NO_x and other oxidants that lead to a large photochemical turnover.

2.6 NIGHT-TIME OXIDATION CHEMISTRY

Though photochemistry does not take place at night, it is important to note, within the context of tropospheric oxidation chemistry, the potential for oxidation chemistry to continue at night. This chemistry does not lead to the production of ozone, in fact the opposite, but has importance owing to the potential for the production of secondary pollutants. In the troposphere, the main night-time oxidant is thought to be the nitrate radical formed by the relatively slow oxidation of NO_2 by O_3, *viz.*

$$NO_2 + O_3 \rightarrow NO_3 + O_2 \qquad (2.52)$$

The time constant for reaction (2.52) is of the order of 15 h at an ozone concentration of 30 ppbv and $T = 290$ K. Other sources include

$$N_2O_5 + M \rightarrow NO_3 + NO_2 + M \qquad (2.53)$$

but as N_2O_5 is formed from

$$NO_3 + NO_2 + M \rightarrow N_2O_5 + M \qquad (2.54)$$

the two species act in a coupled manner. Dinitrogen pentoxide, N_2O_5, is potentially an important product as it can react heterogeneously with water to yield HNO_3. During the daytime the NO_3 radical is rapidly photolysed as it strongly absorbs in the visible, *viz*

$$NO_3 + hv \rightarrow NO + O_2 \qquad (2.55)$$

$$NO_3 + h\nu \rightarrow NO_2 + O(^3P) \qquad (2.56)$$

having a lifetime in the region of 5 s for overhead sun and clear sky conditions. Further, NO_3 will react rapidly with NO

$$NO_3 + NO \rightarrow NO_2 + NO_2 \qquad (2.57)$$

which can have significant daytime concentrations in contrast to the night time, where away from strong source regions, the NO concentrations should be near zero.

The nitrate radical has a range of reactivity towards VOCs. The nitrate radical is highly reactive towards certain unsaturated hydrocarbons such as isoprene, a variety of butenes and monoterpenes, as well as reduced sulfur compounds such as dimethylsulfide (DMS). In the case of DMS, if the NO_2 concentration is 60% that of DMS then NO_3 is a more important oxidant than OH for DMS in the MBL.[27] In general, NO_3 abstraction reactions of the type

$$NO_3 + RH \rightarrow HNO_3 + R \qquad (2.58)$$

are relatively slow, with the alkyl radical reacting with oxygen under atmospheric conditions to form a peroxy radical. In the case of RH being an aldehyde, acyl products will form acylperoxy radicals ($R \cdot CO \cdot O_2$), potential sources of peroxyacylnitrates. In contrast, the reaction of NO_3 with alkenes occurs by an addition mechanism, initiating a complex chemistry involving nitrooxy substituted organic radicals, which can either regenerate NO_2 or produce comparatively stable bifunctional organic nitrate products.[28] For example, the products derived from the reaction of NO_3 with propene in the presence of O_2 and NO_x include CH_3CHO, HCHO, 1,2-propanedioldinitrate (PDDN), nitroxyperoxypropylnitrate (NPPN) and α-(nitrooxy)acetone (See Figure 15).[3] The reaction channel that produces the nitrated acetones also yields peroxy radicals, leading to the potential for a night-time source of OH, either by reaction ($HO_2 + O_3$) or by the direct reaction of the peroxy radical with NO_3. For the reaction of NO_3 with propene the initial addition can take place at either end of the double bond, *viz*

$$NO_3 + CH_2{=}CHCH_2 + M \rightarrow CH_2CHCH_2(ONO_2) + M \qquad (2.59)$$

$$NO_3 + CH_2{=}CHCH_2 + M \rightarrow CH_2CH(ONO_2)CH_2 + M \qquad (2.60)$$

The reaction can then proceed by the mechanism shown schematically in Figure 15. The ratio of final products is dependent on the structure of the individual alkenes. In general, for branched alkenes, there is

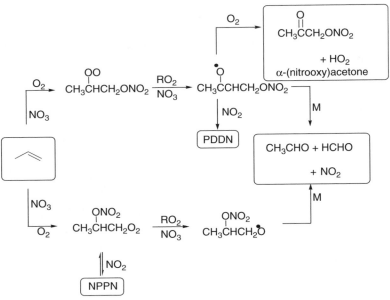

Figure 15 *A schematic representation of the chain propagation reactions in the NO$_3$ radical initiated oxidation of propene*[28]

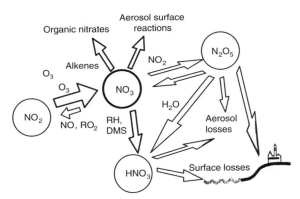

Figure 16 *A simplified reaction scheme for night-time chemistry involving the nitrate radical*[27]

significant regeneration of NO$_x$ and production of unsubstituted carbonyl products while comparatively they are a minor source of HO$_x$.[28] For the less alkyl substituted alkenes, there is a greater yield of HO$_x$ and bifunctional organic nitrate products but a lesser regeneration of NO$_x$. Therefore, depending on the mix of hydrocarbons, NO$_3$ chemistry can act to either recycle NO$_x$, therefore inhibiting the formation of nitrate aerosol or HNO_3 at night and potentially lead to the generation of a night-time source of HO_x. Figure 16 provides a simplified summary of the relevant night-time chemistry involving the nitrate radical.

One important difference between NO_3 chemistry and daytime OH chemistry is that NO_3 can initiate, but not catalyse, the removal of organic compounds. Therefore its concentration can be suppressed by the presence of fast-reacting, with respect to NO_3, organic compounds.

Reactive lifetime with respect to different oxidants
Different oxidants react at different rates in the atmosphere. Further, the average concentrations of the main oxidants can vary. For the following table of data, calculate the relative lifetime with respect to reaction with OH and NO_3, given that the average $[OH] \approx 10^6$ molecule cm^{-3} and the $[NO_3] \approx 10^9$ molecule cm^{-3}.

Compound	$k(OH)$ (cm^3 molecule^{-1} s^{-1})	τ_{OH} (days)	$k(NO_3)$ (cm^3 molecule^{-1} s^{-1})	τ_{NO_3} (days)
CH_4	8.5×10^{-15}	1361	$<1 \times 10^{-19}$	115,740
C_2H_6	2.7×10^{-13}	43	8×10^{-18}	1446
C_2H_4	8.5×10^{-12}	1.3	2×10^{-16}	58
$(CH_3)_2C{=}C(CH_3)_2$	1.1×10^{-10}	0.1	4.5×10^{-11}	2×10^{-4}

The data nicely illustrate that lifetime is a product of rate and the concentration of the radical species.

Case Study III – *Evidence for the role of NO_3 in night-time oxidation chemistry from experimental based studies.*[27,28] Significant NO_3 concentrations have been detected over a wide range of atmospheric conditions, indicating a potential role for NO_3 over large regions of the atmosphere.[28] The atmospheric lifetime of NO_3 can be estimated using the steady-state approximation (*cf.* PSS) to be

$$\tau(NO_3) = \frac{[NO_3]}{k_{52}[NO_2][O_3]} \tag{2.61}$$

A useful quantity with which to compare this with is the reciprocal of the lifetime calculated from the sum of the first order loss processes involving NO_3 and N_2O_5

$$\tau(NO_3)^{-1} \geq \sum_i k_{(NO_3+HC_i)}[HC_i] + k_{(NO_3+DMS)}[DMS] + k_{het}(NO_3)$$
$$+ \left(k^I[H_2O] + k^{II}[H_2O]^2 + k_{het}(N_2O_5)\right).K_{2.54}[NO_2] \tag{2.62}$$

where the pseudo first-order loss rates over the i reactive hydrocarbons are summed and k^I and k^{II} are the first- and second-order components with respect to H_2O of the reaction

$$N_2O_5 + H_2O \rightarrow 2HNO_3 \tag{2.63}$$

$k_{het}(NO_3)$ and $k_{het}(N_2O_5)$ are the heterogeneous loss rates for these species and $K_{2.54}$ is the equilibrium constant for reactions (2.53 and 2.54). Figure 17 shows the observed lifetime of NO_3 in the MBL at Mace Head, Ireland[29] segregated by arrival wind sector. Mace Head experiences a range of air masses from *clean* marine air to European continental outflow. The measured NO_3 lifetime varies from 2 min to 4 h (see Figure 17). An assessment of the parameters controlling NO_3 atmospheric lifetime (*cf*. Equation (2.61)) highlights that, under the conditions encountered the lifetime of NO_3 chemistry is very sensitive to DMS and NMHC chemistry in *clean* marine air. However, in more polluted air the terms involving the indirect loss of N_2O_5 either in the gas-phase with H_2O or through uptake on aerosol tend to dominate.

From the preceding discussion it can be seen that the involvement of NO_3 chemistry in gas-phase tropospheric chemistry has potentially six significant consequences.[28]

(i) The radical can control NO_y speciation in the atmosphere at night (*via* reaction (2.54)).

(ii) Nitric acid can be formed, by hydrolysis of N_2O_5, as a product of a hydrogen abstraction process or indirectly *via* the NO_3-mediated production of OH, which can react with NO_2 (reaction (2.23)) to produce nitric acid.

(iii) Primary organic pollutants can be oxidised and removed at night.

(iv) Radicals (HO_x and RO_2) produced by NO_3 chemistry can act as initiators for chain oxidation chemistry.

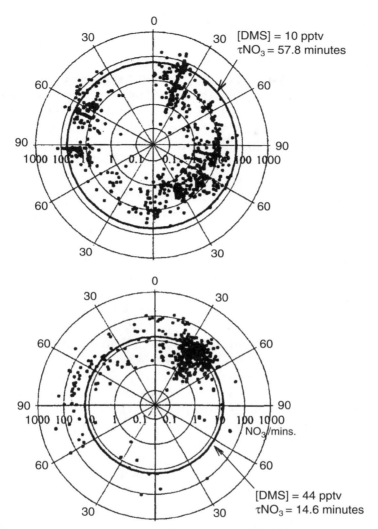

Figure 17 *Polar plots showing the observed lifetime of NO_3 vs. arrival wind direction for Mace Head, Ireland (top) and Tenerife (bottom).[29] The black circle represents the calculated lifetime of NO_3 from reaction with the campaign average [DMS] only*

(v) Toxic or otherwise noxious compounds such as peroxyacylnitrates, other nitrates and partially oxidised compounds, may be formed.

(vi) Nitrate products or NO_3 itself may act as temporary reservoirs in the presence of NO_x.

2.7 OZONE-ALKENE CHEMISTRY

Another potential *dark* source of HO_x in the atmosphere, more particularly in the boundary layer, is from the reactions between ozone and alkenes. The ozonolysis of alkenes can lead to the direct production of the OH radical at varying yields (between 7 and 100%) depending on the structure of the alkene, normally accompanied by the co-production of an (organic) peroxy radical. As compared to both the reactions of OH and NO_3 with alkenes the initial rate of the reaction of ozone with an alkene is relatively slow, this can be offset under regimes where there are high concentrations of alkenes and/or ozone. For example, under typical rural conditions the atmospheric lifetimes for the reaction of ethene with OH, O_3 and NO_3 are 20 h, 9.7 days and 5.2 months, respectively in contrast, for the same reactants with 2-methyl-2-butene the atmospheric lifetimes are 2.0 h, 0.9 h and 0.09 h.

The mechanism for the reaction of ozone with alkenes was first suggested by Criegee[30] in the late 1940s and involves the addition of ozone to form a primary ozonide, which rapidly decomposes to form a vibrationally excited carbonyl oxide (*Criegee intermediate*) and carbonyl products. The Criegee intermediate can then either be collisionally stabilised by a third body (M), or undergo unimolecular decomposition to products (see also Section 2.8). It is now widely believed that alkyl-substituted Criegee intermediates can decompose *via* a vibrationally hot hydroperoxide intermediate to yield an OH radical, along with another radical species of the general form $R_1R_2CC(O)R_3$, which is expected to react rapidly with O_2 to form a peroxy radical (RO_2) in the atmosphere.[31] Figure 18 shows a schematic representation of the ozone-alkene reaction mechanism. The OH and peroxy radical yield is dependent on the structure and mechanism of the individual alkene-ozone reaction.[32] Table 5 shows typical OH yields for the reaction of a range of anthropogenic and biogenic alkenes with ozone. There is growing experimental evidence[33] of the importance of ozone-alkene reactions as significant oxidative sinks for alkenes and a general source of HO_x.

2.8 SULFUR CHEMISTRY

Sulfur chemistry is an integral part of life, owing to its role in plant and human metabolism. Sulfur compounds have both natural and anthropogenic sources. In modern times, the atmospheric sulfur budget has become dominated by anthropogenic emissions, particularly from fossil fuel burning. It is estimated that 75% of the total sulfur emission budget is dominated by anthropogenic sources with 90% of it occurring in the

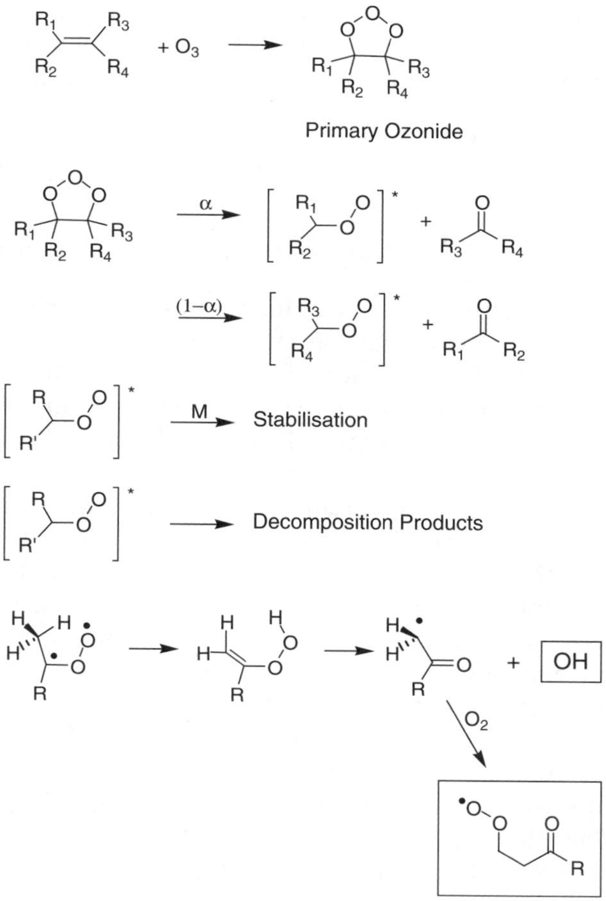

Figure 18 *A schematic representation of the oxidation of an alkene initiated by reaction with ozone (after ref. 14)*

Table 5 *Range of OH yields from the reaction of ozone with alkenes*

Alkene	OH yield
Ethene	0.18 ± 0.06
Propene	0.35 ± 0.07
Methylpropene	0.72 ± 0.12
Δ^3-carene	1.00

Source: All data taken from ref. 32.

Northern Hemisphere. The natural sources include volcanoes, plants, soil and biogenic activity in the oceans.[5] In terms of photochemistry the major sulfur oxide, sulfur dioxide (SO_2) does not photodissociate in the troposphere (*cf.* NO_2), *i.e.*

$$SO_2(X^1A_1) + h\nu(240 < \lambda < 330 \text{ nm}) \rightarrow SO_2(^1A_2, {}^1B_1) \quad (2.64)$$

$$SO_2(X^1A_1) + h\nu(340 < \lambda < 400 \text{ nm}) \rightarrow SO_2(^3B_1) \quad (2.65)$$

The oxidation of sulfur compounds in the atmosphere has implications in a number of different atmospheric problems such as acidification, climate balance and the formation of a sulfate layer in the stratosphere, the so-called Junge layer. By far the largest sulfur component emitted into the atmosphere is SO_2. Figure 19 shows the spatial distribution of SO_2 emissions in 1980 and 2000 from EMEP.[34] In Europe, the source regions for SO_2 are quite apparent, the so-called black triangle region (southern Poland, eastern Germany and the northern part of the Czech Republic) are the largest sources of anthropogenic sulfur pollution. There are a number of other large emission sources including central UK and the Kola Peninsula, also apparent in Figure 19. The absolute maximum in emissions is in southern Italy around Sicily where the largest single source of both natural, the volcano Mt. Etna, and anthropogenic SO_2 is found. SO_2 can be detected from space-borne sensors[35] as a product of volcanic activity and fossil fuel burning. Figure 19 also illustrates another interesting point in that over much of Europe between 1980 and 2000 there has been a decrease of 1000 ton or more owing to legislative limits. Interestingly, with decreasing land emissions the importance of ship emissions has increased (from 5% to 16% of total European SO_2 emissions).

The atmospheric oxidation of SO_2 can take place by a number of different mechanisms, both homogeneously and heterogeneously in the liquid and gas phases (see Figure 20). The gas-phase oxidation of SO_2, *viz*

$$SO_2 + OH + M \rightarrow HSO_3 + M \quad (2.66)$$

$$HSO_3 + O_2 \rightarrow HO_2 + SO_3 \quad (2.67)$$

$$SO_2 + H_2O + M \rightarrow H_2SO_4 \quad (2.68)$$

can lead to the formation of sulfuric acid, which owing to its relatively low vapour pressure can rapidly attach to the condensed phase such as

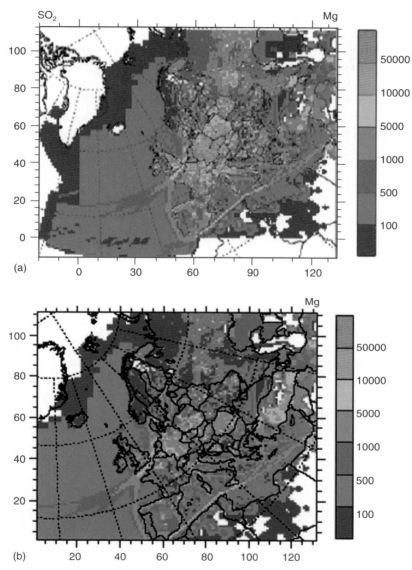

Figure 19 *Spatial distributions of sulfur dioxide emissions in (a) 1980 and (b) 2000.[34] The units are ton year^{-1} grid sq^{-1}*

aerosol particles. The bulk of the H_2SO_4 is lost *via* wet deposition mechanisms in cloud droplets and precipitation. There is another potential gas-phase loss route for SO_2 that can lead to the formation of sulfuric acid in the presence of H_2O, *i.e.* the reaction of SO_2 with Criegee intermediates (see Section 2.7). The aqueous-phase oxidation of SO_2 is more complex, depending on a number of factors such as the nature of the aqueous phase (*e.g.* clouds and fogs), the availabiliity of oxidants (*e.g.* O_3 and H_2O_2) and the availability of light. An overview of the mechanism is given in Figure 20. The key steps include the transport of the gas to the surface of a droplet, transfer across the gas–liquid interface, the formation of aqueous-phase equilibria, the transport from the surface into the bulk aqueous phase and subsequent reaction. In brief, the SO_2 gas is dissolved in the liquid phase, establishing a set of equilibria for a series of S(IV) species, *i.e.* $SO_2 H_2O$, HSO_3^- and SO_3^{2-}.

$$SO_2(g) + H_2O \rightleftharpoons SO_2{\cdot}H_2O(aq) \tag{2.69}$$

$$SO_2{\cdot}H_2O(aq) \rightleftharpoons HSO_3^- + H^+ \tag{2.70}$$

$$HSO_3^- \rightleftharpoons SO_3^{2-} + H^+ \tag{2.71}$$

The solubility of SO_2 is related to the pH of the aqueous phase, decreasing at lower values of pH. The oxidation of sulfur (IV) to sulfur (VI) is a complex process dependent on many physical and chemical factors. The main oxidants seem to be O_2 (catalysed/uncatalysed), O_3, H_2O_2, the oxides of nitrogen and free-radical reactions in clouds and fogs. For example, H_2O_2 is highly soluble in solution so even at relatively low gas-phase concentrations (typically *ca.* 1 ppbv) there is a

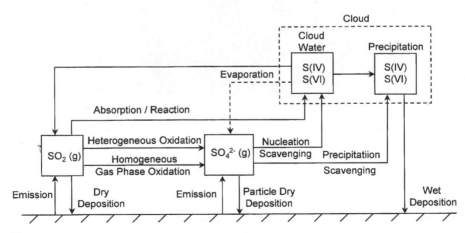

Figure 20 *Summary of emission, oxidation and deposition of S(IV) and S(VI) (after ref. 63)*

significant concentration of H_2O_2 present in solution. The oxidation proceeds as

$$HSO_3^- + H_2O_2 \quad \rightleftharpoons \quad \begin{matrix} O^- \\ \diagdown \\ \diagup \\ O \end{matrix} S-OOH \ + \ H_2O \qquad (2.72)$$

$$\begin{matrix} O^- \\ \diagdown \\ \diagup \\ O \end{matrix} S-OOH \ + HA \quad \rightleftharpoons \quad H_2SO_4 + A^- \qquad (2.73)$$

where HA is an acid. The ubiquitous occurrence of H_2O_2, its solubility, its high reactivity and pH independence (under atmospheric conditions) of the rate constant for the reaction with SO_2 makes H_2O_2 one of the most important oxidants for SO_2 in the troposphere. A more detailed description of aqueous-phase oxidation of SO_2 is given in ref. 36.

2.9 HALOGEN CHEMISTRY

In comparison to the atmospheric chemistry taking place in the stratosphere where halogen chemistry is well known and characterised (see Section 2.10), there has been much debate as to the role of halogen species in the oxidative chemistry of the troposphere. There is growing experimental evidence as to the prevalence of halogen chemistry as part of tropospheric photochemistry (see Table 6).[14,37] Much of the proposed halogen chemistry is propagated through the reactions of a series of halogen atoms and radicals.

Table 6 *Sources of reactive halogen species found in various parts of the troposphere*

Species, site	Likely source mechanism
ClO_x in the polar boundary layer	By-product of the "Bromine Explosion"
ClO_x by salt flats	By-product of the "Bromine Explosion"
BrO_x in the polar boundary layer	Autocatalytic release from sea-salt on ice, "Bromine explosion" mechanism
BrO_x in the Dead sea basin	"Bromine Explosion" (salt pans)
BrO_x in the free troposphere	(1) Photo-degradation of hydrogen-containing organo-halogen species (*e.g.* CH_3Br), (2) "Spill-out" from the boundary layer, (3) Transport from the stratosphere?
BrO_x in the MBL	"Bromine Explosion" mechanism
IO_x in the MBL	(1) Photo-degradation of short-lived organo-halogen species (*e.g.* CH_2I_2) (2) Photolysis of I_2?

Source: After ref. 37.

Bromine oxide species can be formed in the polar boundary layer[38,39] and areas with high salt levels such as the Dead Sea.[40] The major source of gas-phase bromine in the lower troposphere is thought to be the release of species such as IBr, ICl, Br_2 and BrCl from sea-salt aerosol, following the uptake from the gas phase and subsequent reactions of hypohalous acids (HOX, where X = Br, Cl, I).[41]

$$HOBr + (Br^-)_{aq} + H^+ \rightarrow Br_2 + H_2O \qquad (2.74)$$

The halogen release mechanism is autocatalytic[38] and has become known as the "Bromine explosion". The Br_2 produced in reaction (2.74) is rapidly photolysed, producing bromine atoms that can be oxidised to BrO by O_3 (reaction (2.76))

$$BrX + h\nu \rightarrow Br + X \qquad (2.75)$$

$$Br + O_3 \rightarrow BrO + O_2 \qquad (2.76)$$

the resultant BrO reacting with HO_2 (reaction (2.77)) to reform HOBr.

$$BrO + HO_2 \rightarrow HOBr + O_2 \qquad (2.77)$$

Thus, the complete cycle has the form

$$BrO + O_3 + (Br^-)_{aq} + (H^+)_{aq} \overset{surface, HO_x}{\longrightarrow} 2BrO + products \qquad (2.78)$$

where effectively one BrO molecule is converted to two by oxidation of bromide from a suitable surface. Figure 21 shows a schematic representation of the Bromine-explosion mechanism.[37] It is worth noting that the bromine explosion mechanism only occurs from sea salt with a pH < 6.5 therefore requiring acidification of the aerosol potentially caused by the uptake of strong acids likely to be of anthropogenic origins or naturally occurring acids.[42,43] The wide-spread occurrence of these heterogeneous mechanisms are supported by the observed depletion of bromide ions in sea-salt aerosol.[42,43]

Case Study IV – *BrO in the springtime Arctic* – In the spring time in both the Arctic and Antarctic large clouds of BrO-enriched air masses are observable from space (see Figure 22).[44,45] These clouds cover several thousand square kilometres over the polar sea ice with BrO levels up to 30 pptv. The BrO is always coincident with low levels of ozone in the MBL.[38] In order to observe these events, there is a requirement for meteorological conditions that stop mixing between the boundary layer and the free troposphere and sunlight to drive the required photolysis of gaseous bromine (see reaction (2.75)) released heterogeneously through chemical processes on the ice. Because of the prerequisites for strong surface inversions to confine the air and sunlight, episodes of bromine explosion events and boundary layer ozone depletion tend to be confined to spring. It is thought that frost flowers in the Antarctic, which are formed on a liquid layer with high salinity on top of the sea ice surface, are likely to provide a large surface area for the heterogeneous release of reactive bromine but the evidence is not unequivocal.

With respect to iodine chemistry, the major sources of iodine are thought to be from macroalgal sources releasing organoiodine compounds.[46] Photolysis of the organoiodine compounds releases the iodine.

$$RI_x + hv \rightarrow R + I \tag{2.79}$$

$$I + O_3 \rightarrow IO + O_2 \tag{2.80}$$

During daylight hours iodine monoxide, IO exists in a fast photochemical equilibrium with I, *viz.*

$$IO + hv \rightarrow I + O \tag{2.81}$$

The Bromine explosion

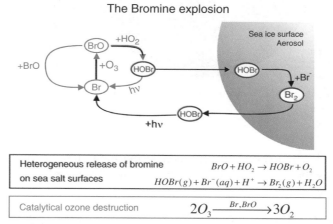

Heterogeneous release of bromine on sea salt surfaces	$BrO + HO_2 \rightarrow HOBr + O_2$ $HOBr(g) + Br^-(aq) + H^+ \rightarrow Br_2(g) + H_2O$
Catalytical ozone destruction	$2O_3 \xrightarrow{Br,BrO} 3O_2$

Figure 21 *A schematic representation of the so-called Bromine explosion mechanism where effectively one BrO molecule is converted to two by oxidation of bromide from a suitable aerosol surface*[64]

GOME BrO, Mar 2000

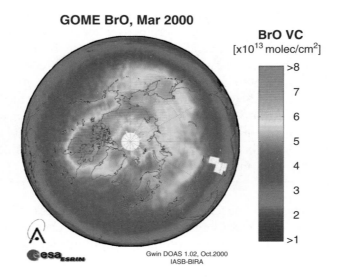

BrO VC
$[\times 10^{13}\,molec/cm^2]$

Gwin DOAS 1.02, Oct.2000
IASB-BIRA

GOME BrO, Oct 1999

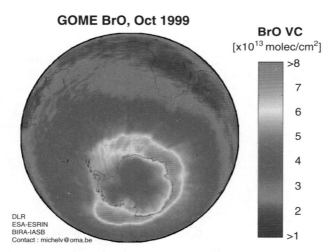

BrO VC
$[\times 10^{13}\,molec/cm^2]$

DLR
ESA-ESRIN
BIRA-IASB
Contact : michelv@oma.be

Figure 22 *[BrO] in the Antarctic and Artic during spring derived from GOME satellite measurements[45]*

The aerosol "explosion" mechanism, previously described for bromine, acts effectively to recycle the iodine back to the gas phase.

Case Study V – *Iodine monoxide chemistry in the coastal margins –* Figure 23 shows the measured concentration of iodine monoxide (IO) at Mace Head in Ireland.[47] The data in Figure 23 show not only a clear requirement for radiation for photochemical production of IO but also a strong correlation with low tide conditions. The correlation with low tidal conditions is indicative of the likely sources of IO, in that the photolysis of organoiodine compounds (reaction (2.79)) such as CH_2I_2 and CH_2IBr emitted from macroalgae in the intertidal zone are possibly the candidate iodine sources. The photolysis lifetime of a molecule such as diodomethane is only a few minutes at midday. Recent experimental observations have suggested the I_2 can be emitted directly from the macroalgae at low tide making this a substantial source of iodine.[48] The chemistry can also lead to an enrichment of iodine in the aerosol in the form of iodate providing a route to move biogenic iodine from ocean to land.[47]

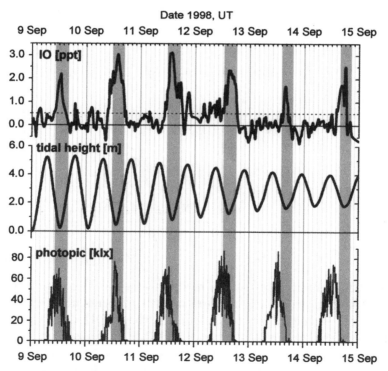

Figure 23 *[IO], tidal height and solar radiation measured at Mace Head in Ireland. The grey areas represent the low-tide areas during the day. The dotted line in the upper panel is the IO detection limit[46,47]*

2.9.1 Tropospheric Halogens and Catalytic Destruction of Ozone

Potentially, the most important effect of reactive halogen species maybe that their chemistry may lead to the catalytic destruction of ozone *via* two distinct cycles

Cycle I:

$$XO + YO \rightarrow X + Y + O_2 \text{ (or } XY + O_2) \tag{2.82}$$

$$\underline{X + O_3 \rightarrow XO + O_2} \tag{2.83}$$

Net:

$$O_3 + O_3 \rightarrow 3O_2 \tag{2.84}$$

In Cycle I, the rate-limiting step involves reaction (2.82), the self- or cross-reaction of the halogen monoxide radicals. Cycle I has been identified to be the prime cause for polar boundary layer ozone destruction.[38] The second-ozone destruction cycle, which is more prevalent at low halogen levels, has the form

Cycle II:

$$XO + HO_2 \rightarrow HOX + O_2 \tag{2.85}$$

$$HOX + h\nu \rightarrow X + OH \tag{2.86}$$

$$X + O_3 \rightarrow XO + O_2 \tag{2.83}$$

$$\underline{OH + CO \rightarrow H + CO_2(+M) \rightarrow HO_2} \tag{2.11, 2.12}$$

$$\text{Net: } O_3 + CO + h\nu \rightarrow CO_2 + O_2 \tag{2.87}$$

The rate-determining step in this reaction sequence is reaction (reaction (2.85)) making ozone destruction linearly dependent on [XO]. The fraction of HOX that photolyses to give back OH depends critically on the accommodation coefficient of HOX on aerosols. Currently, there is a large uncertainty in this parameter. An important side effect of Cycle II is the potential for the reduction of the $[HO_2]/[OH]$ ratio by consumption of HO_2. The inorganic halogen chemistry described is summarised in Figure 24.

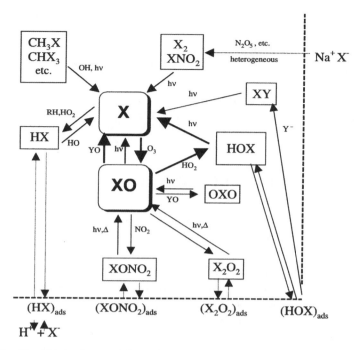

Figure 24 *Simplified scheme of the inorganic halogen reactions in the boundary layer ($X = Cl$, Br, I). The reactive halogen sources are release from sea salt or photochemical degradation of organohalogen species[37]*

Table 7 *Rates of ozone depletion by various catalytic cycles involving XO using measured data typical of MBL just after sunrise*

Rate-determining reaction	Cycle (see text)	O_3 removal rate (ppbv h^{-1})
BrO + IO	I	0.13
IO + HO2	II	0.08
IO + IO	I	0.06
BrO + HO2	II	0.03
BrO + BrO	I	0.01

Source: After ref. 48.
Note: All rates calculated at $T = 295$ K for BrO = 3 pptv, IO = 2 pptv and HO_2 = 3 pptv.

Case Study VI – *The potential effect of tropospheric halogen chemistry on ozone* – An example of the potential effect of the ozone depletion cycles (Cycles I and II) can be assessed using data from the MBL.[49] Table 7 lists the rates of ozone depletion (from Cycles I and II) in the first hour after sunrise at Mace Head in Ireland. Measurements of

[BrO] show a pulse in the first two hours after dawn ascribed to the photolysis of inorganic bromine compounds produced either by the bromine explosion mechanism[41] or the photolysis of mixed bromo/iodo-organohalogens[46] built up overnight. Using measured concentrations of BrO, IO and HO_2, the data in Table 7 show the ozone depletion cycle (Cycle type I) involving the BrO and IO cross-reaction is the most important with an O_3 depletion rate of 0.3 ppbv h^{-1}.

2.10 STRATOSPHERIC CHEMISTRY

The stratosphere is characterised by increasing temperatures with increasing height (see Figure 1). The presence of ozone and oxygen that absorb the UV light and emit infrared radiation heats this region. Approximately, 90% of ozone molecules are found in the stratosphere (see Figure 6). The chemistry of the stratosphere, as compared to that of the troposphere, maybe described as being more chemically simple but initiated by shorter wavelength (more energetic) light.

A British scientist Sydney Chapman[50] suggested the basic ideas of stratospheric ozone in the 1930s, which have become known as the Chapman cycle. Short wavelength UV (hv) can dissociate molecular oxygen and the atomic oxygen fragments produced react with oxygen molecules to make ozone.

$$O_2 + hv \rightarrow O + O \qquad (2.88)$$

$$O + O_2 \xrightarrow{M} O_3 \qquad (2.20)$$

There are some further reactions that complete the picture, which are involved in the interconversion and removal of ozone and atomic oxygen

$$O_3 + hv \rightarrow O + O_2 \qquad (2.89)$$

$$O + O_3 \rightarrow O_2 + O_2 \qquad (2.90)$$

As O and O_3 can rapidly interconvert (reactions (2.89 and 2.90)) they are often referred to as odd oxygen.

Even from this simple chemistry, it is possible to see why there is an ozone *layer* (see Figure 6). For the production of ozone, both UV radiation (hv) and molecular oxygen are required (O_2). High up in the atmosphere there is a plentiful supply of short-wavelength UV radiation but little oxygen, lower down in the atmosphere the opposite situation pertains. Thus, the point where there is both enough molecular oxygen and UV light is where the maximum in the ozone concentrations are going to be found. In the terrestrial atmosphere this point is at a height of about 15–40 km, as shown in Figure 6. When the Chapman oxygen-only mechanism is applied to real atmospheric measurements, it is found that about five times as much ozone is predicted as is measured. This therefore follows that there must be some other, faster, way of removing ozone in the stratosphere. One idea put forward by Bates and Nicolet[51] is that a catalytic process could be speeding up the destruction of ozone. A catalyst is something that can promote a chemical reaction without itself being consumed. In the stratosphere, a catalytic cycle like

$$X + O_3 \rightarrow XO + O_2 \tag{2.83}$$

$$O + XO \rightarrow X + O_2 \tag{2.91}$$

$$\text{Net } O + O_3 \rightarrow O_2 + O_2 \tag{2.90}$$

can lead to the net destruction of ozone (reaction (2.90)). In this case the catalyst is the molecule X, which by destroying ozone is converted into XO, but the XO then reacts with an oxygen reforming X. This, the so-called *chain* process, can proceed many hundreds of thousands of times before termination. As to the identity of X, it is molecules like chlorine atoms (Cl), nitrogen oxide (NO) and the hydroxyl radical (OH) and therefore the corresponding XO is ClO, NO_2 and HO_2, *i.e.*

$$Cl + O_3 \rightarrow ClO + O_2 \tag{2.92}$$

$$O + ClO \rightarrow Cl + O_2 \tag{2.93}$$

These cycles are often referred to as the HO_x, NO_x and ClO_x cycles and the groups of species as families.

On the surface there is now an apparent contradiction; with only the Chapman cycle we produce too much ozone, whereas the catalytic cycles could destroy all the ozone! There are a number of cycles that can interconvert the catalytic cycles without odd oxygen removal these will be in competition with the catalytic cycles, *e.g.*

$$NO + O_3 \rightarrow NO_2 + O_2 \tag{2.24}$$

$$NO_2 + h\nu \rightarrow NO + O \qquad (2.2)$$

$$\text{Net } O_3 + h\nu \rightarrow O_2 + O \qquad (2.89)$$

The products of reaction (2.2) recombining *via* reaction (2.20) to reform ozone. This form of cycle (2.24, 2.2, 2.89) is often referred to as a null cycle. The fraction of the XO_x tied up in null cycles is ineffective as a catalyst.

Another key feature with respect to the effectiveness of catalytic cycles is the formation of reservoir species *via* holding cycles. At any given time about 99% of active Cl is held as reservoir species.

$$Cl + CH_4 \rightarrow HCl + CH_3 \qquad (2.94)$$

$$Cl + H_2 \rightarrow HCl + H \qquad (2.95)$$

$$ClO + NO_2 \xrightarrow{M} ClONO_2 \qquad (2.96)$$

These reservoirs are of great importance to the chemistry of the stratosphere as they act to divert potential catalytic species from active to inactive forms, but they remain available to release the active catalysts again. HCl is the longest-lived and most abundant Cl reservoir species having a lifetime of about one month in the lower stratosphere. It is returned to active Cl largely *via* reaction with OH

$$OH + HCl \rightarrow Cl + H_2O \qquad (2.97)$$

$ClONO_2$ formed *via* reaction (2.96) is destroyed mainly by photolysis or *via* reaction with O atoms, leading to regeneration of active Cl species.

$$ClONO_2 + h\nu \rightarrow Cl + NO_3 \qquad (2.98a)$$

$$ClONO_2 + h\nu \rightarrow ClO + NO_2 \qquad (2.98b)$$

$$O + ClONO_2 \rightarrow \text{Products} \qquad (2.99)$$

The lifetime of $ClONO_2$ is approximately 6 h in the lower stratosphere (< 30 km) decreasing to about an hour at 40 km owing to the increase in UV light. Figure 25 shows the key chemical interconversions of the chlorine chemistry in the stratosphere, delineating the molecules into source, active and reservoir species.

Where do these catalytic species come from and what effect does man have in increasing the levels of these catalysts? The catalytic families of

Chlorine Chemistry in Stratosphere

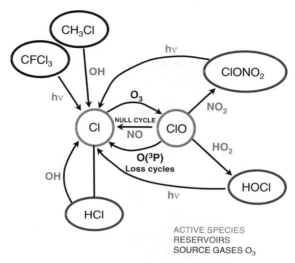

Figure 25 *Inorganic chemistry involved in the interconversion chlorine species in the stratosphere. The relationship between source gases, active species and reservoir species is illustrated*

HO_x, NO_x and ClO_x are present in the natural atmosphere but in the contemporary atmosphere these have been supplemented by anthropogenic sources. For example, with the chlorine species the most abundant natural precursor is methyl chloride, CH_3Cl. There is little interhemispheric difference in the concentration of CH_3Cl indicative of a strong oceanic source, though there is a rising contribution for biomass burning as well as small contribution from volcanic emission.[52] As CH_3Cl has a tropospheric lifetime of a year, it can be transported to the stratosphere where on reaction with OH or *via* photolysis it can release the chlorine. Figure 26 shows the primary sources of chlorine species for the stratosphere separating the anthropogenic and biogenic source categories.

There are a number of potential sources of anthropogenic pollution that could find there way to the stratosphere. One of the first problems that scientists were worried about was Concorde or stratospheric supersonic transport (SST). Aircraft engines produce large amounts of nitrogen oxides (*e.g.* NO) on combustion of the fuel. Injection of these nitrogen oxides directly into the atmosphere could supplement the relevant catalytic cycle plus they can be very long-lived at high altitudes. Initial estimates suggested that SST would lead to substantial losses of

Primary Sources of Chlorine and Bromine for the Stratosphere in 1999

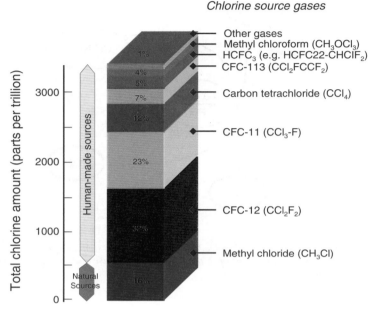

Figure 26 *Stratospheric source gases – A variety of gases transport chlorine into the stratosphere. These gases are emitted from natural sources and human activities. For chlorine, human activities account for most that reaches the stratosphere. The CFCs are the most abundant of the chlorine-containing gases released in human activities. Methyl chloride is the most important natural source of chlorine*[58]

ozone. For two reasons it became apparent that this would not be the case. First, the number of aeroplanes required to observe a loss of ozone would be 1500, flying 7 h a day, 7 days a week. This number of aeroplanes would be able to transport the population of London to New York, or *vice versa* in a matter of weeks. Second, refinements to the understanding of the fundamental chemistry used to make the initial predictions have shown the impact is likely to be less. A warning note should be interposed at this point because the rise in international air travel and the new generation of aircraft means that this problem is back on the scientific agenda.[53,54] The potential of subsonic aircraft to pollute the upper atmosphere is now being studied extensively. Other areas of potential stratospheric pollution have been identified such as NASA space shuttle launches and the extensive use of agricultural fertilisers.

A more serious culprit with respect to the depletion of the atmospheric ozone layer was identified from about 1974 onwards. US scientists postulated[55] that some industrial chemical starting to reach the stratosphere might interfere with the mechanisms of ozone formation and destruction. The chemicals were chlorofluorocarbons, or CFCs, and had been introduced by General Motors in the 1930s mainly as coolants for refrigerators and air conditioning systems, aerosol propellants and blowing agents, *etc.* The perceived virtue of these compounds was their physical properties, chemical inertness and lack of toxicity. The potential problem becomes apparent at higher altitudes where the CFCs (*e.g.* CF_2Cl_2) are not so inert and can be broken down by UV light

$$CF_2Cl_2 + hv \rightarrow Cl + CF_2Cl \tag{2.100}$$

releasing a chlorine atom that can enter the previously described catalytic cycles (reactions (2.92 and 2.93)) and destroy ozone. Fortuitously, as described previously, most of the active chlorine, including that provided by CFC degradation is tied up in the reservoir compounds. There are many myths about CFCs and chlorine in the atmosphere, such as if CFCs are heavier than air how can they make it high up into the atmosphere? The truth is that thousands of measurements have found CFCs high in the atmosphere and though heavier than air, the atmosphere is not still and the winds can carry and mix gases high into the stratosphere.[52] Figure 26 shows the amount and variety of CFC compounds that can act as source gases for chlorine in the stratosphere.

Based on predictions of the effect of CFCs on the ozone layer, in 1987 a previously unprecedented step was taken when many countries signed the UN Montreal protocol specifying the control and phase-out of these ozone-depleting chemicals.[52] Since that time the protocol has been modified in order to speed up the schedule and extend the range of chemical covered to further lessen the effect of these chemicals (see Figure 27). One of the factors that lead to more rapid world action on CFCs was the discovery of the so-called Antarctic ozone *hole*.

2.10.1 The Antarctic Ozone Hole

British scientists from the British Antarctic Survey (BAS) had been making measurements of ozone from their base at Halley Bay (76°S) for many years. They detected a decline in the springtime ozone since 1977, and by October 1984, they had detected a 30% decline in the total ozone.[56] Today, this value has reached 60%. The ozone *hole* begins to develop in August, is fully developed by early October and has normally broken up by early December (see Figure 28).

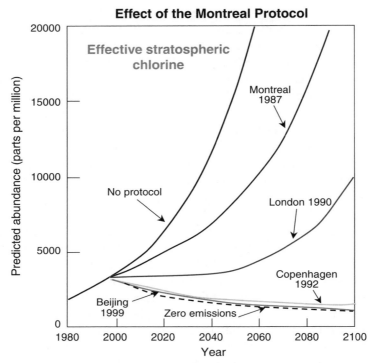

Figure 27   *Predictions for the future abundance of effective stratospheric chlorine are shown in (1) the assuming no Protocol regulations, (2) only the regulations in the original 1987 Montreal Protocol and (3) additional regulations from the subsequent Amendments and Adjustments. The city names and years indicate where and when changes to the original 1987 Protocol provisions were agreed upon. Effective stratospheric chlorine as used here accounts for the combined effect of chlorine and bromine gases. The "zero emissions" line shows stratospheric abundances if all emissions were reduced to zero beginning in 2003*[58]

Figure 29 shows a recent picture of the ozone *hole* taken from space. Strictly speaking the use of the word "*hole*" to describe what happens to ozone in the Antarctic is an exaggeration. There is undoubtedly a massive depletion of ozone, particularly between 12 and 20 km in the Antarctic stratosphere (up to 100%) but the total column of ozone is depleted rather than removed altogether (see Figure 28). The exact location and size of the *hole* varies with meteorological conditions, but the area covered has increased over the past 10 years or so (see Figure 30). Currently, in the austral spring the *hole* extends over the entire Antarctic continent, occasionally including the tip of South America, covering an area equivalent to the North American continent (*ca.* 22 million km^2) (see Figure 31).

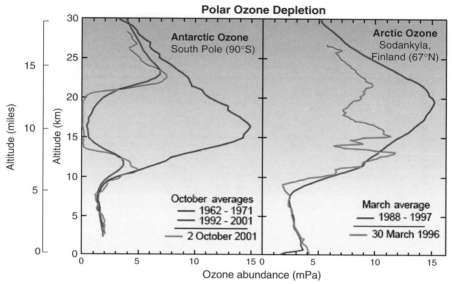

Figure 28 *Antarctic and Artic ozone distributions in the SH and NH spring periods.[58]*
The figure shows the substantial decrease in ozone in the 14–20 km range after
polar sunrise. For the Antarctic the data show there was little depletion before
the 1980s. For the Arctic, there is substantial interannual variability owing to
the temperature of the stratosphere

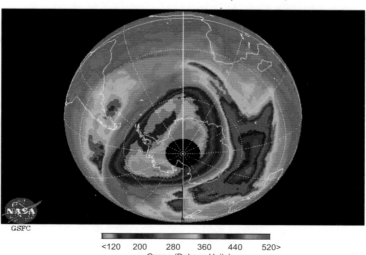

Figure 29 *The 2000 ozone hole over the Antarctic seen by the satellite EP-TOMS (Image*
courtesy of NASA)

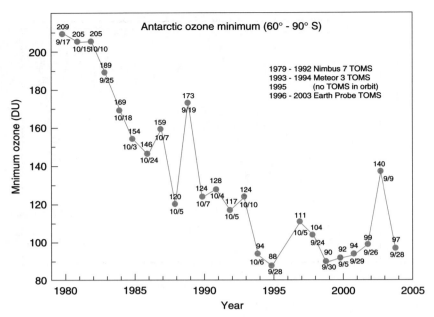

Figure 30 *Lowest values of ozone measured from satellite each year in the ozone hole. Global average ozone is about 300 Dobson Units. Before 1980 ozone less than 200 Dobson Units was rarely seen. In recent years ozone near 100 Dobson Units has become normal in the ozone hole. Ozone in the year 2002 ozone hole was higher than we have come to expect because of unusually high temperatures in the Antarctic stratosphere (from NASA data)*

What happens in Antarctic stratosphere to make it different from many other places? It is colder than the rest of the earth, ozone depletion occurs in the Austral spring when it is particularly cold ($-80°C$). Polar meteorology leads to very strong westerly winds (up to 100 ms^{-1}) forming a stable vortex, which prevents mixing of air from lower latitudes (see Figure 32). The vortex develops a core of very cold air and it is these temperatures that allow polar stratospheric clouds (PSCs) to form. When the sun returns in September, the temperatures begin to rise, the winds weaken and typically this drives the vortex to break down by November. In a sense the air over the pole is more or less confined into what can be thought of as a giant chemical reactor. There is a slow downward circulation (see Figure 32), which draws air down through the vortex modifying this to a notional flow reactor.

In 1986 and 1987, a huge international expedition was mounted to examine the chemistry and meteorology that leads to the polar ozone loss. Many different measurements were made using satellites, aeroplanes and ground-based instruments.[57] The results showed some stark

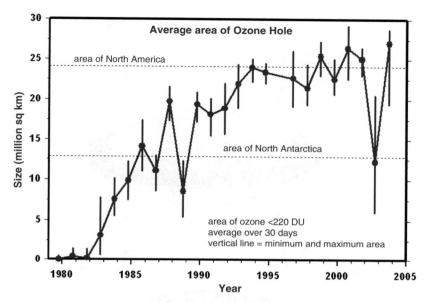

Figure 31   *Growth of the average area of the ozone hole from 1979 to 2003. The ozone hole is defined as the area for which ozone is less than 220 Dobson Units, a value rarely seen under normal conditions. It shows that the ozone in this area hardly occurred at all in 1980, but by year 2000 covered an area larger than North America, 26.5 million km² (from NASA data)*

contrasts between the chemistry inside and outside the vortex before and after sunrise in the Austral spring (see Figure 33). Within the vortex concentrations of *e.g.* ClO were a factor of 10 larger than outside. These measurements also showed that the stratosphere within the disturbed region was abnormally dry and highly deficient in nitrogen oxides.

The reason for the dehydration and denitrification of the Antarctic stratosphere is the formation of the PSCs, whose chemistry perturbs the composition in the Antarctic stratosphere. Polar stratospheric clouds can be composed of small (<1 μm diameter) particles rich in HNO_3 or at lower temperatures (<190 K) larger (10 μm) mainly ice particles. These are often split into two categories, the so-called Type I PSC, which contains the nitric acid either in the form of liquid ternary solutions with water and sulfuric acid or as solid hydrates of nitric acid, or Type II PSCs made of ice particles. The ice crystals on these clouds provide a surface for reactions such as

$$ClONO_2 \text{ (g)} + HCl \text{ (ice)} \rightarrow Cl_2 \text{ (g)} + HNO_3 \text{ (ice)} \qquad (2.101)$$

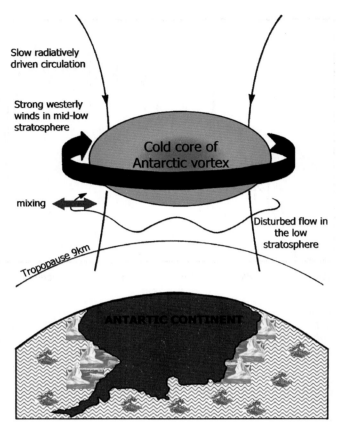

Figure 32 *The Winter vortex over Antarctica. The cold core is almost isolates from the rest of the atmosphere, and acts as a reaction vessel in which chlorine activation may take place in the polar night (after ref. 3)*

$$ClONO_2 \text{ (g)} + H_2O \text{ (ice)} \rightarrow HOCl \text{ (g)} + HNO_3 \text{ (ice)} \quad (2.102)$$

$$HOCl \text{ (g)} + HCl \text{ (ice)} \rightarrow Cl_2 \text{ (g)} + H_2O \text{ (ice)} \quad (2.103)$$

$$N_2O_5 \text{ (g)} + H_2O \text{ (ice)} \rightarrow 2HNO_3 \text{ (ice)} \quad (2.104)$$

In reactions (2.101 and 2.102), the normally inactive reservoir for Cl and ClO is converted into molecular chlorine. It is worth noting that the HNO_3 is left in the ice-phase. These reactions can continue all winter long without light converting inactive chlorine to active chlorine. The two main effects therefore of the heterogeneous chemistry involving

Observations of stratospheric constituents from ER2 aircraft

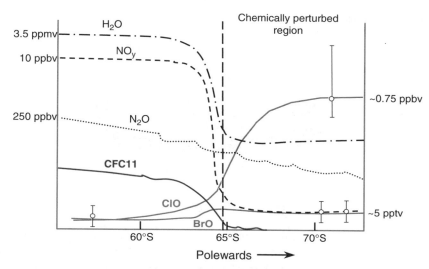

Figure 33 *Schematic representations of the changes in the concentration of some strato-spheric species on entering the chemically perturbed region of the Antarctic vortex (after ref. 57)*

PSCs are the release of active chlorine from its usual reservoirs and the removal of NO_x from the system by the production of HNO_3.

When the sun rises in the spring it triggers the depletion chemistry.

$$Cl_2 + hv \rightarrow Cl + Cl \qquad (2.105)$$

$$Cl + O_3 \rightarrow ClO + O_2 \qquad (2.92)$$

But there is one further unforeseen modification as the concentration of oxygen atoms is too low first thing in spring to regenerate Cl *via* the normal reaction (2.93), so there is instead a further low-temperature reaction that regenerates the chlorine atoms

$$ClO + ClO + M \rightarrow (ClO)_2 + M \qquad (2.106)$$

$$(ClO)_2 + hv \rightarrow Cl + ClOO \qquad (2.107)$$

$$ClOO + M \rightarrow Cl + O_2 + M \qquad (2.108)$$

that can re-enter the catalytic cycle and destroy large amounts of ozone. The absence of significant concentrations of NO_x stops rapid reformation of the ClO_x reservoir species (*e.g. via* reaction (2.96)). The ClO self-reaction ozone destruction cycle is supplemented by a coupled ClO/BrO cycle.

$$ClO + BrO \rightarrow Cl + Br + O_2 \qquad (2.109)$$

or

$$\begin{cases} ClO + BrO \rightarrow BrCl + O_2 \\ BrCl + hv \rightarrow Cl + Br \end{cases} \qquad (2.110, 2.111)$$

$$Cl + O_3 \rightarrow ClO + O_2 \qquad (2.92)$$

$$Br + O_3 \rightarrow BrO + O_2 \qquad (2.76)$$

The stratospheric bromine comes from a range of source gases including CH_3Br.[58] The contribution of bromine to ozone loss in the polar regions has increased faster than that of chlorine because abundances of bromine continue to increase at a time when those of chlorine are levelling off (see Figure 27).[52] Cycles (2.105, 2.92, 2.106–2.108) and (2.76, 2.92, 2.109–2.111) account for most of the ozone loss observed in the Antarctic stratosphere in the late winter/spring season. At high ClO abundances, the rate of ozone destruction can reach 2–3% day^{-1} in late winter/spring.

Figure 34 shows a summary of the photochemistry and dynamics in the polar stratosphere, illustrating the time profiles of the key chlorine species coupled to the temperature requirements in the vortex. Figure 35 shows measurements of a range of chemical species and temperature in the 2004 ozone hole. The data reflects the main chemical and physical features on the ozone hole.

In summary, the key features of Antarctic ozone loss are

   (i) The circulating winds in the polar regions enable the formation of a stable vortex which provides a gigantic "reaction vessel" for ozone depletion to occur within.
  (ii) The low temperatures in the vortex encourage the formation of PSCs, which enhance the production of active chlorine species.
 (iii) Pre-conditioning of the atmosphere takes place during the polar winter releasing chlorine from reservoir molecules.
 (iv) At sunrise, molecular chlorine is dissociated into free atoms that can destroy the ozone.
  (v) Chlorine atoms are regenerated by the dimer reactions.

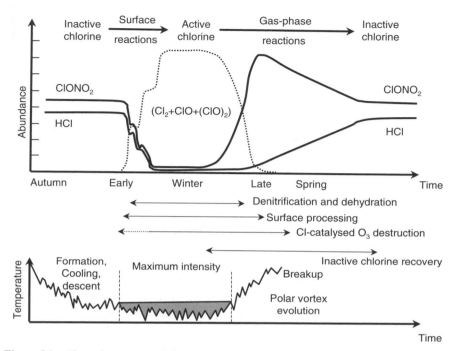

Figure 34 *Photochemistry and dynamics in the polar stratosphere (after ref. 65)*

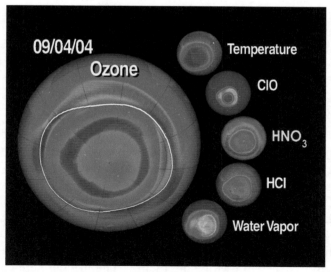

Figure 35 *Chemical species and temperature as measured by satellite in the 2004 ozone hole (image courtesy of NASA). The white line on the ozone figure gives the extent of the vortex*

(vi) The removal of nitric acid *(and other oxides of nitrogen)*, by slow, gravitational settling means that the reservoir species cannot be regenerated. The HNO_3 is liberated within the particle.

In more recent times, there have been discoveries of ozone depletion in the Arctic that occur by similar mechanisms as the ones described here (see Figure 28). The Arctic equivalent does not tend to be as dramatic owing to the fact the Artic stratosphere does not get as cold as the Antarctic, mainly owing to a less well-formed vortex, largely owing to northern hemisphere topography.[52]

2.11 SUMMARY

The chemistry of the atmosphere is diverse, driven in the main by the interaction of light with a few molecules that drives a complex array of chemistry. The type and impact of atmospheric chemistry varies in concert with the physical and biological change throughout the atmosphere. An integral understanding of atmospheric chemistry within the earth system context underpins many contemporary global environmental problems and is therefore vital to sustainable development.

This chapter has outlined the main features of the chemistry of the troposphere and stratosphere. In some senses it must be remembered that our understanding of atmospheric chemistry is still evolving, but the science expressed gives entry into the rich and changing world of atmospheric science. In the coming years we need to understand the interaction of atmospheric chemistry with climate as well as better understand the multi-phase impacts of the chemistry. There is still much to discover, rationalise and understand in atmospheric chemistry.

QUESTIONS

(i) The earth system is made up of the atmosphere, hydrosphere, lithosphere and biosphere. Describe, with examples, the chemical interactions between these different elements of the earth system.

(ii) The amount of any chemical substance in the atmosphere is dependent on the source–transport–sink relationship for that compound. Describe the different types of sources of trace gases in the atmosphere.

(iii) Discuss the following statement: "The atmosphere can be split into different regions according to different physical or chemical properties".

(iv) Explain the following statement: "Ozone can be photolysed both in the troposphere and stratosphere leading to quite different chemistry".

(v) Why is the stratosphere separate from the troposphere?

(vi) Discuss the chemistry of ozone in the stratosphere in terms of the relationships between catalytic species and the different types of reservoir chemistry.

(vii) Explain the main chemical and physical features of the so-called ozone "hole".

(viii) The hydroxyl radical (OH) has different role in the stratosphere and troposphere. Give an example of the reactions of OH in troposphere and stratosphere and describe their significance.

(ix) What role does the boundary layer play in the mixing of gases?

(x) What are the main reactions that contribute to the formation of the so-called photochemical smog?

(xi) What physical, chemical and social effects can aggravate the formation of photochemical smog?

(xii) Peroxyacetylnitrate (PAN) is of interest as a characteristic product of tropospheric photochemistry.

(i) How is PAN formed in the atmosphere?

(ii) The unimolecular decomposition of PAN to peroxy acetyl radicals and NO_2 is strongly temperature dependent. The following expression gives the temperature dependence of the unimolecular rate constant:

$$k = 1.0 \times 10^{17} \exp\left(-\frac{14,000}{T}\right).$$

If the lifetime is given by $1/k$, calculate the atmospheric lifetime of PAN at $T=310, 298, 290$ and $280\,\mathrm{K}$.

(iii) In the light of your answer to part (b), comment on the potential of PAN as a reservoir for reactive nitrogen in remote regions.

(xiii) Explain the following:

(i) In the shadow of a thick cloud the concentration of OH falls to nearly zero.

(ii) On a global scale the concentration distribution of CO_2 is about the same whereas the concentration of isoprene varies from one location to the next.

(iii) There is more than one chemical pathway to the initiation of night-time chemistry.

(iv) Both HONO and HCHO are detected in abundance in areas suffering from photochemical smog and may lead to the acceleration of the production of smog.

(v) The effect that NO_x from aeroplanes and lightning can have on ozone can be quite different from NO_x emitted from the surface of the earth.

(xiv) The figure below shows two sets of ozone profiles taken over Halley Bay in Antarctica during Winter/Spring.

(i) What measurement platforms are normally used to take an ozone profile?

(ii) Why under normal conditions (August 15th) is the ozone maximum at 20 km?

(iii) Describe the key processes that have contributed to the loss of ozone by October 13th.

(iv) What molecules are thought to be responsible for the ozone loss shown in the figure?

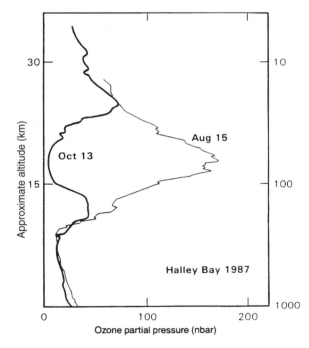

(xv) The ozone continuity equation below is a reasonable representation of the processes that controls ozone in the boundary layer, *viz*

$$\frac{d[O_3]}{dt} = C + \frac{E_v([O_3]_{ft} - [O_3])}{H} + \frac{v_d[O_3]}{H}$$

C is a term representative of the photochemistry (production or destruction), E_v the entrainment velocity, $[O_3]_{ft}$ the concentration of free tropospheric ozone, v_d the dry deposition velocity and H the height of the boundary layer. The ozone budget shown in Table 8 has been calculated both in the summer and winter using the ozone continuity equation for a site in the marine boundary layer.

(i) Describe the chemistry involved in calculating the term C in the ozone continuity equation.

(ii) Given the magnitude of the photochemical terms in Table 8, comment on the amount of available NO_x.

(iii) How does the ability to make or destroy ozone change with season?

(iv) What factors affect the magnitude of a deposition velocity?

(v) What role does entrainment play in the boundary layer?

(vi) What are the weaknesses of using this approach for the calculation of chemical budgets?

(vii) Predict the likely diurnal cycle of ozone and hydrogen peroxide in the remote marine boundary layer.

(xvi) Nitric acid has many potential loss processes in the troposphere. Given that the reaction

$$OH + HNO_3 \rightarrow H_2O + NO_3$$

has a rate constant of 2.0×10^{-13} cm^3 molecule^{-1} s^{-1} and the hydroxyl radical concentration is 4×10^6 molecule cm^{-3}. The

Table 8 *Calculated average ozone removal and addition rates according to pathway, ppbv day^{-1} (upper part) and fractional contributions to overall production or destruction patyways (lower part)*

| Pathway | O_3 removal | | O_3 addition | |
	Photochemistry (%)	Deposition (%)	Photochemistry (%)	Entrainment (%)
Summer	1.19	0.18	0.56	2.1
Winter	0.61	0.35	0.29	0.1
Summer	87	13	21	79
Winter	64	36	74	26

Table 9   *Photoprocess data for HNO$_3$*

Wavelength (nm)	F(λ) (photon cm^{-2} s^{-1})	σ (cm^2 molecule^{-1})	ϕ
295–305	2.66×10^{13}	0.409×10^{-20}	1
305–315	4.20×10^{14}	0.146×10^{-20}	1
315–325	1.04×10^{15}	0.032×10^{-20}	1
325–335	1.77×10^{15}	0.005×10^{-20}	1
335–345	1.89×10^{15}	0	1
345–355	2.09×10^{15}	0	1

deposition velocity for HNO$_3$ is 0.005 cm s^{-1} over a 1 km well-mixed boundary layer and the photoprocess data given in Table 9. What is the atmospheric lifetime of HNO$_3$?

REFERENCES

1. A.K. Smith, *J. Atmos. Sol.-Terr. Phy.*, 2004, **66**, 839.
2. J.M.C. Plane, *Chem. Rev.*, 2003, **103**, 4963.
3. R.P. Wayne, *Chemistry of Atmospheres*, 3rd edn, Oxford University Press, UK, 2000.
4. R.A. Cox and R.G. Derwent, *Specialist Periodical Reports Chem. Soc.*, 1981, **4**, 189.
5. A.V. Jackson, in *Handbook of Atmospheric Science*, Blackwell publishing, Oxford, 2003, Chapter 5.
6. R.R. Stull, *An Introduction to Boundary Layer Meteorology*, Kluwer Academic Publishers, London, UK, 1988.
7. C.E. Junge, *Tellus*, 1962, **14**, 363.
8. P. Fabian and P.G. Pruchniewz, *J. Geophys. Res.*, 1977, **82**, 2063.
9. P.J. Crutzen, *Pure Appl. Geophys.*, 1973, **106–108**, 1399.
10. W.L. Chameides and J.C.G. Walker, *J. Geophys. Res.*, 1973, **78**, 8760.
11. A. Volz and D. Kley, *Nature*, 1988, **332**, 240.
12. IPCC, *Climate Change 2001: The Scientific Basis*, Cambridge University Press, Cambridge, UK, 2001.
13. A.M. Thompson, *Science*, 1992, **256**, 1157.
14. P.S. Monks, *Chem. Soc. Rev.*, 2005, **34**, 376.
15. P.A. Leighton, *The Photochemistry of Air Pollution*, Academic Press, New York, 1961.
16. J.M. Roberts, Reactive odd-nitrogen (NOy) in the atmosphere, in *Composition and Climate of the Atmosphere*, H.B. Singh (ed), Van Nostrand Reinhold, New York, 1995.
17. J. Lelieveld and P.J. Crutzen, *Nature*, 1990, **343**, 227.
18. S. Sillman, *Atmos. Environ.*, 1999, **33**, 1821.

19. L. Jaeglé, D.J. Jacob, W.H. Brune, I.C. Faloona, D. Tan, Y. Kondo, G.W. Sasche, B. Anderson, G.L. Gregory, S. Vay, H.B. Singh, D.R. Blake and R. Shetter, *Geophys. Res. Lett.*, 1999, **26**, 3081.

20. (a) G.P. Ayers, S.A. Penkett, R.W. Gillett, B.J. Bandy, I.E. Galbally, C.P. Meyer, C.M. Elsworth, S.T. Bentley and B.W. Forgan, *Nature*, 1992, **360**, 446–448; (b) P.S. Monks, G. Salisbury, G. Holland, S.A. Penkett and G.P. Ayers, *Atmos. Environ.*, 2000, **34**, 2547–2561.

21. C.N. Hewitt (ed), *Reactive Hydrocarbons in the Atmosphere*, Academic Press, London, 1999.

22. N. Poisson, M. Kanakidou and P.J. Crutzen, *J. Atmos. Chem.*, 2000, **36**, 157.

23. S. Houweling, F. Dentener and J. Lelieveld, *J. Geophys. Res. (Atmos.)*, 1998, **103**, 10673.

24. A.J. Haagen-Smit, *Ind. Eng. Chem.*, 1952, **44**, 1342.

25. G. Lammel and J.N. Cape, *Chem. Soc. Rev.*, 1996, **25**, 361.

26. H.B. Singh, D. O'Hara, D. Herlth, J.D. Bradshaw, S. Sandholm, G.L. Gregory, G.W. Sachse, D.R. Blake, P.J. Crutzen and M. Kanakidou, *J. Geophys. Res.*, 1992, **97**, 16511.

27. B.J. Allan, N. Carslaw, H. Coe, R.A. Burgess and J.M.C. Plane, *J. Atmos. Chem.*, 1999, **33**, 129.

28. R.P. Wayne, I. Barnes, P. Biggs, J.P. Burrows, C.E. Canosa-Mas, J. Hjorth, G. LeBras, G.K. Moortgat, D. Perner, G. Poulet, G. Restelli and H. Sidebottom, *Atmos. Environ.*, 1991, **25**, 1.

29. B.J. Allan, G. McFiggans, J.M.C. Plane, H. Coe and G.G. McFadyen, *J. Geophys. Res.*, 2000, **105**, 24191.

30. R. Criegee, *Agnew. Chem. Internat. Edit.*, 1975, **14**, 745.

31. J.G. Calvert, R. Atkinson, J.A. Kerr, S. Madronich, G.K. Moortgat, T.J. Wallington and G. Yarwood, *The Mechanisms of Atmospheric Oxidation of the Alkenes*, Oxford University Press, Oxford, 2000.

32. S.E. Paulson, M.Y. Chung and A.S. Hasson, *J. Phys. Chem. A*, 1999, **103**, 8125.

33. G. Salisbury, A.R. Rickard, P.S. Monks, B.J. Allan, S. Bauguitte, S.A. Penkett, N. Carslaw, A.C. Lewis, D.J. Creasey, D.E. Heard, P.J. Jacobs and J.D. Lee, *J. Geophys. Res. (Atmos.)*, 2001, **106**, 12669.

34. EMEP, in *EMEP Assessment, Part I: European Perspective*, G. Lövblad, L. Tarrasón, K. Tørseth and S. Dutchak (eds), Oslo, October, 2004.

35. M. Eisinger and J.P. Burrows, *Geophys. Res. Lett.*, 1998, **22**, 4177.

36. B.J. Finlayson-Pitts and J.N. Pitts Jr., *Chemistry of the Upper and Lower Atmosphere*, Academic Press, San Diego, 2000.
37. U. Platt and G. Honniger, *Chemosphere*, 2003, **52**, 325–338.
38. L.A. Barrie, J.W. Bottenheim, R.C. Schnell, P.J. Crutzen and R.A. Rasmussen, *Nature*, 1988, **334**, 138.
39. S.M. Fan and D.J. Jacob, *Nature*, 1992, **359**, 522.
40. K. Heibestreit, J. Stutz, D. Rosen, V. Matveiv, M. Peleg, M. Luria and U. Platt, *Science*, 1990, **283**, 55.
41. R. Vogt, P.J. Crutzen and R. Sander, *Nature*, 1996, **383**, 327.
42. G.P. Ayers, R.W. Gillett, J.M. Cainey and A.L. Dick, *J. Atmos. Chem.*, 1999, **33**, 299.
43. R. Sander, W.C. Keene, A.A.P. Pszenny, R. Arimoto, G.P. Ayers, E. Baboukas, J.M. Cainey, P.J. Crutzen, R.A. Duce, G. Hönninger, B.J. Huebert, W. Maenhaut, N. Mihalopoulos, V.C. Turekian and R. Can Dingenen, *Atmos. Chem. Phys.*, 2003, **3**, 1301.
44. (a) T. Wagner and U. Platt, *Nature*, 1998, **395**, 486–490; (b) T. Wagner, C. Leue, M. Wenig, K. Pfeilsticker and U. Platt, *J. Geophys. Res.*, 2001, **106**, 24225–24235.
45. M. Van Roozendael, C. Fayt, J.-C. Lambert, I. Pundt, T. Wagner, A. Richter and K. Chance, Development of a bromine oxide product from GOME, *Proceedings of ESAMS' 99-European Symposium on Atmospheric Measurements from Space*, ESTEC, Noordwijk, The Netherlands, 18–22 January 1999, WPP-161, 1999, 543–547.
46. L.J. Carpenter, *Chem. Rev.*, 2003, **103**, 4953.
47. L.J. Carpenter, K. Hebestreit, U. Platt and P.S. Liss, *Atmos. Chem. Phys.*, 2001, **1**, 9–18.
48. A. Saiz-Lopez and J.M.C. Plane, *Geophys. Res. Lett.*, 2004, **31**, L04112.
49. A. Saiz-Lopez, J.M.C. Plane and J.A. Shillito, *Geophys. Res. Lett.*, 2004, **31**, L03111.
50. S. Chapman, *Mem. Roy. Meterol. Soc.*, 1930, **3**, 103.
51. D.R. Bates and M. Nicolet, *J. Geophys. Res.*, 1950, **55**, 301.
52. WMO, Scientific assessment of ozone depletion: 2002, Global Ozone Research and Monitoring Project, Report No. 47, Geneva, 2003.
53. IPCC, Aviation and the Global Atmosphere, Special Report, Intergovernmental Panel on Climate Change, 1999.
54. S.R. Kawa, J.G. Anderson, S.L. Baughcum, C.A. Brock, W.H. Brune, R.C. Cohen, D.E. Kinnison, P.A. Newman, J.M. Rodriguez, R.S. Stolarski, D. Waugh and S.C. Wofsy, Assessment of the effects of high-speed aircraft in the stratosphere: 1998, NASA Tech. Pub. 1999-209236, NASA Goddard Space Flight Center, Greenbelt, MD, 1999.

55. M.J. Molina and F.S. Rowland, *Nature*, 1974, **249**, 810.
56. J.C. Farman, B.G. Gardiner and J.D. Shanklin, *Nature*, 1985, **315**, 207.
57. J.G. Anderson, D.W. Toohey and W.H. Brune, *Science*, 1991, **251**, 39.
58. WMO, Twenty questions and answers about the ozone layer, in *WMO, Scientific Assessment of Ozone Depletion: 2002*, Global Ozone Research and Monitoring Project, Report No. 47, Geneva, 2003.
59. D. Ehhalt, *Phys. Chem. Chem. Phys.*, 1999, **24**, 5401.
60. J.N. Howard, J.I.F. King and P.R. Gast, Thermal radiation, *Handbook of Geophysics*, Macmillan, New York, 1960.
61. P.S. Monks, A.R. Rickard, S.L. Hall and N.A.D. Richards, *J. Geophys. Res.*, 2004, **109**, D17206.
62. P.D. Lightfoot, R.A. Cox, J.N. Crowley, M. Destriau, G.D. Hayman, M.E. Jenkin, G.K. Moortgat and F. Zabel, *Atmos. Environ.*, 1992, **10**, 1805.
63. D. Lamb, D.F. Miller, N.F. Robinson and A.W. Gertler, *Atmos. Environ.*, 1987, **21**, 2333.
64. U. Frieß, Spectroscopic measurements of atmospheric trace gases at Neumayer station, Antarctica, Ph.D. Thesis, University of Heidelberg, Germany, 2001.
65. WMO, Scientific assessment of ozone depletion: 1998, Global Ozone Research and Monitoring Project, Report No. 44, Geneva, 1999.

CHAPTER 3

Chemistry of Freshwaters

MARGARET C. GRAHAM AND JOHN G. FARMER

School of GeoSciences, University of Edinburgh, Crew Building, King's Buildings, West Mains Road, Edinburgh EH9 3JN

3.1 INTRODUCTION

Water is essential for all life on Earth. As the Earth's population continues to increase rapidly, the growing human need for freshwater (*e.g.* for drinking, cooking, washing, carrying wastes, cooling machines, irrigating crops, receiving sewage and agricultural runoff, recreation, and industrial purposes) is leading to a global water resources crisis. Rees[1] has commented that 'there is a growing consensus that if current trends continue, water scarcity and deteriorating water quality will become the critical factors limiting future economic development, the expansion of food production, and the provision of basic health and hygiene services to millions of disadvantaged people in developing countries'. In 2003, the UN International Year of Freshwater, 20% of the world's people lacked good drinking water and 40% lacked adequate sanitation facilities.[2]

The inventory of water at the Earth's surface (Table 1)[3] shows that the oceans, ice caps, and glaciers contain 98.93% of the total, with groundwater (1.05%) accounting for most of the rest, and lakes and rivers amounting to only 0.009% and 0.0001%, respectively.[3] The average annual water withdrawal for use by humans is currently about 4000 km^3, of which 69% is used by agriculture, 21% by industry/power, and 10% for domestic purposes.[2] Figures vary widely between continents and countries. On a per capita basis, North America withdraws seven times as much freshwater as Africa. The USA (41% agriculture, 46% industry/power, 13% domestic) withdraws almost 20% of its renewable freshwater resources each year, in comparison with 7% for the UK (3%, 75%, 22%), 22% for China (68%, 26%, 7%), 33% for Poland (8%, 79%, 13%), 51%

Table 1 *Inventory of water at the Earth's surface*

Reservoir	Volume 10^6 km^3 (10^{18} kg)	Per cent of total
Oceans	1400	95.96
Mixed layer	50	
Thermocline	460	
Abyssal	890	
Ice caps and glaciers	43.4	2.97
Groundwater	15.3	1.05
Lakes	0.125	0.009
Rivers	0.0017	0.0001
Soil moisture	0.065	0.0045
Atmosphere total[a]	0.0155	0.001
Terrestrial	0.0045	
Oceanic	0.0110	
Biosphere	0.002	0.0001
Approximate total	1459	

Source: Global Environment: Water Air Geochemical Cycles by Berner/Berner, © 1996.
Adapted by permission of Prentice-Hall, Upper Saddle River, NJ.[3]
[a] As liquid volume equivalent of water vapour.

for India (86%, 5%, 8%), and 76% for Bangladesh (96%, 1%, 3%).[2] Many developing countries are using up to 40% of their renewable freshwater for irrigation.[2] As a consequence of population growth, pollution and climate change, the average supply of water per person is likely to drop by one-third over the next two decades,[4] with a global supply crisis projected to occur between 2025 and 2050, although much earlier for some individual countries.[1,2,4]

The quality of freshwater is also of concern. For example, about 75% of Europe's drinking water is taken from groundwater,[2] a resource described as overexploited in almost 60% of European industrial and urban centres and threatened by pollutants.[5] River and lake eutrophication caused by excessive phosphorus and nitrogen from agricultural, domestic, and industrial effluents is a problem across most of Europe. The Eastern European countries in particular fare badly when judged against the criteria of pathogenic agents, organic matter, salinization, nitrate, phosphorus, heavy metals, pesticides, acidification, and radioactivity, and even the Nordic countries of Norway, Sweden, and Finland, despite having generally better freshwater conditions than the rest of Europe, suffer from acidification.[5] In Asia, 86% of all urban wastewater and 65% of all wastewater is discharged, untreated, into the aquatic environment.[6] There is also the unexpected and insidious chemical contamination of apparently clean drinking water supplies, exemplified most notably by the arsenic contamination of groundwater that affects millions in Bangladesh and other Asian countries.[2]

The effects of contaminants or pollutants on freshwater depend upon their chemical, physical, and biological properties, their concentrations and duration of exposure. Aquatic life may be affected indirectly (*e.g.* through depletion of oxygen caused by biodegradation of organic matter) or directly, through exposure to toxic or carcinogenic chemicals, some of which may accumulate in organisms. Such toxicity may, however, be modified by the presence of other substances and the characteristics of the particular water body. For example, metal toxicity is affected by pH, which influences speciation; dissolved organic carbon has been shown to reduce bioavailability by forming metal complexes and by subsequent adsorption to particulate matter in freshwater.[7] The toxicity of many heavy metals to fish is also inversely related to water hardness, with Ca^{2+} competing with free metal ions for binding sites in biological systems.

In biologically productive lakes which also develop a thermocline in summer, the bottom water may become depleted in oxygen, leading to a change from oxidizing to reducing conditions. As a result, the potential exists for remobilization of nutrients and metals from bottom sediments to the water column, one more example of how alterations in master variables in the hydrological cycle, in this case the redox status, can affect the fate and influence of pollutants.

A further demonstration of the importance of fundamental properties of both pollutants and water bodies is provided by the behaviour of chemicals upon reaching a groundwater aquifer. Soluble chemicals, such as nitrate, move in the same direction as groundwater flow. A poorly soluble liquid which is less dense than water, such as petrol, spreads out over the surface of the water table and flows in the direction of the groundwater. Poorly soluble liquids which are denser than water, such as various chlorinated solvents, sink below the water table and may flow separately along low permeability layers encountered at depth in the aquifer and not necessarily in the same direction as that of the overlying groundwater.[7]

The rest of this chapter therefore consists of two major sections, first on fundamental aquatic chemistry of relevance to the understanding of pollutant behaviour in the aquatic environment and second on associated case studies and examples drawn from around the world, including reference to water treatment methods where appropriate.

3.2 FUNDAMENTALS OF AQUATIC CHEMISTRY

3.2.1 Introduction

3.2.1.1 Concentration and Activity. All natural waters contain dissolved solutes. In order to understand their chemical behaviour in aqueous systems, it is first necessary to consider the relationship between solute concentration and solute activity.

The activity of a solute is a measure of its observed chemical behaviour in (aqueous) solution. Interactions between the solute and other species in solution lead to deviations between solute activity {i} and concentration [i]. An activity coefficient, γ_i, is therefore defined as a correction factor, which interrelates solute activity and concentration.

$$\{i\} = \gamma_i \, [i] \tag{3.1}$$

Concentration and activity of a solute are only the same for very dilute solutions, *i.e.* γ_i approaches unity as the concentration of all solutes approaches zero. For non-dilute solutions, activity coefficients must be used in chemical expressions involving solute concentrations.[†] Although freshwaters are sufficiently dilute to be potable (containing less than about 1000 mg L^{-1} total dissolved solids (TDS)),[8] it cannot be assumed that activity coefficients are close to unity.

Calculation of activities using the expression in Equation (3.1) would appear to be straightforward, requiring only a value for the activity coefficient and the concentration of the solute, i. The activity coefficients for individual ions, γ_i, cannot, however, be measured. The presence of both cations and anions in solution means that laboratory experiments quantifying either the direct (*e.g.* salt solubility) or indirect (*e.g.* elevation of boiling point, depression of freezing point, *etc.*) effects of the presence of solutes in aqueous solution generally lead to the mean ion activity coefficients, $\gamma_\pm$. Traditionally, single-ion activity coefficients have been obtained from the mean values, *e.g.* using the mean-salt method (Example 3.1),[9,10] but this approach has been shown to be inadequate because it ignores ion pairing effects at higher solute concentrations.[11,12] This is generally more significant for anions than for cations because the extent of anion association at higher solute concentration is much greater.[13]

Example 3.1: Determination of single-ion activity coefficients for Ca^{2+} from mean values for $CaCl_2$ using the mean-salt method.

The solubility of calcium chloride in water can be expressed as
$$CaCl_{2(s)} \rightleftharpoons Ca^{2+}_{(aq)} + 2Cl^-_{(aq)} \qquad K_{SP}$$

[†] Concentration is expressed on the molar scale in terms of mol L^{-1} of solvent or on the molal scale in terms of mol kg^{-1} of solvent. The molal scale gives concentrations that are independent of temperature and pressure. In this chapter, the molar scale will be used on the basis that molarity and molality are almost identical for the low ionic strengths commonly associated with freshwaters.

where the solubility product,

$$K_{SP} = \{Ca^{2+}\}\{Cl^-\}^2 = \gamma_{Ca^{2+}}\gamma_{Cl^-}^2 [Ca^{2+}][Cl^-]^2$$

We define the mean ion activity coefficient as the geometric mean of the single-ion activity coefficients

$$\gamma_{\pm CaCl_2} = (\gamma_{Ca^{2+}}\gamma_{Cl^-}\gamma_{Cl^-})^{1/3}$$

The mean ion activity coefficient values can be obtained from experiments where the effect of electrolyte concentration on the K_{SP} value for a salt is determined. The mean values are then compared with those for KCl under the same solution conditions. The single-ion activity coefficient for Ca^{2+} can then be computed if an assumption is made about the individual values for K^+ and Cl^-. These ions have the same magnitude of charge and similar electronic configuration, ionic radii, and ionic mobilities. On the basis of these properties, the MacInnes convention (1919) states that

$$\gamma_{K^+} = \gamma_{Cl^-} = \gamma_{\pm KCl}$$

In general: $\gamma_{cation} = (\gamma_{\pm cation Cl_{x-1}})^x / (\gamma_{\pm KCl})^{x-1}$

Thus $\gamma_{\pm CaCl_2} = (\gamma_{Ca^{2+}}\gamma_{Cl^-}^2)^{1/3} = (\gamma_{Ca^{2+}}\gamma_{\pm KCl}^2)^{1/3}$

and $\gamma_{Ca^{2+}} = (\gamma_{\pm CaCl_2})^3 / (\gamma_{\pm KCl})^2$

For $I = 0.1$ mol l^{-1}: $\gamma_{Ca^{2+}} = (0.518)^3 / (0.770)^2 = 0.234$

(Mean salt data from Robinson and Stokes.[9])

Although other methods using such experimental data have been documented,[14] theoretical and/or empirical considerations have led to various expressions involving a dependence on ionic strength, I, which is a major factor influencing the activity of solutes in aqueous solution.

3.2.1.2 Ionic Strength. Ionic strength, I, is a measure of the charge density in solution and is defined as

$$I = 0.5 \sum_i ([i]z_i^2) \tag{3.2}$$

where [i] is the concentration of solute ion, i; z_i is the integral charge associated with solute ion, i.

Ionic strength, expressed in terms of mol L^{-1}, is most accurately determined by carrying out a total water analysis, *i.e.* quantification of the concentrations of all ionic species in solution and calculation of I using Equation (3.2). It may, however, be estimated from measurements of either the TDS in solution or, preferably, the specific conductance (SpC) of the solution, *e.g.* $I = 2.5 \times 10^{-5}$ TDS (mg L^{-1}) or $I = 1.7 \times 10^{-5}$ SpC (μS cm^{-1}).[12]

The simplest theoretical relationship between I and γ_i is expressed in the Debye–Hückel (DH) Equation (3.3).[‡] The DH model assumes that ions can be represented as point charges, *i.e.* of infinitely small radius, and that long-range coulombic forces between ions of opposite charge are responsible for the differences between the observed chemical behaviour, *i.e.* activity, and the predicted behaviour on the basis of solute concentration.

$$\log \gamma_i = -0.5 z_i^2 \sqrt{I} \quad \text{DH equation} \tag{3.3}$$

$$\log \gamma_i = -0.5 z_i^2 \left(\sqrt{I} / \left(1 + 0.33\, a_i \sqrt{I} \right) \right) \quad \text{Extended DH equation} \tag{3.4}$$

$$\log \gamma_i = -0.5 z_i^2 \left(\sqrt{I} / \left(1 + \sqrt{I} \right) \right) \text{Güntelberg approximation} \tag{3.5}$$

An infinitely dilute solution is defined as having an ionic strength $< 10^{-5}$ mol L^{-1} and activity coefficients calculated by any of the above equations would have a value of unity. For freshwaters where I is in the range 10^{-5}–$10^{-2.3}$ mol L^{-1}, activity coefficients can be calculated using the DH Equation (3.3). Where I is in the range $10^{-2.3}$–10^{-1} mol L^{-1}, however, the values calculated using Equation (3.3) differ from experimental data and this stems from the assumption that the solute ions are point charges. The extended DH expression (3.4) includes a parameter, a_i, which is related to ion size. Example 3.2 illustrates the use of both DH and extended DH expressions to calculate single-ion activity coefficients.

[‡] The general forms of the DH and extended DH equations are $\log \gamma_i = -A z_i^2 \sqrt{I}$ and $\log \gamma_i = -A z_i^2 (\sqrt{I}/(1 + B a_i \sqrt{I}))$, respectively, where A and B are temperature dependent constants. A $= 1.824928 \times 10^6 \rho_0^{0.5} (\varepsilon T)^{-1.5} \sim 0.5$ and B $= 50.3 (\varepsilon T)^{-0.5} \sim 0.33$ at 298 K, where ρ_0 is the density and ε is the dielectric constant of water at 298 K.

Example 3.2: Calculation of single-ion activity coefficients using the DH and extended DH equations.

If a solution contains 0.01 mol L^{-1} $MgSO_4$, 0.006 mol L^{-1} Na_2CO_3, and 0.002 mol L^{-1} $CaCl_2$ then the ionic strength of solution is

$$I = 0.5(0.01 \times (+2)^2 + 0.01 \times (-2)^2 + 0.006 \times 2 \times (+1)^2$$
$$+ 0.006 \times (-2)^2 + 0.002 \times (+2)^2 + 0.002 \times 2 \times (-1)^2)$$
$$= 0.5(0.04 + 0.04 + 0.012 + 0.024 + 0.008 + 0.004)$$
$$= 0.064 \, mol \, L^{-1}$$

Calculating the single-ion activity coefficient for Na^+ using both the DH and the extended DH equations (use a value of 4.5 for the ion size parameter, a_i)

DH $\log \gamma_{Na^+} = -0.5(+1)^2 \sqrt{(0.064)} = -0.126$ $\gamma_{Na^+} = 0.747$

Ex-DH $\log \gamma_{Na^+} = -0.5(+1)^2 \sqrt{(0.064)}$
$$/(1 + 0.33 \times 4.5 \times \sqrt{(0.064)})$$
$$= -0.0919 \qquad\qquad \gamma_{Na^+} = 0.809$$

At $I = 0.064$ mol L^{-1}, the single-ion activity coefficient calculated by the DH model is $\sim 8\%$ lower than that calculated by the extended DH model. Now consider a solution which has an ionic strength of only 0.001 mol L^{-1}.

DH $\log \gamma_{Na^+} = -0.0158$ $\gamma_{Na^+} = 0.964$

Ex $-$ DH $\log \gamma_{Na^+} = -0.0151$ $\gamma_{Na^+} = 0.966$

Clearly, at lower ionic strength, the single-ion activity coefficient is much closer to unity. Also, the DH and the extended DH models give almost exactly the same value. This is because the denominator $(1+0.33a_i\sqrt{I})$ of the extended DH equation approaches a value of unity, *i.e.* $0.33a_i\sqrt{I}$ approaches zero, as the ionic strength decreases towards zero. In other words, the two equations become identical at very low ionic strength.

It should be noted that, for solutions of ionic strength $<10^{-1}$ mol L^{-1}, either Equation (3.4) or the Güntelberg approximation (3.5), which incorporates an average value of 3 for a_i, can be used.[13] The Güntelberg approximation is particularly useful in calculations where a number of ions are present in solution or when values of the ion size parameter are poorly defined.

For solutes in higher ionic strength solutions, however, Equations (3.3)–(3.5) tend to underestimate activity coefficients in part because of the assumptions that are inherent in the DH and extended DH Equations, *e.g.* only long-range coulombic interactions, ion size parameters are independent of ionic strength, ions of like charge do not interact, each ion can construct its own atmosphere of ions of opposite charge.[8,12,15,16] Additional factors are that (i) short-range interactions (including those between ions of like charge and also between ions and neutral molecules),[12,15–17] rather than the long-range coulombic interactions, between ions become increasingly important and (ii) the number of water molecules involved in ion hydration spheres means that the activity of water decreases to a value significantly less than unity.[12] In order to compensate particularly for the effect of (ii), various empirical correction factors have subsequently been added to the extended DH expression.

$$\log \gamma_i = -0.5\, z_i^2 \left(\sqrt{I}/(1 + 0.33\, a_i \sqrt{I}) - bI \right) \quad \text{Truesdell – Jones equation} \quad (3.6)$$

$$\log \gamma_i = -0.5\, z_i^2 \left(\sqrt{I}/(1 + \sqrt{I}) - bI \right) \quad \text{Davies equation} \quad (3.7)$$

The Truesdell–Jones[18] equation contains an empirical term, bI, as well as the ion size parameter, a_i. Values of both a_i and b are often obtained by fitting this equation to the individual activity coefficients obtained from mean salt data using the MacInnes convention (see Example 3.1).[9,10] This equation is used in some geochemical computer models, *e.g.* PHREEQE.[19] The Davies equation, which also incorporates an empirical term bI but uses an average ion size parameter ($a_i = 3$), is more commonly used because the ion size parameter is not well defined for all ions. The Davies equation can be used to calculate activity coefficients for monovalent and divalent ions in aqueous solutions, where $I = 0.1$–0.7 mol L^{-1} and $I = 0.1$–$\sim$$0.3$ mol L^{-1}, respectively. Originally, Davies used a value of 0.2 for b but later suggested that the value be increased to 0.3.[20,21] The geochemical computer model MINTEQA2 employs the Davies equation to calculate single-ion activity coefficients but uses a value of 0.24 for b.[22]

Specific Ion Interaction Theory (SIT) is a more advanced empirical approach that was developed by Guggenheim[23] and extended by Scatchard and others.[24–26] The SIT equation has an extended DH term that accounts for the non-specific long-range interactions but also has an ion- and electrolyte-specific correction factor, ε_{ij} [j], for each pair of ions, i and j, of opposite charge. Brønsted[27] also recognized the importance of short-range interactions in higher ionic strength solutions and Equation (3.8) is often referred to as the Brønsted–Guggenheim equation.

$$\log \gamma_i = \text{`Extended DH'} + \Sigma_j \varepsilon_{ij} \text{ [j]} \quad (3.8)$$

Originally the extended DH term was similar to that in Equation (3.5), *i.e.* with an average ion size parameter, $a_i = 3$. Scatchard[28] showed that a better fit with experimental data was obtained using an average value of $a_i = 4.6$, *i.e.* $0.33a_i = 1.5$, and so the extended DH expression used in the SIT model is often

$$\log \gamma_i = -0.510 \, z_i^2 \, \sqrt{I}/(1 + 1.5\sqrt{I}) \qquad (3.9)$$

Overall, however, the SIT approach is limited in application to intermediate ionic strength solutions (0.1–3.5 mol L^{-1}).

The Pitzer model can be used to obtain activity coefficients for solutes in low (<0.1 mol L^{-1}), intermediate (0.1–3.5 mol L^{-1}) and high (>3.5 mol L^{-1}) ionic strength solutions. The Pitzer equations[29] include terms for binary and ternary interactions between solute species as well as a modified DH expression. The general formula is

$$\ln \gamma_i = z_i^2 f^{\gamma} + \sum_j D_{ij} \, [j] + \sum_{jk} E_{ijk} \, [j][k] + \dots \qquad (3.10)$$

where f^{γ} is the modified DH term

$$f^{\gamma} = -0.392 \, (\sqrt{I}/(1 + 1.2\sqrt{I}) + (2/1.2) \ln(1 + 1.2 \, \sqrt{I})) \quad (3.11)$$

and the D_{ij} terms describe interactions between pairs of ions, i and j, while the E_{ijk} terms describe interactions among three ions, i, j, and k. Higher-order terms can also be added to this general formula. A key difference between the SIT and Pitzer models is that interactions between ions of like charge are also included. For ionic strength solutions of up to ~ 3.5 mol L^{-1}, there is good agreement between the two models but, at >3.5 mol L^{-1}, the ternary and higher-order terms in the Pitzer equations become more important and this model gives the more accurate ion activity coefficients.[12]

Although naturally occurring brines and some high ionic strength contaminated waters may require the more complicated expressions developed in the Davies, SIT, or Pitzer models, the use of Equations (3.3)–(3.5) is justified for the ionic strengths of many freshwaters.

Finally, this section has focused primarily on ions in aqueous solution but it should be remembered that aqueous solutions also contain important dissolved neutral species, *e.g.* O_2, CO_2, H_2CO_3, $Si(OH)_4$. For non-ideal solutions, the Setschenow Equation (1899) has traditionally been used to describe the ionic strength dependence of the activity coefficients for dissolved neutral species and can be written as

$$\log \gamma_i = k_i \, [i] \qquad (3.12)$$

where k_i is the temperature dependent salting out coefficient for molecular species, i.

At higher ionic strength values, an additional dependence on $[i]^2$ is often required to fit the observed solubility data.[17] Alternatively, the Pitzer model can be used with a high degree of accuracy to describe the short-range binary (neutral–neutral, neutral–cation, neutral–anion) and ternary (neutral–neutral–neutral, neutral–cation–anion) interactions between ions and neutral species in single and mixed electrolyte solutions.[17]

3.2.1.3 Equilibria and Equilibrium Constants. Many of the important reactions involving solutes in freshwaters can be described by equilibria. This approach means that an equilibrium constant, K, relating the activities of the solutes, can be defined for each stoichiometric equilibrium expression.

$$a\,A + b\,B \rightleftharpoons c\,C + d\,D \tag{3.13}$$

$$K = \{C\}^c\,\{D\}^d\,/\,\{A\}^a\,\{B\}^b \tag{3.14}$$

The equilibrium constant can also be defined in terms of the concentrations of the solutes.

$$K = (\gamma_c\,[C])^c\,(\gamma_d\,[D])^d/(\gamma_a\,[A])^a\,(\gamma_b\,[B])^b \tag{3.15}$$

For infinitely dilute solutions, $\{\,\} = [\,]$ and the equilibrium constant can be written as

$$K = [C]^c\,[D]^d/[A]^a\,[B]^b \tag{3.16}$$

For solutions of fixed ionic strength, or, for example, where major ions in solution, *e.g.* conservative cations and anions, are present at concentrations several orders of magnitude greater than the species involved in the chemical equilibrium, *e.g.* A, B, C, and D in Equation (3.13), it can be assumed that the solute activity coefficients are also constants and can be incorporated into the equilibrium constant. The equilibrium constant for a fixed ionic strength aqueous solution is termed a constant ionic strength equilibrium constant, ^cK.

$$^cK = [C]^c\,[D]^d/[A]^a\,[B]^b \text{ where } ^cK = (\gamma_a^a\,\gamma_b^b\,/\,\gamma_c^c\,\gamma_d^d)\,K \tag{3.17}$$

This again enables the use of concentrations rather than activities in equilibrium calculations.

It is sometimes advantageous to use a mixture of activities and concentrations and a mixed equilibrium constant, K', is defined as

$$K' = \{C\}^c\,[D]^d\,/\,[A]^a\,[B]^b \quad \text{where } K' = (\gamma_a^a\,\gamma_b^b\,/\,\gamma_d^d)\,K \tag{3.18}$$

For example, a mixed acidity constant is frequently used where pH has been measured according to the IUPAC convention as the activity of hydrogen ions but the concentrations of the conjugate acid-base pair are used.

The relationship between K and ^cK or K and K' as defined above can be used to calculate the effect of ionic strength of solution on the true equilibrium constant, K. ^cK and K' can be calculated using the Equations (3.3)–(3.5) together with experimentally determined values of K (Example 3.3).

Example 3.3: Calculation of the mixed acidity constant, K', using the Güntelberg approximation and the Davies equation.

For an acid-base equilibrium:
$$HA \rightleftharpoons H^+ + A^- \qquad K = \{H^+\}\{A^-\}/\{HA\}$$

$$K' = (\gamma_{HA}/\gamma_{A^-})\, K$$

Taking logs:

$$\log K' = \log(\gamma_{HA}/\gamma_{A^-}) + \log K$$

$$pK' = pK + \log \gamma_{A^-} - \log \gamma_{HA}$$

Using the Güntelberg approximation:

$$pK' = pK + 0.5\left(z_{HA}^2 - z_{A^-}^2\right)\left(\sqrt{I}/(1 + \sqrt{I})\right)$$

Using the Davies equation:

$$pK' = pK + 0.5\left(z_{HA}^2 - z_{A^-}^2\right)\left(\sqrt{I}/(1 + \sqrt{I}) - 0.3I\right)$$

Now calculate the mixed acidity constant, K', at ionic strength values of (i) 0.05 mol L^{-1} using the Güntelberg approximation and (ii) 0.4 mol L^{-1} using the Davies equation for a monoprotic acid, HA, with $K = 6.8 \times 10^{-8}$:

(i)
$$pK' = 7.17 + 0.5\,(0 - (-1)^2)\,(\sqrt{0.05}/(1 + \sqrt{0.05}))$$
$$= 7.17 - 0.0914 = 7.08$$
$$K' = 8.34 \times 10^{-8}$$

(ii)
$$pK' = 7.17 + 0.5(0 - (-1)^2)\,(\sqrt{0.4}/(1 + \sqrt{0.4}) - 0.3 \times 0.4)$$
$$= 7.17 - 0.5\,(0.387 - 0.12) = 7.04$$
$$K' = 9.19 \times 10^{-8}$$

The values for the corrected constants, K', are 22% and 35% greater, respectively, than the infinite dilution constant, K.

At higher ionic strengths, the activity coefficients could be calculated using the SIT or Pitzer equations and thus a more accurate correction made to the infinite dilution constant, K.

Finally, it should be noted that all values of the equilibrium constants apply only at a specified temperature. In Sections 3.2.2–3.2.4, the values of the equilibrium constants apply at 298 K unless stated otherwise.

3.2.2 Dissolution/Precipitation Reactions

Section 3.2.1 has provided relationships between solute activities and concentrations in solution in order that solute behaviour can be quantified. This section discusses dissolution and precipitation reactions that impart or remove solutes to/from natural waters, and therefore modify the chemical composition of natural waters.

3.2.2.1 Physical and Chemical Weathering Processes. Initial steps involving physical weathering by thermal expansion and contraction or abrasion lead to the disintegration of rock. Disintegration increases the surface area of the rock, which, in the presence of water, can undergo chemical weathering. Water acts not only as a reactant but also as a transporting agent of dissolved and particulate components, and so weathering processes are extremely important in the hydrogeochemical cycling of elements. Mineral dissolution reactions often involve hydrogen ions from mineral or organic acids, *e.g.* acid hydrolysis of Na-feldspar in Equation (3.21). Alternatively, the transfer of electrons (sometimes simultaneously with proton transfer) may promote the dissolution of minerals, *e.g.* reductive dissolution of biotite.

Chemical weathering of minerals results not only in the introduction of solutes to the aqueous phase but often in the formation of new solid phases. Dissolution is described as congruent, where aqueous phase solutes are the only products, or incongruent, where new solid phase(s) in addition to aqueous phase solutes are the products. These reactions can be represented by equilibria, where the equilibrium constant is related only to the activities of the aqueous solutes, *i.e.* an assumption is made that the activities of solid phases have the value of unity. Examples of weathering of primary minerals, those formed at the same time as the parent rock, are shown in (3.19)–(3.22):

Congruent dissolution of quartz

$$SiO_{2(s)} + 2\,H_2O \rightleftharpoons Si(OH)^0_{4(aq)} \tag{3.19}$$
$$\text{quartz} \qquad\qquad \text{silicic acid}$$

$$K = \{Si(OH)_4^0\} \tag{3.20}$$

At pH < 9, the equilibrium between quartz and silicic acid can be represented as in (3.19) and the value of K at 298 K is 1.05×10^{-4} mol L^{-1},[30] indicating that the solubility of quartz is low. At higher pH values, the dissociation of silicic acid results in the increased solubility of quartz (see Section 3.2.4.1).

Incongruent dissolution of Na-feldspar

$$2\,NaAlSi_3O_{8(s)} + 2\,H^+ + 9\,H_2O \rightleftharpoons Al_2Si_2O_5(OH)_{4(s)} + 2\,Na^+_{(aq)}$$
$$\text{Na-feldspar} \qquad\qquad\qquad \text{kaolinite}$$
$$+\,4\,Si(OH)^0_{4(aq)} \tag{3.21}$$
$$\text{silicic acid}$$

$$K = \{Si(OH)^0_4\}^4\{Na^+\}^2/\{H^+\}^2 \tag{3.22}$$

Alteration of primary minerals such as the feldspars gives rise to secondary minerals such as kaolinite ($Al_2Si_2O_5(OH)_4$), smectites (*e.g.* $Na_{0.33}Al_{2.33}Si_{3.67}O_{10}(OH)_2$), and gibbsite ($Al(OH)_3$). As the composition of the new mineral phases and the potential for further alteration of secondary minerals are a function of the prevailing geochemical conditions, many equilibrium expressions must be used to fully describe chemical weathering processes in natural systems. For example, reactions leading to the formation of kaolinite from the primary mineral, Na-feldspar, as well as alteration of the secondary mineral, *e.g.* to gibbsite, must be considered. The composition of the aqueous phase is of major importance in determining the nature of the solid products formed during chemical weathering. In particular, the solution activities of silicic acid, metal ions, and hydrogen ions are key parameters influencing the formation and alteration processes of new solid phases (Example 3.4).[8]

Example 3.4: Investigation of the stability relationship between gibbsite and kaolinite in natural waters at pH < 7.

Consider the acid hydrolysis weathering reaction undergone by Na-feldspars (pH < 7):

$$NaAlSi_3O_{8(s)} + H^+_{(aq)} + 7H_2O \rightleftharpoons Na^+_{(aq)} + Al^{3+}_{(aq)} + 3OH^-_{(aq)} \\ + 3Si(OH)_{4(aq)}$$

This equilibrium expression represents congruent dissolution of Na-feldspar but the solution quickly becomes saturated with respect to the solid phase $Al(OH)_3$ (gibbsite $K_{SP} = 10^{-33.9}$) and so the secondary mineral, gibbsite, is formed.

$$NaAlSi_3O_{8(s)} + H^+_{(aq)} + 7H_2O \rightleftharpoons Al(OH)_{3(s)} + Na^+_{(aq)} + 3Si(OH)_{4(aq)}$$

The concentration of silicic acid in solution is a key parameter influencing the stability of gibbsite. In the presence of even low concentrations of silicic acid, *e.g.* 10^{-4} mol L^{-1}, kaolinite would be formed instead.

$$2NaAlSi_3O_{8(s)} + 2H^+_{(aq)} + 9H_2O \rightleftharpoons Al_2Si_2O_5(OH)_{4(s)} + 2Na^+_{(aq)} \\ + 4Si(OH)_{4(aq)}$$

In natural systems, the role of water as a transporting agent is important. Geochemical conditions promoting removal of silicic acid from the solid-water interface (*e.g.* water flow) favour the formation of gibbsite over kaolinite.
(Adapted from Drever.[8])

3.2.2.2 Solubility. Dissolution of minerals during chemical weathering releases species into solution but aqueous phase concentrations are limited by the solubility of solid phases, such as amorphous silica, gibbsite, and metal salts. For example, the solubility of Al^{III} and Ca^{II} may be limited by the formation of gibbsite and gypsum, respectively.

$$Al(OH)_{3 \text{ gibbsite}} \rightleftharpoons Al^{3+}_{(aq)} + 3OH^-_{(aq)} \tag{3.23}$$

$$CaSO_4.2H_2O_{\text{gypsum}} \rightleftharpoons Ca^{2+}_{(aq)} + SO^{2-}_{4(aq)} + 2H_2O \tag{3.24}$$

The equilibrium solubilities are defined by the respective solubility products, K_{SP} (values from ref. 30).

$$K_{SP} = \{Al^{3+}\}\{OH^-\}^3 = 10^{-33.9} \tag{3.25}$$

$$K_{SP} = \{Ca^{2+}\} \{SO_4^{2-}\} = 10^{-4.58} \tag{3.26}$$

A solution is considered to be undersaturated, saturated, or oversaturated with respect to a solid phase, for example gypsum, if $K_{SP} > \{Ca^{2+}\}\{SO_4^{2-}\}_{observed}$, $K_{SP} = \{Ca^{2+}\}\{SO_4^{2-}\}_{observed}$, or $K_{SP} < \{Ca^{2+}\} \{SO_4^{2-}\}_{observed}$, respectively.

3.2.2.3 Influence of Organic Matter. Natural organic matter is present in most natural systems. With respect to weathering processes, its importance in freshwaters and indeed soils and sediments in contact with freshwaters can be attributed to the presence of organic acids, which include low molecular weight compounds, *e.g.* oxalic acid, and extremely complex, high molecular weight coloured compounds described as humic substances – a heterogeneous mixture of polyfunctional macromolecules ranging in size from a few thousand to several hundred thousand Daltons (Da). In addition to inorganic acids, *e.g.* H_2CO_3, these provide hydrogen ions for the acid hydrolysis of minerals. As well as promoting dissolution, natural organic matter can influence the formation of new mineral phases.[31]

3.2.3 Complexation Reactions in Freshwaters

In this section, as a starting point, it is assumed that all species in solution are in hydrated form, *e.g.* $M^{III}(H_2O)_6^{3+}$, where the six water molecules form the first co-ordination sphere of the metal ion, M^{III}. The hydrated form is often represented as, for example, $M_{(aq)}^{3+}$.

3.2.3.1 Outer and Inner Sphere Complexes. Outer sphere complexation involves interactions between metal ions and other solute species in which the co-ordinated water of the metal ion and/or the other solute species are retained. For example, the initial step in the formation of ion pairs, where ions of opposite charge approach within a critical distance and are then held together by coulombic attractive forces, is described as outer sphere complex formation.

$$Mg_{(aq)}^{2+} + SO_{4(aq)}^{2-} \rightleftharpoons (Mg^{2+}(H_2O)(H_2O)SO_{4(aq)}^{2-}) \tag{3.27}$$

The formation of ion pairs is influenced by the nature of the oppositely charged ions, the ionic strength of solution, and ion charge. Ion pairs are generally formed between hard (low polarizability) metal cations, *e.g.* Mg^{2+}, Fe^{3+}, and hard anions, *e.g.* CO_3^{2-}, SO_4^{2-}, and ion pair formation is generally most significant in high ionic strength aqueous phases. In freshwaters, which are frequently of low ionic strength, ion charge is a key parameter. Greater coulombic interactions occur between oppositely charged ions with high charge.

Inner sphere complexation involves interactions between metal ions and other species in solution which possess lone pairs of electrons. Inner sphere complexation involves the transfer of at least one lone pair of electrons. Those species which possess electron lone pairs are termed ligands and reactions may involve inorganic or organic ligands.

3.2.3.2 Hydrolysis. In aqueous systems, hydrolysis reactions are an important example of inner sphere complexation for many metal ions. The interaction between hydrated metal ions and water can be written as a series of equilibria for which the equilibrium constants are denoted $*K_n$.[§]

$$Al^{3+}_{(aq)} + H_2O_{(1)} \rightleftharpoons Al(OH)^{2+}_{(aq)} + H^+_{(aq)} \quad *K_1 = 10^{-4.95} \quad (3.28)$$

$$Al(OH)^{2+}_{(aq)} + H_2O_{(1)} \rightleftharpoons Al(OH)^+_{2(aq)} + H^+_{(aq)} \quad *K_2 = 10^{-5.6} \quad (3.29)$$

$$Al(OH)^+_{2(aq)} + H_2O_{(1)} \rightleftharpoons Al(OH)_{3(aq)} + H^+_{(aq)} \quad *K_3 = 10^{-6.7} \quad (3.30)$$

$$Al(OH)_{3(aq)} + H_2O_{(1)} \rightleftharpoons Al(OH)^-_{4(aq)} + H^+_{(aq)} \quad *K_4 = 10^{-5.6} \quad (3.31)$$

Hydrolysis and, more generally, complex formation equilibria may be described by cumulative stability constants, β_n. Using the Equations (3.28)–(3.31)

$$Al^{3+}_{(aq)} + H_2O_{(1)} \rightleftharpoons Al(OH)^{2+}_{(aq)} + H^+_{(aq)} \quad *\beta_1 = *K_1 = 10^{-4.95} \quad (3.32)$$

[§]The hydrolysis of hydrated metal ions can also be written as the interaction between the hydrated metal ion and the hydroxyl ligand. For example

$$Al^{3+}_{(aq)} + OH^-_{(aq)} \rightleftharpoons Al(OH)^{2+}_{(aq)} \quad K_1$$

where K_1 and $*K_1$ are related as follows

$$Al^{3+}_{(aq)} + OH^-_{(aq)} \rightleftharpoons Al(OH)^{2+}_{(aq)} \quad K_1$$

$$H_2O_{(1)} \rightleftharpoons H^+_{(aq)} + OH^-_{(aq)} \quad K_w$$

so

$$Al^{3+}_{(aq)} + H_2O_{(1)} \rightleftharpoons Al(OH)^{2+}_{(aq)} + H^+_{(aq)} \quad *K = K_1 K_w$$

The values for $*K_n$ are those obtained by Wesolowski and Palmer;[32] all values are given in unitless form but fundamentally are defined in terms of the products of the activities of ions on the molar or molal scales.

$$Al^{3+}_{(aq)} + 2H_2O_{(l)} \rightleftharpoons Al(OH)^+_{2(aq)} + 2H^+_{(aq)}$$
$$*\beta_2 = *K_1 *K_2 = 10^{-10.55} \quad (3.33)$$

$$Al^{3+}_{(aq)} + 3H_2O_{(l)} \rightleftharpoons Al(OH)_{3(aq)} + 3H^+_{(aq)}$$
$$*\beta_3 = *K_1 *K_2 *K_3 = 10^{-17.25} \quad (3.34)$$

$$Al^{3+}_{(aq)} + 4H_2O_{(l)} \rightleftharpoons Al(OH)^-_{4(aq)} + 4H^+_{(aq)}$$
$$*\beta_4 = *K_1 *K_2 *K_3 *K_4 = 10^{-22.85} \quad (3.35)$$

Clearly, these hydrolysis reactions are dependent on hydrogen ion activity. The relationship between speciation and pH, which is influenced by the characteristics of the metal ion, will be discussed further in Section 3.2.4.1.

3.2.3.3 Inorganic Complexes. The main inorganic ligands in oxygenated freshwaters, in addition to OH^-, are HCO_3^-, CO_3^{2-}, Cl^-, SO_4^{2-}, and F^-, and, under anoxic conditions, HS^- and S^{2-} (see Section 3.2.4.4). The stability of complexes formed between metal ions and inorganic ligands depends on the nature of the metal ion as well as the properties of the ligand. Table 2 shows some of the major complexed metal species involving inorganic ligands in oxygenated freshwaters at pH 8.

3.2.3.4 Surface Complex Formation. Metal ions form both outer and inner sphere complexes with solid surfaces, *e.g.* hydrous oxides of iron, manganese, and aluminium. In addition, metal ions, attracted to charged surfaces, may be held in a diffuse layer, which, depending upon ionic strength, extends several nanometres from the surface into solution.

Table 2   *Major species in freshwaters*

Metal ion	Major species	$[M^{x+}_{(aq)}]/[M_{TOT\,(aq)}]$
Mg^{II}	Mg^{2+}	0.94
Ca^{II}	Ca^{2+}	0.94
Al^{III}	$Al(OH)_3^0$, $Al(OH)_2^+$, $Al(OH)_4^-$	1×10^{-9}
Mn^{IV}	MnO_2^0	–
Fe^{III}	$Fe(OH)_3^0$, $Fe(OH)_2^+$, $Fe(OH)_4^-$	2×10^{-11}
Ni^{II}	Ni^{2+}, $NiCO_3^0$	0.4
Cu^{II}	$CuCO_3^0$, $Cu(OH)_2^0$	0.01
Zn^{II}	Zn^{2+}, $ZnCO_3^0$	0.4
Pb^{II}	$PbCO_3^0$	0.05

Source: Adapted from Sigg and Xue,[33] and reproduced with permission from W. Stumm and J.J. Morgan, *Aquatic Chemistry* 3rd Edn, © Wiley, 1996.

Diffuse layer metal retention and outer sphere complex formation involve electrostatic attractive forces, which are characteristically weaker than co-ordinative interactions leading to inner sphere surface complex formation. A number of factors influence metal interactions with surfaces, including the chemical composition of the surface, surface charge, and the nature and speciation of the metal ion. The importance of the pH of the aqueous phase in these interactions will be discussed further in Section 3.2.4.1.

3.2.3.5 Organic Complexes. Dissolved organic matter consists of a highly heterogeneous mixture of compounds, including low molecular weight acids and sugars as well as the high molecular weight coloured compounds termed humic substances. Humic substances are secondary synthetic products derived from decaying organic debris. Although they are structurally poorly defined, it is accepted that large numbers of mainly oxygen-containing functional groups are attached to a flexible, predominantly carbon backbone. Individual organic molecules, in particular those of humic substances, can provide more than one functional group for complex formation with a hydrated metal ion. Ligands which provide two and three functional groups are termed bidentate and tridentate, respectively. Complexes where more than one functional group from the same organic molecule is involved are more stable than those where functional groups are from discrete organic molecules. The concentration of dissolved organic matter in freshwaters is generally low (2–6 mg C L^{-1}) and humic substances, comprising molecules which possess large numbers of functional groups in numerous different chemical environments, are implicated as the component of natural organic matter most important in metal binding.[31]

3.2.4 Species Distribution in Freshwaters

3.2.4.1 pH as a Master Variable. pH is one of the key parameters which influences species distribution in aqueous systems. Many equilibrium expressions contain a hydrogen ion activity term, *e.g.* acidity constants, complexation constants. It is therefore useful to consider the relationship between the activity of the species of interest (*e.g.* contaminant metal ions, organic pollutants, naturally occurring inorganic and organic solutes, and weakly acidic functional groups on mineral surfaces) and the activity of hydrogen ions. For example, an acid – conjugate base pair can be represented as HA and A$^-$, respectively.

$$HA \rightleftharpoons A^- + H^+ \qquad (3.36)$$

$$K = \{A^-\}\{H^+\}/\{HA\} \tag{3.37}$$

The activity of acid initially in solution, C, termed the analytical activity, is equal to the sum of the equilibrium activities of the acid and its conjugate base.

$$C = \{HA\} + \{A^-\} \tag{3.38}$$

Equations (3.37) and (3.38) can be combined to give an expression for the activity of the conjugate base in terms of the equilibrium constant, the analytical activity, which is also a constant, and the hydrogen ion activity.

$$K = \{A^-\}\{H^+\}/(C-\{A^-\}) \tag{3.39}$$

$$\{A^-\} = KC/(K + \{H^+\}) \tag{3.40}$$

Similarly an expression for the activity of the acid can be written as

$$\{HA\} = \{H^+\}C/(K + \{H^+\}) \tag{3.41}$$

Alternatively, the activities of acid and conjugate base can be represented as a fraction of the analytical activity.

$$\alpha_0 = \{HA\}/C = \{H^+\}/(K + \{H^+\}) \tag{3.42}$$

$$\alpha_1 = \{A^-\}/C = K/(K + \{H^+\}) \tag{3.43}$$

$$\alpha_0 + \alpha_1 = 1 \tag{3.44}$$

α_0 and α_1 are termed dissociation fractions for the acid and conjugate base, respectively. The expressions for the dissociation fractions are now independent of the analytical activity.

A plot of dissociation fraction against pH for an acid-conjugate base pair is shown in Figure 1.

More frequently, log{ } against pH is plotted using (3.45)–(3.46).

$$\log \{HA\} = \log\{H^+\} + \log C - \log(K + \{H^+\}) \tag{3.45}$$

$$\log \{A^-\} = \log K + \log C - \log(K + \{H^+\}) \tag{3.46}$$

The equilibrium pH can be obtained directly from the log{ } *vs.* pH plot by adding lines showing log $\{H^+\}$ and log $\{OH^-\}$ and by using an expression for charge balance.

$$\{H^+\} = \{A^-\} + \{OH^-\} \tag{3.47}$$

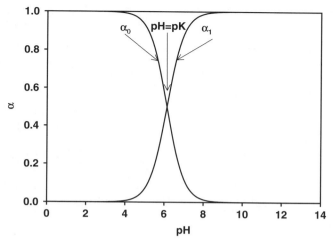

Figure 1 *The influence of pH on α_0 and α_1 for a monoprotic acid*

For a solution containing a weak acid, the charge balance relationship (3.47) is satisfied at the equilibrium pH (Example 3.5). The same log{ } *vs.* pH plot can also be used to determine the equilibrium pH of a solution containing the salt of a weak acid, *e.g.* NaA, by using the appropriate charge balance expression.

$$\{Na^+\} + \{H^+\} = \{A^-\} + \{OH^-\} \tag{3.48}$$

Using $C = \{Na^+\} = \{HA\} + \{A^-\}$ gives

$$\{HA\} + \{H^+\} = \{OH^-\} \tag{3.49}$$

Example 3.5: Graphical illustration of the relationship between (i) log{2,4,6-TCP} and pH; (ii) log{2,4,6-TCP$^-$} and pH for an organic pollutant, 2,4,6-trichlorophenol (2,4,6-TCP), that has been released into a natural water (assume $I = 0$ mol L^{-1}). Use $C = \{2,4,6\text{-}TCP_{TOTAL}\} = 4 \times 10^{-4}$ mol L^{-1} and $K = 10^{-6.13}$.

Graphical representation is best approached by dividing the pH range into two regions, $pH < pK$ and $pH > pK$. The slope of the lines in each of these two regions can be determined by differentiating (3.45) and (3.46) with respect to pH.

For $pH < pK$, $(K + \{H^+\}) \rightarrow \{H^+\}$ and
$$\text{d log } \{2, 4, 6\text{-}TCP\}/\text{d pH} = \text{d log}\{H^+\}/\text{d pH} + \text{d log } C/\text{d pH}$$
$$- \text{d log}\{H^+\}/\text{d pH} = 0$$

d log {2, 4, 6-TCP$^-$}/d pH = d log K/d pH + d log C/d pH
$$- \text{d log}\{H^+\}/\text{d pH} = +1$$

For pH > pK, $(K+\{H^+\}) \rightarrow K$ and
d log {2, 4, 6-TCP}/d pH = d log {H$^+$}/d pH + d log C/d pH
$$- \text{d log } K/\text{d pH} = -1$$

d log {2, 4, 6-TCP$^-$}/d pH = d log K/d pH + d log C/d pH
$$- \text{d log } K/\text{d pH} = 0$$

To construct the straight line sections of the graph, note that log{2,4,6-TCP} = log C for pH < pK and log{2,4,6-TCP$^-$} = log C for pH > pK. Then using the full expressions for log{2,4,6-TCP} and log{2,4,6-TCP$^-$}, calculate several points on lines of slope +1 and −1 to complete the graph.

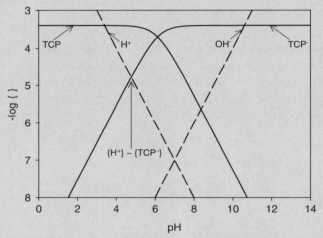

The equilibrium pH for a solution containing the weak acid is obtained by approximating the charge balance expression by {H$^+$} ∼ {TCP$^-$} (because the solution will be acidic and thus {OH$^-$} is very small). Conversely, the equilibrium pH for a solution of the salt of the weak acid is obtained by approximating the appropriate charge balance expression by {TCP} ∼ {OH$^-$} (because the solution will be basic and so {H$^+$} is very small).
Note also that, as shown in Example 3.4, the equilibrium constant, K, can be corrected for ionic strength of solution. In general, the equilibrium pH value decreases with increasing ionic strength.

Carbonate Equilibria. Another example of the importance of acid-base behaviour, not only in aerated freshwaters but also in seawater (see

Chapter 4), is the dissociation of diprotic acids such as carbonic acid, H_2CO_3. The hydration of dissolved $CO_{2(aq)}$ in natural waters gives rise to carbonic acid. The dissociation of carbonic acid not only influences the pH of the water but also provides ligands which can complex trace metals. Example 3.6 illustrates the relationship between log { } and pH for a closed aqueous system which contains dissolved CO_2 (the presence of mineral phases has been ignored). By assuming that a system is closed to the atmosphere, it is possible to treat carbonic acid as a non-volatile acid and to consider only the hydration reaction which converts $CO_{2(aq)}$ to H_2CO_3. H_2CO_3* is used to represent the sum of the activities of $CO_{2(aq)}$ and H_2CO_3, *i.e.* to take account of the presence of $CO_{2(aq)}$, which is in equilibrium with H_2CO_3. The dissociation of carbonic acid can then be described by (3.50)–(3.51) and the equilibrium constants, K_1 and K_2, are defined by (3.52)–(3.53).

$$H_2CO_3^* \rightleftharpoons HCO_3^- + H^+ \tag{3.50}$$

$$HCO_3^- \rightleftharpoons CO_3^{2-} + H^+ \tag{3.51}$$

$$K_1 = \{HCO_3^-\}\{H^+\}/\{H_2CO_3^*\} \tag{3.52}$$

$$K_2 = \{CO_3^{2-}\}\{H^+\}/\{HCO_3^-\} \tag{3.53}$$

The analytical activity of dissolved inorganic carbon species can be written as the sum of undissociated and dissociated acid species (3.54). For a closed system, the value of C is constant over the entire pH range.

$$C = \{H_2CO_3^*\} + \{HCO_3^-\} + \{CO_3^{2-}\} \tag{3.54}$$

Combining (3.52)–(3.54), an expression for each of these (H_2CO_3*, HCO_3^-, and CO_3^{2-}) is obtained.

$$\{H_2CO_3^*\} = C\,(\{H^+\}^2/(\{H^+\}^2 + K_1\{H^+\} + K_1K_2)) \tag{3.55}$$

$$\{HCO_3^-\} = C\,(K_1\{H^+\}/(\{H^+\}^2 + K_1\{H^+\} + K_1K_2)) \tag{3.56}$$

$$\{CO_3^{2-}\} = C\,(K_1K_2/(\{H^+\}^2 + K_1\{H^+\} + K_1K_2)) \tag{3.57}$$

The dissociation fractions, α_0, α_1, and α_2, are obtained by dividing each of (3.55)–(3.57) by the analytical activity, C (*cf.* (3.42)–(3.43)).

$$\alpha_0 = \{H^+\}^2/(\{H^+\}^2 + K_1\{H^+\} + K_1K_2) \tag{3.58}$$

$$\alpha_1 = K_1\{H^+\}/(\{H^+\}^2 + K_1\{H^+\} + K_1K_2) \qquad (3.59)$$

$$\alpha_2 = K_1K_2/(\{H^+\}^2 + K_1\{H^+\} + K_1K_2) \qquad (3.60)$$

where

$$\alpha_0 + \alpha_1 + \alpha_2 = 1 \qquad (3.61)$$

Graphical representation of log { } *vs.* pH can again be used to obtain the equilibrium pH (Example 3.6).

Example 3.6: Graphical illustration of the relationship between (i) log{H₂CO₃*}, (ii) log{H₂CO₃}, (iii) log{HCO₃⁻}, and (iv) log{CO₃²⁻} *and pH for a closed aqueous system. Use* $C = 2 \times 10^{-5}$ *mol* L^{-1}, $K_1 =$ 5.1×10^{-7} *and* $K_2 = 5.1 \times 10^{-11}$.

The presence of dissolved CO_2 is taken into account by the following equilibrium:

$$CO_{2(aq)} + H_2O \rightleftharpoons H_2CO_{3(aq)}$$

$$K_{hydration} = \{H_2CO_3\}/\{CO_2\} = 1.54 \times 10^{-3}$$

We defined $\{H_2CO_3^*\} = \{CO_2\} + \{H_2CO_3\}$ but $\{H_2CO_3^*\} \sim \{CO_2\}$ because the hydration equilibrium lies far to the left. So $K_{hydration} = \{H_2CO_3\}/\{H_2CO_3^*\}$ and $\{H_2CO_3\} = \{H_2CO_3^*\} \times 1.54 \times 10^{-3}$. In this way the presence of dissolved CO_2 is taken into account while the true concentration of carbonic acid can still be determined.

By assuming a closed system, the total analytical activity of carbonic acid, C, is constant [this is not true for an open system (Example 3.7)]. Thus (3.52)–(3.54) can be combined to give (3.55)–(3.57).

For a diprotic acid such as $H_2CO_3^*$, the pH range can be divided into three regions, $pH < pK_1$, $pK_1 < pH < pK_2$, and $pH > pK_2$, and the procedure outlined in Example 3.5, *i.e.* obtaining logarithmic expressions for $\{H_2CO_3^*\}$, $\{HCO_3^-\}$, and $\{CO_3^{2-}\}$ and differentiating each with respect to pH, can then be utilized to construct the plot of $-\log\{ \ \}$ against pH.

For $\{H_2CO_3^*\}$

$$pH < pK_1: \quad \log\{H_2CO_3^*\} = 2\log\{H^+\} + \log C - 2\log\{H^+\}$$
$$= \log C$$
$$d\log\{H_2CO_3^*\}/d\,pH = 0$$

$$pK_1 < pH < pK_2 : \quad \log\{H_2CO_3^*\} = \log\{H^+\} + \log C - \log K_1$$
$$d\log\{H_2CO_3^*\}/d\,pH = -1$$

$$pH > pK_2 : \quad \log\{H_2CO_3^*\} = 2\log\{H^+\} + \log C - \log K_1 K_2$$
$$d\log\{H_2CO_3^*\}/d\,pH = -2$$

The slopes for the other lines are obtained in the same way. Approximating the charge balance expression (3.62) by $\{H^+\} \sim \{HCO_3^-\}$ gives the equilibrium pH for a closed aqueous system containing carbonic acid.

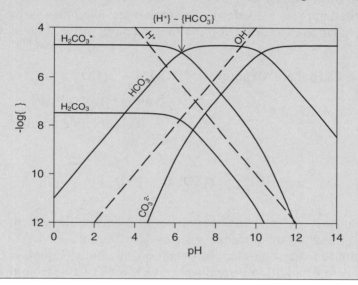

Alkalinity. An important definition arising from the considerations of carbonate equilibria is that of alkalinity, which is derived with reference to the charge balance expression for carbonic acid.

$$\{H^+\} = \{HCO_3^-\} + 2\{CO_3^{2-}\} + \{OH^-\} \tag{3.62}$$

Alkalinity is described as the acid neutralizing capacity of an aqueous system or equivalently as the amount of base possessed by the system.

$$\text{Alkalinity} = \{HCO_3^-\} + 2\{CO_3^{2-}\} + \{OH^-\} - \{H^+\} \tag{3.63}$$

Species other than carbonate can contribute to alkalinity and an alternative definition is

$$\begin{aligned}
\text{Alkalinity} = &\{HCO_3^-\} + 2\{CO_3^{2-}\} + \{NH_3\} + \{HS^-\} + 2\{S^{2-}\} \\
&+ \{H_3SiO_4^-\} + 2\{H_2SiO_4^{2-}\} + \{Org^-\} + \{HPO_4^{2-}\} \\
&+ 2\{PO_4^{3-}\} + \{OH^-\} - \{H^+\} - \{H_3PO_4\}
\end{aligned} \tag{3.64}$$

In addition to the charge balance expression for H_2CO_3, Equation (3.64) includes appropriate expressions for NH_4^+, H_2S, H_4SiO_4, HOrg, and $H_2PO_4^-$. In each case, the reference point has been selected on the basis of the pK values for these species. For example, phosphoric acid, H_3PO_4, has p$K_1 = 2.15$ and p$K_2 = 7.20$ and so $H_2PO_4^-$ will be the main 'P' species in solution at the reference point, *i.e.* the equilibrium pH of a solution of carbonic acid and water (pH ~ 5.65). The contribution of 'P' species to alkalinity is thus determined by using the charge balance expression for NaH_2PO_4.

Another approach to determine carbonate alkalinity is by using a charge balance expression for the major conservative ions in aqueous solution.

$$\begin{aligned}
\text{Carbonate Alkalinity} &= \{HCO_3^-\} + 2\{CO_3^{2-}\} \\
&= \sum(\text{conservative cations}) \\
&\quad - \sum(\text{conservative anions})
\end{aligned}$$

$$\begin{aligned}
&= \{Na^+\} + \{K^+\} + 2\{Ca^{2+}\} + 2\{Mg^{2+}\} \\
&\quad - \{Cl^-\} - 2\{SO_4^{2-}\} - \{NO_3^-\}
\end{aligned} \tag{3.65}$$

Alkalinity is an important parameter in assessing the effects of environmental change on aqueous systems (see Section 3.3.4.1). It is also important to understand that, by definition, alkalinity (Equation (3.61)) is independent of addition or removal of CO_2 (or H_2CO_3) from the system (*cf.* Equation (3.62) – H_2CO_3 does not appear in the charge balance expression). This can be very useful in the determination of the concentration of dissolved inorganic carbon species in aqueous systems that are in equilibrium with an atmosphere containing $CO_{2(g)}$ (Example 3.7).

Example 3.7: Graphical illustration of the relationship between (i) $log\{H_2CO_3{}^*\}$, (ii) $log\{H_2CO_3\}$, (iii) $log\{HCO_3{}^-\}$, and (iv) $log\{CO_3{}^{2-}\}$ and pH for an open aqueous system. Use $p_{CO2} = 3.5 \times 10^{-4}$ atm, $K_H = 3.2 \times 10^{-2}$, $K_1 = 5.1 \times 10^{-7}$ and $K_2 = 5.1 \times 10^{-11}$.

$$CO_{2(g)} + H_2O \rightleftharpoons H_2CO_3{}^* \quad K_H = \{H_2CO_3{}^*\}/p_{CO2}$$
$$= 3.2 \times 10^{-2} \text{ mol L}^{-1} \text{ atm}^{-1}$$

where K_H is the Henry's Law constant for atmosphere-aqueous phase equilibria involving gases.

Clearly, $\{H_2CO_3^*\}$ is now a constant for a fixed partial pressure of CO_2 and is independent of pH, *i.e.* $\{H_2CO_3^*\} = K_H p_{CO2}$. This means that $\{H_2CO_3\}$ is also constant over the entire pH range.

Combining this new expression for $\{H_2CO_3^*\}$ with (3.52)–(3.53) gives expressions for $\{HCO_3^-\}$ and $\{CO_3^{2-}\}$:

$$\{HCO_3^-\} = K_1 K_H p_{CO2}/\{H^+\} \text{ and } \{CO_3^{2-}\} = K_1 K_2 K_H p_{CO2}/\{H^+\}^2$$

The analytical activity, C, is not a constant but can be expressed as

$$C = K_H p_{CO2}/\alpha_0$$

which increases with increasing pH according to the function $(1 + K_1/\{H^+\} + K_1 K_2/\{H^+\}^2)$ as shown by the heavy black line on the accompanying graph.

Plotting $-\log\{\ \}$ *vs.* pH for $\{HCO_3^-\}$ and $\{CO_3^{2-}\}$ gives lines of slope $+1$ and $+2$, respectively. The charge balance expression, which is the same as that for the closed system, can again be used to determine the equilibrium pH.

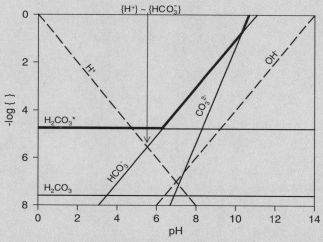

Finally, it is useful to note that the expressions describing the pH dependence of carbonate species concentration for both closed (Example 3.6) and open systems can also be used in mass balance expressions for metal complexation.

pH Dependence of Complex Formation. Other equilibria such as metal complexation reactions can be considered as acid–base reactions and plots of log $\{\ \}$ against pH also provide information about the dominant species present in solution under different geochemical conditions. Hydrolysis of metal cations occurs progressively with increasing pH, *e.g.*

$M^{3+}_{(aq)}$, $M(OH)^{2+}_{(aq)}$, $M(OH)_2^{+}_{(aq)}$, $M(OH)_3^{0}_{(aq)}$, and $M(OH)_4^{-}_{(aq)}$ are the hydrolysis products for many M^{III} cations (see Section 3.2.3.2). $M^{3+}_{(aq)}$ is generally found at low pH and only at extremely high pH is $M(OH)_4^{-}$ formed. The dependence of Cr^{III} hydrolysis reactions on pH is illustrated in Example 3.8.

Example 3.8: Determination of Cr^{III} speciation in an aqueous solution at pH 1, 6, and 11. Use $\{Cr^{III}_T\} = 10^{-6}$ mol L^{-1} and log*$\beta_1 = -4.00$, log*$\beta_2 = -9.62$, log*$\beta_3 = -16.75$, and log*$\beta_4 = -27.77$. Assume that no $Cr(OH)_3$ is precipitated, i.e. homogeneous solution, and that inorganic and organic ligands are absent.

$$Cr^{3+} + H_2O \rightleftharpoons CrOH^{2+} + H^+ \qquad {}^*\beta_1$$

$$Cr^{3+} + 2H_2O \rightleftharpoons Cr(OH)_2^+ + 2H^+ \qquad {}^*\beta_2$$

$$Cr^{3+} + 3H_2O \rightleftharpoons Cr(OH)_3^0 + 3H^+ \qquad {}^*\beta_3$$

$$Cr^{3+} + 4H_2O \rightleftharpoons Cr(OH)_4^- + 4H^+ \qquad {}^*\beta_4$$

$$\{Cr_T^{III}\} = \{Cr^{3+}\} + \{CrOH^{2+}\} + \{Cr(OH)_2^+\} + \{Cr(OH)_3^0\}$$
$$+ \{Cr(OH)_4^-\} = 10^{-6} \text{ mol } L^{-1}$$
$$= \{Cr^{3+}\}(1 + {}^*\beta_1/\{H^+\} + {}^*\beta_2/\{H^+\}^2 + {}^*\beta_3/\{H^+\}^3$$
$$+ {}^*\beta_4/\{H^+\}^4)$$

This equation can be solved for $\{Cr^{3+}\}$ at various pH values by using the equilibrium constants for each of the four equilibria above, e.g. at pH 2, $\{Cr^{3+}\} = 9.90 \times 10^{-7}$ mol L^{-1}.

The concentrations of all other species can be calculated using $\{Cr^{3+}\}$ at each selected pH value, e.g. at pH 2, $\{CrOH^{2+}\} = (*\beta_1/\{H^+\})\{Cr^{3+}\} = 0.01 \times 9.90 \times 10^{-7} = 9.90 \times 10^{-9}$ mol L^{-1}.

Plotting $-\log\{\ \}$ against pH illustrates that the dominant species of Cr^{III} are Cr^{3+}, $Cr(OH)_2^+$, and $Cr(OH)_4^-$ at pH 1, 6, and 11, respectively. Alternatively, the same information may be presented in a plot of mole fraction, χ, against pH. The mole fractions can be obtained by dividing species activity by the total activity of Cr^{III}, e.g. $\chi_{Cr3+} = \{Cr^{3+}\}/\{Cr^{III}_T\}$.

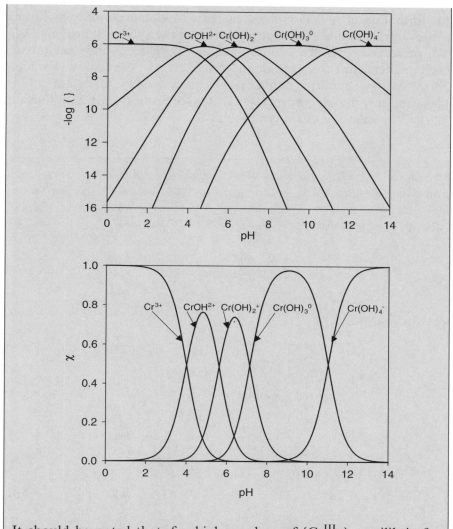

It should be noted that, for higher values of $\{Cr^{III}_T\}$, equilibria for polymeric species should be included and that $Cr(OH)_3$ may precipitate. In the presence of iron, the mixed precipitate $Fe_xCr_{1-x}(OH)_3$ may be formed.

Example 3.9 shows, however, that, for Al^{III}, ligands such as F^- and SO_4^{2-} may compete successfully with the hydroxyl ion at low pH values (see Section 3.3.1.2). For aqueous systems in contact with solid phases, the formation of such complexes increases mineral solubility by several orders of magnitude over that predicted by the solubility product, K_{SP}. The presence of even trace concentrations of organic ligands can have an even greater effect over a wide pH range on Al^{III} speciation (Example 3.9).[12] More generally, complexation clearly has an important role to play, not only in reducing metal toxicity (see Section 3.3.1.2) but also in markedly increasing metal ion mobility.

Example 3.9: Determination of Al^{III} speciation at pH 2, 6, and 10 in an aqueous solution where $\{Al^{III}_T\} = 1 \times 10^{-7}$ mol L^{-1}. Use log $*\beta_1 = -4.95$, log $*\beta_2 = -10.55$, log $*\beta_3 = -17.25$, log $*\beta_4 = -22.85$, log $\beta_{1\,fluoride} = 7.00$, log $\beta_{2\,fluoride} = 12.70$, log $\beta_{1\,sulfate} = -3.50$, log $\beta_{2\,sulfate} = 5.00$, and log $\beta_{1\,org} = 5.9$. Assume $\Sigma F^- = 5 \times 10^{-8}$ mol L^{-1}, $\Sigma SO_4^{2-} = 1 \times 10^{-4}$ mol L^{-1}, $\Sigma Org = 1 \times 10^{-7}$ mol L^{-1} and that no precipitation of $Al(OH)_3$ occurs.

Hydrolysis only:

$$\{Al^{III}_T\} = \{Al^{3+}\} + \{Al(OH)^{2+}\} + \{Al(OH)_2^+\}$$
$$+ \{Al(OH)_3\} + \{Al(OH)_4^-\}$$
$$= 10^{-7} \, mol \, L^{-1}$$
$$= \{Al^{3+}\} \left(1 + {}^*\beta_1/\{H^+\} + {}^*\beta_2/\{H^+\}^2 \right.$$
$$\left. + {}^*\beta_3/\{H^+\}^3 + {}^*\beta_4/\{H^+\}^4\right)$$

Main species: pH 2: Al^{3+}; pH 6: $Al(OH)^{2+}$, $Al(OH)_2^+$; pH 10: $Al(OH)_4^-$.

Hydrolysis and one inorganic ligand, F^-:

$$\{Al^{III}_T\} = \{Al^{3+}\} + \{Al(OH)^{2+}\} + \{Al(OH)_2^+\} + \{Al(OH)_3\}$$
$$+ \{Al(OH)_4^-\} + \{AlF^{2+}\} + \{AlF_2^+\} = 10^{-7} \, mol \, L^{-1}$$
$$= \{Al^{3+}\} (1 + {}^*\beta_1/\{H^+\} + {}^*\beta_2/\{H^+\}^2 + {}^*\beta_3/\{H^+\}^3$$
$$+ {}^*\beta_4/\{H^+\}^4 + \beta_{1\,fluoride}\{F^-\} + \beta_{2\,fluoride}\{F^-\}^2)$$

Main species: pH 2: Al^{3+}, AlF^{2+}; pH 6: $Al(OH)^{2+}$, $Al(OH)_2^+$; pH 10: $Al(OH)_4^-$.

Hydrolysis and two inorganic ligands, F^- and SO_4^{2-}:

$$\begin{aligned}
\{Al_T^{III}\} =& \{Al^{3+}\} + \{Al(OH)^{2+}\} + \{Al(OH)_2^+\} \\
& + \{Al(OH)_3\} + \{Al(OH)_4^-\} + \{AlF^{2+}\} \\
& + \{AlF_2^+\} + \{AlSO_4^+\} + \{Al(SO_4)_2^-\} = 10^{-7}\,mol\,L^{-1} \\
=& \{Al^{3+}\}\,(1 + {}^*\beta_1/\{H^+\} + {}^*\beta_2/\{H^+\}^2 \\
& + {}^*\beta_3/\{H^+\}^3 + {}^*\beta_4/\{H^+\}^4 \\
& + \beta_{1\,fluoride}\{F^-\} + \beta_{2\,fluoride}\{F^-\}^2 \\
& + \beta_{1\,sulfate}\{SO_4^{2-}\} + \beta_{2\,sulfate}\{SO_4^{2-}\}^2)
\end{aligned}$$

Main species: pH 2: Al^{3+}, AlF^{2+}, $(AlSO_4^+)$; pH 6: $Al(OH)^{2+}$, $Al(OH)_2^+$; pH 10: $Al(OH)_4^-$.

Hydrolysis, two inorganic ligands, F^- and SO_4^{2-}, and organic ligand, Org^-:

$$\begin{aligned}
\{Al_T^{III}\} =& \{Al^{3+}\} + \{Al(OH)^{2+}\} + \{Al(OH)_2^+\} \\
& + \{Al(OH)_3\} + \{Al(OH)_4^-\} + \{AlF^{2+}\} \\
& + \{AlF_2^+\} + \{AlSO_4^+\} + \{Al(SO_4)_2^-\} + \{AlOrg^{2+}\} \\
& = 10^{-7}\,mol\,L^{-1} \\
=& \{Al^{3+}\}\,(1 + {}^*\beta_1/\{H^+\} + {}^*\beta_2/\{H^+\}^2 \\
& + {}^*\beta_3/\{H^+\}^3 + {}^*\beta_4/\{H^+\}^4 \\
& + \beta_{1\,fluoride}\{F^-\} + \beta_{2\,fluoride}\{F^-\}^2 \\
& + \beta_{1\,sulfate}\{SO_4^{2-}\} \\
& + \beta_{2\,sulfate}\{SO_4^{2-}\}^2 + \beta_{1\,org}\{Org^-\})
\end{aligned}$$

Main species: pH 2: Al^{3+}, AlF^{2+}, $(AlSO_4^+)$; pH 6: $AlOrg^{2+}$, $Al(OH)_2^+$; pH 10: $Al(OH)_4^-$.

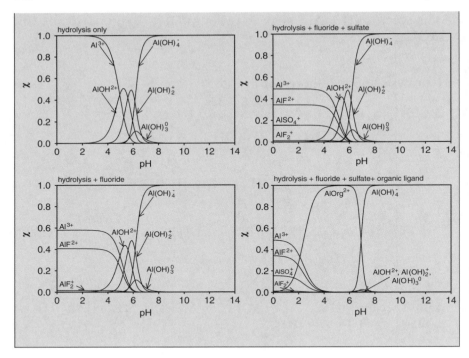

Influence of Ionic Strength. It should be noted that corrections to take account of ionic strength as discussed in Section 3.2.1.2 apply not only to the acid–base equilibrium constants but also to the stability constants for complex formation.

3.2.4.2 pε as a Master Variable. Chemical speciation is also influenced by the redox conditions prevailing in natural waters. Although redox reactions are often slow, and therefore species are present at activities far from equilibrium, they are commonly represented by thermodynamic equilibrium expressions, which can provide the boundary conditions towards which a system is proceeding.

pε, a parameter describing redox intensity, gives the hypothetical electron activity at equilibrium. It measures the relative tendency of a solution to accept or donate electrons with a high pε being indicative of a tendency for oxidation, *i.e.* accepting electrons, while a low pε is indicative of a tendency for reduction, *i.e.* donating electrons. It is defined as

$$p\varepsilon = -\log\{e^-\} \qquad (3.66)$$

The pε scale is thus analogous to the pH scale (pH = $-\log\{H^+\}$), since a low value for pε is obtained where the hypothetical $\{e^-\}$ is large (pH is low where $\{H^+\}$ is large) and conversely a high value of pε is obtained where $\{e^-\}$ is small (pH is high where $\{H^+\}$ is small). $p\varepsilon$ is related to the electrode potential, E_H, by the expression

$$p\varepsilon = 16.9\ E_H\ (V) \tag{3.67}$$

This relationship can be derived from the Nernst Equation

$$E_H = E_H^0 + (2.303\ RT\ n/F)\ \log(\Pi\ (\text{oxidized species})/$$
$$\Pi\ (\text{reduced species})) \tag{3.68}$$

where R is the universal gas constant, T the absolute temperature, F the Faraday constant, and n the number of electrons.

Starting with the equilibrium expression

$$aA + bB + ne^- \rightleftharpoons cC + dD \tag{3.69}$$

we can write

$$K = \{C\}^c\{D\}^d/\{A\}^a\{B\}^b\{e^-\}^n \tag{3.70}$$

Logarithmic expressions are then

$$\log K = \log\ (\Pi(\text{reduced species})/\Pi(\text{oxidized species})) - n\ \log\{e^-\}$$
$$p\varepsilon = (1/n)\ \log K + (1/n)\ \log\ (\Pi(\text{oxidized species})/\Pi(\text{reduced species}))$$

Multiply by 2.303 RT/F to give

$$(2.303\ RT/F)\ p\varepsilon = (2.303\ RT\ n/F)\ \log K$$
$$+ (2.303\ RT\ n/F)\ \log\ (\Pi(\text{oxidized species})/\Pi(\text{reduced species}))$$

Now use the standard Gibbs free energy, $\Delta G^0 = -2.303\ RT\ \log K = -nFE_H^0$ to give

$$(2.303\ RT/F)\ p\varepsilon = E_H^0$$
$$+ (2.303\ RT\ n/F)\ \log\ (\Pi(\text{oxidized species})/\Pi(\text{reduced species}))$$
$$\tag{3.71}$$

The right hand side of Equation (3.71) is the same as that of the Nernst Equation (3.68) and so the left hand side of Equation (3.71) must be equal to the left hand side of the Nernst Equation (3.68), thus leading to Equation (3.67). An additional definition is

$$p\varepsilon^0 = (1/n)\ \log K \tag{3.72}$$

and so

$$p\varepsilon^0 = (2.303\ RT/F)\ E_H^0 = 16.9\ E_H^0 \tag{3.73}$$

The usefulness of Equation (3.72) will be demonstrated further in Example 3.10.

Although the electron activity is a hypothetical phenomenon, pε is a useful parameter to describe the redox intensity of natural systems and hence the species distribution under prevailing redox conditions (Example 3.10).

Example 3.10: Calculate the pε values for the following solutions (298 K, $I = 0$) (i) a solution at pH 2 containing $\{Fe^{3+}\} = 10^{-4.5}$ mol L^{-1} and $\{Fe^{2+}\} = 10^{-2.7}$ mol L^{-1} where log $K = 13$; (ii) a neutral solution containing $\{Mn^{2+}\} = 10^{-6}$ mol L^{-1} in equilibrium with the solid phase, $Mn^{IV}O_2$ and log $K = 40.84$.

  (i) At pH 2, the hydrated metal ion has not undergone hydrolysis to any significant extent:

$$Fe^{3+} + e^- \rightleftharpoons Fe^{2+} \quad K = \{Fe^{2+}\}/\{Fe^{3+}\}\{e^-\}$$

$$p\varepsilon = \log K + \log(\{Fe^{3+}\}/\{Fe^{2+}\})$$

$$= 13 + \log(10^{-4.5}/10^{-2.7})$$

$$= 13 - 1.8 = 11.2$$

(ii)

$$MnO_2 + 4H^+ + 2e^- \rightleftharpoons Mn^{2+} + 2H_2O$$

$$K = \{Mn^{2+}\}/\{H^+\}^4\{e^-\}^2$$

$$p\varepsilon = 0.5 \log K + 0.5 \log(\{H^+\}^4/\{Mn^{2+}\})$$

$$= 20.42 + 0.5 \log(10^{-28}/10^{-6})$$

$$= 20.42 - 11 = 9.42$$

For a fixed pH value (*e.g.* pH = 2), the species distribution (*e.g.* where $\{Fe_T\} = 1 \times 10^{-3}$ mol L^{-1}) at different pε values can be illustrated by plotting log{ } against pε. This is analogous to the treatment of pH as a master variable. The pε range is split into two parts, $p\varepsilon < p\varepsilon^0$ and $p\varepsilon > p\varepsilon^0$. Using $p\varepsilon^0 = (1/n) \log K$, this becomes $\{e^-\} > K^{-1}$ and $\{e^-\} < K^{-1}$ (because n = 1 in this Example). Thereafter, construction of the graph is achieved by the same method used in Example 3.5.

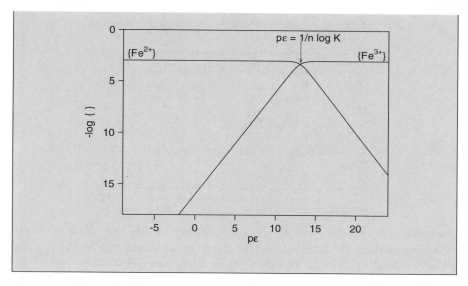

Redox intensity or electron activity in natural waters is usually determined by the balance between those processes which introduce oxygen (*e.g.* dissolution of atmospheric oxygen, photosynthesis) and those which remove oxygen (*e.g.* microbial decomposition of organic matter). Often these processes are controlled by the availability of inorganic nutrients such as phosphate and nitrate, *e.g.* as utilized in the formation of organic matter during photosynthesis (see Section 3.3.4).

$$106CO_2 + 16NO_3^- + HPO_4^{2-} + 122H_2O + 18H^+$$
$$\rightleftharpoons C_{106}H_{263}O_{110}N_{16}P_1 + 138O_2 \qquad (3.74)$$

The decay of organic matter produced in this manner leads to the subsequent consumption of oxygen, *e.g.* by respiration.

$$C_{106}H_{263}O_{110}N_{16}P_1 + 138O_2 \rightleftharpoons 106CO_2 + 16NO_3^- + HPO_4^{2-}$$
$$+ 122H_2O + 18H^+ \qquad (3.75)$$

The decay of organic matter requires the presence of a terminal electron acceptor and in Equation (3.75) molecular oxygen is reduced to water. Other terminal electron acceptors present in natural waters include NO_3^-, Mn^{IV}, Fe^{III}, SO_4^{2-}, and CO_2. Once all molecular oxygen has been consumed, organic matter is decomposed *via* reactions involving other terminal electron acceptors in a series determined by the $p\varepsilon$ intensity as shown in Table 3.

The sequence of redox reactions involving organic matter, all of which are microbially mediated, can be thought of as progressing through

Table 3   *Redox processes – terminal electron acceptors*

Terminal electron acceptor	$p\varepsilon$
O_2	13.75
NO_3^- (reduction to N_2)	12.65
Mn^{IV}	8.9
NO_3^- (reduction to NH_4^+)	6.15
Fe^{III}	−0.8
Reducible organic matter	−3.01
SO_4^{2-}	−3.75
CO_2	−4.13

Source: Adapted with permission from W. Stumm and J.J. Morgan, *Aquatic Chemistry* 3rd Edn,[15] © Wiley, 1996.

decreasing levels of $p\varepsilon$ and so, for example, NO_3^- in denitrification reactions would be utilized as the electron acceptor before Mn^{IV} and SO_4^{2-} would be utilized before CO_2 (see Section 3.3.3.2).

Redox processes are important for elements which can exist in more than one oxidation state in natural waters, *e.g.* Fe^{II} and Fe^{III}, Mn^{II}, and Mn^{IV}. These are termed redox-sensitive elements. The redox conditions in natural waters often affect the mobility of these elements since the inherent solubility of different oxidation states of an element may vary considerably. For example, Mn^{II} is soluble whereas Mn^{IV} is highly insoluble. In oxic systems, Mn^{IV} is precipitated in the form of oxyhydroxides. In anoxic systems, Mn^{II} predominates and is able to diffuse along concentration gradients both upwards and downwards in a water column. This behaviour gives rise to the classic concentration profiles observed for Mn (and Fe) at oxic-anoxic interfaces as illustrated in Figure 2.

Other examples of redox-sensitive elements include heavy elements such as uranium, plutonium, and neptunium, all of which can exist in multiple oxidation states in natural waters. Redox conditions in natural waters are also indirectly important for solute species associated with redox-sensitive elements. For example, dissolution of iron (hydr)oxides under reducing conditions may lead to the solubilization and hence mobilization of associated solid phase species, *e.g.* arsenate, phosphate (see Sections 3.3.2.1, 3.3.3.2, and 3.3.4.1).

3.2.4.3 $p\varepsilon – pH$ Relationships.   Many redox equilibria also involve the transfer of protons and so chemical speciation depends on both $p\varepsilon$ and pH. Stability relationships involving redox processes can be investigated *via* $p\varepsilon$–pH diagrams. The region of interest for natural waters displayed in these diagrams is defined by the stability limits of water, *i.e.* where the total partial pressure of oxygen and hydrogen is no greater than 1 atm.

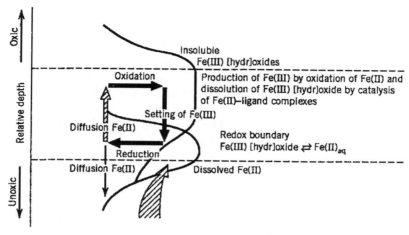

Figure 2 *Oxic-anoxic boundary in the water or sediment column[15]*
(Reproduced with permission from ref 15, © John Wiley & Sons Inc. 1996)

Upper Limit

$$O_{2(g)} + 4e^- + 4H^+ \rightleftharpoons 2H_2O_{(1)} \qquad\qquad \log K = 83.1 \qquad (3.76)$$

Where $p_{O2} = 1$, $\log p_{O2} = 0$ and

$$\log K = -4\log\{e^-\} - 4\log\{H^+\}$$
$$p\varepsilon = 20.78 - pH \qquad\qquad (3.77)$$

Lower Limit

$$2H^+ + 2e^- \rightleftharpoons H_{2(g)} \qquad\qquad \log K = 0 \qquad (3.78)$$

or

$$2H_2O + 2e^- \rightleftharpoons H_{2(g)} + 2OH^- \qquad\qquad \log K = -28 \qquad (3.79)$$

Where $p_{H2} = 1$, $\log p_{H2} = 0$ and

$$\log K - 2\log K_w = 2\log\{e^-\} - 2\log\{H^+\}$$

or

$$\log K = -2\log\{e^-\} - 2\log\{H^+\}$$

In both cases:

$$p\varepsilon = -pH \qquad\qquad (3.80)$$

For any system, stability relationships between solution phase species and solid phases can be used to construct pε–pH diagrams representing species distribution within the upper and lower stability limits of water. In general, the relationships between pε and pH for redox equilibria involving proton and electron transfer give lines, *e.g.* pε = a pH + b (where a and b are constants), with slope = a. Redox equilibria involving only electron transfer give rise to horizontal lines, *i.e.* pε = b, a = 0. Vertical lines arise from equilibria involving only proton transfer and pH = c (where c is a constant). The relationships between pε and pH for As–H_2O and Cr–H_2O systems at 298 K and 1 atm total are shown in Examples 3.11 and 3.12, respectively. The importance of both As and Cr speciation as represented in these pε–pH diagrams is discussed further in Sections 3.3.2.1 and 3.3.2.5, respectively.

Example 3.11: Construction of a pε–pH graph for Cr-H_2O. Assume that no solid phases are formed.

Acid-base behaviour of Cr^{VI}:
$$HCrO_4^- \rightleftharpoons CrO_4^{2-} + H^+ \quad K_1 = 10^{-6.45}$$

Hydrolysis behaviour of Cr^{III}:

$$Cr^{3+} + H_2O \rightleftharpoons Cr(OH)^{2+} + H^+ \qquad K_2 = 10^{-4.00}$$
$$Cr(OH)^{2+} + H_2O \rightleftharpoons Cr(OH)_2^+ + H^+ \quad K_3 = 10^{-5.62}$$
$$Cr(OH)_2^+ + H_2O \rightleftharpoons Cr(OH)_3^0 + H^+ \quad K_4 = 10^{-7.13}$$
$$Cr(OH)_3^0 + H_2O \rightleftharpoons Cr(OH)_4^- + H^+ \quad K_5 = 10^{-11.02}$$

Reduction of Cr^{VI} to Cr^{III}:

$$HCrO_4^- + 7H^+ + 3e^- \rightleftharpoons Cr^{3+} + 4H_2O \qquad K_6 = 10^{-70.55}$$
$$HCrO_4^- + 6H^+ + 3e^- \rightleftharpoons Cr(OH)^{2+} + 3H_2O \quad K_7 = 10^{-66.55}$$
$$HCrO_4^- + 5H^+ + 3e^- \rightleftharpoons Cr(OH)_2^+ + 2H_2O \quad K_8 = 10^{-60.93}$$
$$CrO_4^{2-} + 6H^+ + 3e^- \rightleftharpoons Cr(OH)_2^+ + 2H_2O \quad K_9 = 10^{-67.38}$$
$$CrO_4^{2-} + 5H^+ + 3e^- \rightleftharpoons Cr(OH)_3^0 + H_2O \qquad K_{10} = 10^{-60.25}$$
$$CrO_4^{2-} + 4H^+ + 3e^- \rightleftharpoons Cr(OH)_4^- \qquad K_{11} = 10^{-49.23}$$

Construction of vertical lines:
e.g. $K_1 = \{H^+\}\{CrO_4^{2-}\}/\{HCrO_4^-\}$
On the line, the concentrations of the Cr^{VI} species are equal, so pH = pK_1 = 6.45.
Construction of sloped lines:
e.g. $K_8 = \{Cr(OH)_2^+\}/\{HCrO_4^-\}\{H^+\}^5\{e^-\}^3$
On the line, the concentrations of the Cr^{III} and Cr^{VI} species are equal, so pε = $-\log K_8/3 - 5/3$ pH = 20.31–1.6 pH.
Crossing points:
For the lines defined by K_1 and K_8, substitute pH = 6.45 in pε = 20.31–1.6 pH, so pε = 9.99.

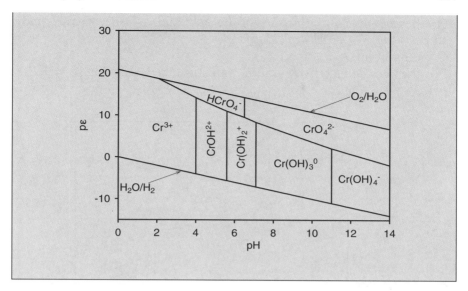

Example 3.12: Construction of a pε–pH diagram for As–H_2O. Assume that no solid phases are formed.

Acid-base behaviour of As^V:

$$H_3AsO_4 \rightleftharpoons H_2AsO_4^- + H^+ \qquad K_1 = 10^{-2.24}$$
$$H_2AsO_4^- \rightleftharpoons HAsO_4^{2-} + H^+ \qquad K_2 = 10^{-6.76}$$
$$HAsO_4^{2-} \rightleftharpoons AsO_4^{3-} + H^+ \qquad K_3 = 10^{-11.60}$$

Acid-base behaviour of As^{III}:

$$H_3AsO_3 \rightleftharpoons H_2AsO_3^- + H^+ \qquad K_4 = 10^{-9.23}$$
$$H_2AsO_3^- \rightleftharpoons HAsO_3^{2-} + H^+ \qquad K_5 = 10^{-12.10}$$

Reduction of As^V to As^{III}:

$$H_3AsO_4 + 2H^+ + 2e^- \rightleftharpoons H_3AsO_3 + H_2O \qquad K_6 = 10^{19.46}$$
$$H_2AsO_4^- + 3H^+ + 2e^- \rightleftharpoons H_3AsO_3 + H_2O \qquad K_7 = 10^{21.70}$$
$$HAsO_4^{2-} + 4H^+ + 2e^- \rightleftharpoons H_3AsO_3 + H_2O \qquad K_8 = 10^{28.46}$$
$$HAsO_4^{2-} + 3H^+ + 2e^- \rightleftharpoons H_2AsO_3^- + H_2O \qquad K_9 = 10^{19.23}$$
$$AsO_4^{3-} + 4H^+ + 2e^- \rightleftharpoons H_2AsO_3^- + H_2O \qquad K_{10} = 10^{30.83}$$
$$AsO_4^{3-} + 3H^+ + 2e^- \rightleftharpoons HAsO_3^{2-} + H_2O \qquad K_{11} = 10^{18.73}$$

Construction of vertical lines:
e.g. $K_1 = \{H^+\}\{H_2AsO_4^-\}/\{H_3AsO_4\}$
On the line, the concentrations of the As^V species are equal,
so pH = pK_1 = 2.24.
Construction of sloped lines:
e.g. $K_6 = \{H_3AsO_3\}/\{H_3AsO_4\}\{H^+\}^2 \{e^-\}^2$

On the line, the concentrations of the As^{III} and As^V species are equal.
so pε = log $K/2$ – pH = 9.73–pH.
Crossing points:
For the lines defined by K_1 and K_6, substitute pH = 2.24 in pε = 9.73–
pH, so pε = 7.49.

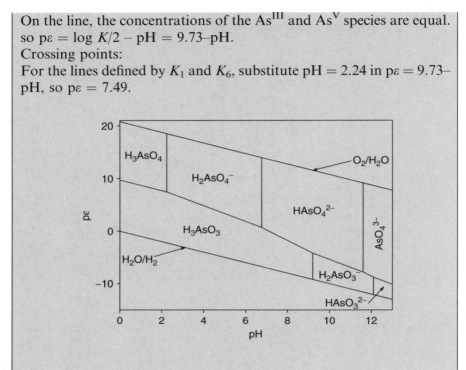

In Examples 3.11 and 3.12 it has been assumed that no solid phases
are formed. At higher solution concentrations, however, the formation
of solid phases such as insoluble hydroxides, carbonates and sulfides
must be considered. The stability of these phases is often pε and/or pH
dependent. Redox equilibria involving both aqueous and solid phase 'S'
species are shown in Example 3.13.

*Example 3.13: Construction of a pε–pH diagram for SO_4–$S_{(s)}$–H_2S
system. Use $S_T = 10^{-2}$ mol L^{-1}.*

Acid-base behaviour of S^{VI}:
$$HSO_4^- \rightleftharpoons SO_4^{2-} + H^+ \qquad\qquad K_1 = 10^{-2.0}$$

Acid-base behaviour of S^{-II}:
$$H_2S_{(aq)} \rightleftharpoons HS^- + H^+ \qquad\qquad K_2 = 10^{-7.0}$$

Reduction of S^{VI} to S^0:
$$HSO_4^- + 7H^+ + 6e^- \rightleftharpoons S_{(s)} + 4H_2O \qquad K_3 = 10^{-34.2}$$

$$SO_4^{2-} + 8H^+ + 6e^- \rightleftharpoons S_{(s)} + 4H_2O \qquad\qquad K_4 = 10^{-36.2}$$

Reduction of S^{VI} to S^{-II}:

$$SO_4^{2-} + 9H^+ + 8e^- \rightleftharpoons HS^- + 4H_2O \qquad\qquad K_5 = 10^{-34.0}$$

Reduction of S^0 to S^{-II}:

$$S_{(s)} + 2H^+ + 2e^- \rightleftharpoons H_2S_{(aq)} \qquad\qquad K_6 = 10^{4.8}$$

Construction of sloped lines involving a solid phase:

$p\varepsilon = -\log K_6/2 - \log\{H_2S_{(aq)}\}/2 - pH$

On the line, assume that $S_T = \log\{H_2S_{(aq)}\} = 10^{-2}$ mol L^{-1} so $p\varepsilon = 3.4–pH$.

The information contained in this $p\varepsilon$–pH diagram is particularly useful in the determination of stability zones for solid phase metal sulfides (Example 3.14). For example, it is the reduction of sulfate to sulfide which determines the stability field of pyrite (FeS_2).

Example 3.14: Construction of a $p\varepsilon$–pH diagram for the Fe–O_2–'S'– CO_2–H_2O open system where $\{Fe_T\}$ is 1×10^{-6} mol L^{-1}, $\{S_{(aq)\,T}\} = 1 \times 10^{-2}$ mol L^{-1} and $p_{CO2} = 10^{-3}$ atm.

Reduction of Fe^{III} to Fe^{II}:

$$Fe^{3+} + e^- \rightleftharpoons Fe^{2+} \qquad\qquad K_1 = 10^{-13.0}$$

$$Fe(OH)^{2+} + e^- + H^+ \rightleftharpoons Fe^{2+} + H_2O \qquad\qquad K_2 = 10^{15.2}$$

$$Fe(OH)_2^+ + e^- + 2H^+ \rightleftharpoons Fe^{2+} + 2H_2O \qquad K_3 = 10^{18.7}$$

$$Fe(OH)_{3(s)} + e^- + 3H^+ \rightleftharpoons Fe^{2+} + 3H_2O \qquad K_4 = 10^{17.9}$$

$$Fe(OH)_{3(s)} + e^- + H^+ \rightleftharpoons Fe(OH)_{2(s)} + H_2O \qquad K_5 = 10^{5.5}$$

$$Fe(OH)_4^- + e^- + 2H^+ \rightleftharpoons Fe(OH)_{2(s)} + 2H_2O \qquad K_6 = 10^{22.2}$$

Reduction of Fe^{II} to Fe^0:

$$Fe^{2+} + 2e^- \rightleftharpoons Fe_{(s)} \qquad K_7 = 10^{-13.8}$$

$$Fe(OH)_{2(s)} + 2e^- + 2H^+ \rightleftharpoons Fe_{(s)} + 2H_2O \qquad K_8 = 10^{1.4}$$

Hydrolysis of Fe^{III}:

$$Fe^{3+} + H_2O \rightleftharpoons Fe(OH)^{2+} + H^+ \qquad K_9 = 10^{-2.2}$$

$$Fe(OH)^{2+} + H_2O \rightleftharpoons Fe(OH)_2^+ + H^+ \qquad K_{10} = 10^{-3.5}$$

$$Fe(OH)_2^+ + H_2O \rightleftharpoons Fe(OH)_{3(s)} + H^+ \qquad K_{11} = 10^{0.8}$$

$$Fe(OH)_{3(s)} + H_2O \rightleftharpoons Fe(OH)_4^- + H^+ \qquad K_{12} = 10^{-16.7}$$

Hydrolysis of Fe^{II}:

$$Fe^{2+} + 2H_2O \rightleftharpoons Fe(OH)_{2(s)} + 2H^+ \qquad K_{13} = 10^{-12.4}$$

Carbonate complexation of Fe^{II} and additional reduction reactions:

$$FeCO_{3(s)} + H^+ \rightleftharpoons Fe^{2+} + HCO_3^- \qquad K_{14} = 10^{5.8}$$

$$Fe(OH)_{3(s)} + HCO_3^- + 2H^+ + e^- \rightleftharpoons FeCO_{3(s)} + 3H_2O \qquad K_{15} = 10^{12.1}$$

$$Fe(OH)_4^- + HCO_3^- + 3H^+ + e^- \rightleftharpoons FeCO_{3(s)} + 4H_2O \qquad K_{16} = 10^{32.4}$$

In the stability field of sulfate:

$$SO_4^{2-} + Fe^{2+} + 16H^+ + 14e^- \rightleftharpoons FeS_{2(s)} + 8H_2O \qquad K_{17} = 10^{86.8}$$

$$FeCO_{3(s)} + 2SO_4^{2-} + 14e^- + 18H^+ \rightleftharpoons FeS_{2(s)} + H_2CO_3 + 8H_2O$$
$$K_{18} = 10^{85.4}$$

At lower $p\varepsilon$ values (in the H_2S and HS^- stability fields):

$$FeS_{2(s)} + 4H^+ + 2e^- \rightleftharpoons Fe^{2+} + 2H_2S_{(aq)} \qquad K_{19} = 10^{-5.0}$$

$$FeS_{2(s)} + H^+ + 2e^- \rightleftharpoons FeS_{(s)} + HS^- \qquad K_{20} = 10^{-14.6}$$

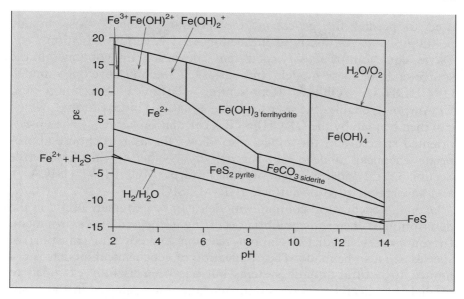

It should be remembered that these diagrams are based on equilibrium conditions and that redox reactions in natural waters may be far from equilibrium. Additional assumptions are often made, *e.g.* the activities of the main components are fixed and that those of all minor species are known, the activity of all solid phases is unity, and only a few reactive components need to be considered.[34] As a consequence, straight line boundaries between species are generated. Importantly, adsorption, ion exchange, and metal-organic interactions are often omitted, limiting the usefulness of classical pε–pH diagrams (*e.g.* Examples 3.11–3.14).[34] Generation of more environmentally representative diagrams requires the use of geochemical computer models (see Section 3.2.5). For example, Kinniburgh and Cooper[34] demonstrate how PHREEQC[35] and ORCHESTRA[36] can be used to obtain predominance diagrams (including pε–pH diagrams) for a wide range of metals, metalloids, and anionic species.

3.2.5 Modelling Aquatic Systems

This chapter has discussed the fundamental importance of the master variables pε and pH in determining species distribution under the conditions prevailing in natural systems (pε -10 to $+17$ and pH 4 to 10). The equilibria contained in Sections 3.2.2–3.2.4 also provide the basis for many of the currently available geochemical computer models (*e.g.* WATEQ4F,[30] PHREEQC,[35] and MINTEQA2[22]), which are being used to predict the geochemical behaviour of metal ions, organic pollutants, *etc.* As discussed in Section 3.2.1.2, certain equations, *e.g.* Davies equation for activity coefficient calculation, are contained in the source code of these models. In contrast, a recently developed model, ORCHESTRA (Objects Representing CHEmical Speciation and TRAnsport), enables the user to define equations in text format, which can then be read by the ORCHESTRA calculation kernel.[36] The object-oriented structure of this model also allows the user to choose appropriate chemical models, *e.g.* for activity coefficient calculation (see Section 3.2.1.2), adsorption at organic surfaces (MODEL V,[37] NICA[38]), and adsorption at oxide surfaces (CD-MUSIC[39]).

Overall, geochemical computer models can be extremely useful in the description of chemical equilibria occurring in the aquatic environment. In some cases, predictions about reaction kinetics and transport of species can also be made. The application of geochemical models is not limited to natural aquatic systems but has been usefully extended to predict the effectiveness of certain remediation strategies in the treatment of waters emanating from contaminated sites.[40]

3.3   CASE STUDIES

3.3.1   Acidification

3.3.1.1   Diatom Records.   The onset and flourishing of the Industrial Revolution, with its dependence for energy upon the combustion of fossils fuels, especially coal, released huge quantities of sulfur dioxide and nitrogen oxides to the atmosphere, which ultimately were returned to the land surface in the form of acid precipitation (H_2SO_4, HNO_3). The effect of this upon freshwater lakes, especially in poorly buffered catchment areas (*e.g.* low-carbonate soils, granitic bedrock) was a significant decrease in pH and a concomitant decline in biological productivity, often leading to clear acid waters devoid of fish. That the acidification of freshwater lakes is a comparatively recent (post-1850) phenomenon linked to acid deposition has been demonstrated by the diatom record in the sediments at the bottom of acidified lakes.[41] Diatoms are algae with silicified cell walls. Depending upon the acidity of the water, some species flourish more than others. Upon deposition, these siliceous remains are preserved in the sediment column. The dramatic change in pH after 1850 inferred from the diatom record in dated sediments from the Round Loch of Glenhead in the Galloway region of south-west Scotland rules out long-term acidification processes as the cause.[42]

3.3.1.2 Aluminium. The effect of such acidification upon the mobilization, behaviour, and subsequent biological impact of aluminium in freshwaters has been spectacular. Constituting 8.13% of the Earth's crust, aluminium is the third most abundant element and is largely associated with crystalline aluminosilicate minerals of low solubility. In the absence of strong acid inputs, processes of soil development normally lead to only a small fraction of aluminium becoming available to participate in biogeochemical reactions. Thus, in the Northern Temperate Region, mobilization of aluminium from upper to lower mineral horizons by organic acids (or H_2CO_3) leached from foliage or forest floors often leads to the formation of $Al(OH)_3$ in a process known as podzolization. Under elevated partial pressures of CO_2, however, dissociation of H_2CO_3 produces H^+, which may solubilize aluminium, while HCO_3^- serves as a counterion for transport of cationic $Al^{III}_{(aq)}$ through soil. Upon discharge to a surface water, CO_2 degasses, resulting in aluminium hydrolysis and precipitation as $Al(OH)_3$. Thus, in most natural waters, concentrations of dissolved aluminium are generally low due to the relatively low solubility of natural aluminium minerals under circumneutral pH values. In the case of strong acid (H_2SO_4, HNO_3) inputs, however, especially to sensitive regions with small pools of basic cations (Ca^{2+}, Mg^{2+}, Na^+, K^+) and an inability to retain inputs of strong acid anions (SO_4^{2-}, NO_3^-, Cl^-), acidic cations (H^+, Al^{III}) are transported along with acid anions from soil to surface water.[43]

The speciation of dissolved aluminium in surface waters is highly pH-dependent (see Section 3.2.3.2), with, in a simple system (considering only monomeric aluminium species) in equilibrium with solid phase $Al(OH)_3$, the aquated ion $Al^{3+}_{(aq)}$ predominating at pH < 3, progressive hydrolysis to $Al(OH)^{2+}$ and $Al(OH)_2^+$ in the pH range 4.5–6.5, followed by $Al(OH)_3^0$ from pH 6.5 to 7, and then $Al(OH)_4^-$ predominating as the pH increases beyond 7. From pH 4.5–7.5, the solubility of aluminium is low and in this range it is often precipitated as $Al(OH)_3$. Below pH 4.5 and above pH 7.5 the concentration of aluminium in solution increases rapidly. In a part experimental fractionation, part speciation modelling study of aluminium in acidified Adirondack surface waters in the north-eastern USA, Driscoll and Schecher[43] found that, in the absence of organic ligands, $Al^{3+}_{(aq)}$ was significant at pH < 5, and that the inorganic monomeric complexes of aluminium predominated at pH > 5. In particular, Al–F complexes were the dominant form at pH 5–6 whereas, at pH > 6, Al–OH species became the major form. In the presence of organic ligands, however, they found that, over a broad pH range of 4.3–7.0, alumino-organic complexes were a major component of the monomeric aluminium. It is the chemically labile inorganic monomeric aluminium that has been identified as the toxic agent for fish which, in acidified lakes and streams at aluminium concentrations of 100–200 μg L^{-1}, are unable to maintain their osmoregulatory balance and are susceptible to respiratory problems from coagulation of mucus and Al(OH)$_3$ on their gills at their physiological pH of 7.2.[44]

In response to a decline in acid deposition as a result of improved atmospheric emission controls, there was a significant decline from the early 1980s to the late 1990s in the concentrations of both inorganic and organic monomeric aluminium in soil solutions draining from mineral soils and in stream water in New Hampshire, USA (Figure 3).[45] These decreases in aluminium concentration were accompanied by decreases in SO$_4^{2-}$ concentrations, but there was only a modest increase in stream water acid-neutralizing capacity and no change in pH. The hydroxylation of aluminium appears to have been critical in buffering stream waters against such increases. In a survey of Adirondack lakes in the north-eastern USA, however, Driscoll and co-workers measured a

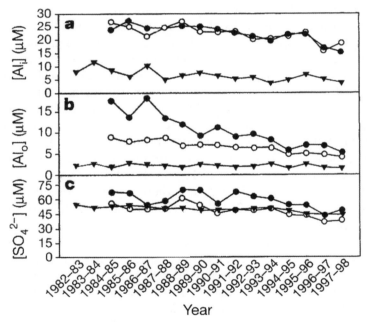

Figure 3 *Long-term declines in sulfate concentrations mitigate the mobilization of aluminium in soil solutions and stream water in the Hubbard Brook Experimental Forest, New Hampshire, USA.[45] **a–c**, annual volume-weighted concentrations of **a**, inorganic monomeric aluminium (Al$_i$); **b**, organic monomeric aluminium (Al$_o$); and **c**, sulfate (SO$_4^{2-}$) in mineral soil solutions at 750 m (●) and 730 m (○), and in stream water (triangles)*
(Reprinted with permission from ref 45, © Macmillan Publishing, 2002)

significant increasing trend in acid-neutralizing capacity of 1.60 µeq L^{-1} per year from 1992 to 2000 for 60% of the monitored lakes.[46] Furthermore, there was a growing concentration of dissolved organic carbon in 15% of the lakes, helping to shift the aluminium from its toxic inorganic form towards less toxic organic forms although likely to delay recovery of acid-neutralizing capacity. Overall, however, by 2000 both USA and European surface waters appeared to be on the way to long-term recovery from acidification, with decreases in SO_4^{2-} concentrations, increases in pH and acid-neutralizing capacity, and decreases in concentrations of labile toxic forms of aluminium.[47–49]

With respect to human exposure to aluminium in drinking water, it must be remembered that aluminium has often been deliberately added, in the form of $Al_2(SO_4)_3$, to water supplies at treatment works to remove coloration by organic compounds in reservoirs in upland catchments. This it achieves by hydrolyzing to a gelatinous, high surface area, precipitate of $Al(OH)_3$, which helps to remove the coloured organic colloids. It has been suggested, partially because of observed water aluminium-related dialysis dementia in some chronic renal patients on artificial kidney dialysis machines in the 1960s and 1970s, that exposure to low concentrations of aluminium in drinking water, for which there is a current EC Maximum Admissible Concentration of 200 µg L^{-1}, might be implicated in Alzheimer's Disease.[50] There is still no proof of this at present,[51] although it is known that the incidence of senile dementia on the remote Pacific island of Guam, where bauxite is mined and the local water is elevated in aluminium, is a hundred-fold greater than in the USA. Ten years after an accident at Camelford in south-west England in 1988, when 20 tonnes of $Al_2(SO_4)_3$ were emptied directly into the main water supply, there did appear to be a lasting effect on the cerebral function of those who had been exposed to greatly enhanced aluminium concentrations in the tap water.[52] It has been mooted that, for humans, a high intake of silicic acid (H_4SiO_4) could reduce the bioavailability of aluminium by two orders of magnitude for, at slightly alkaline intestinal pH, hydroxyaluminosilicates are stable and unavailable for absorption.[53] This would mimic the situation in nature, where, in the absence of a strong acid input, the bioavailability of aluminium is kept low by the formation of hydroxyaluminosilicate species. The latter has been described as a 'geochemical brake on an aluminium juggernaut which would otherwise career into the biotic cycle'.[54] More generally, in addition to aluminium, the role of other metals, including zinc, copper, iron, and manganese, in neurodegenerative disease is now under scrutiny, there being suggestions of adverse effects upon protein folding and subsequent oxidative stress.[55]

3.3.1.3   Acid Mine Drainage and Ochreous Deposits.   One of the major pollution problems affecting freshwaters is that of acid waters discharged from coal and metal mines.[56] In an active underground coal mine, pumping lowers the water table. The deeper strata thus become exposed to air with the result that pyrite (FeS_2) present is subject to oxidation, generating acid conditions.

$$2FeS_{2(s)} + 2H_2O + 7O_{2(g)} \rightleftharpoons 2FeSO_{4(s)} + 2H_2SO_{4(1)} \quad (3.81)$$

When mining and pumping cease, the water table returns to its natural level. While the flooding of the mine stops the direct oxidation of pyrite, it does bring the sulfuric acid and iron sulfates into solution. Some of the ferrous ion (Fe^{2+}) may be oxidized to the ferric ion (Fe^{3+}), a slow process at low pH, but one which can be catalysed by bacteria, and the ferric ion may react further with pyrite.[57]

$$4Fe^{2+} + O_{2(g)} + 4H^+ \rightleftharpoons 4Fe^{3+} + 2H_2O \qquad (3.82)$$

$$FeS_{2(s)} + 14Fe^{3+} + 8H_2O \rightleftharpoons 15Fe^{2+} + 2SO_4^{2-} + 16H^+ \quad (3.83)$$

Underground in the Richmond Mine at Iron Mountain, California, extremely acidic mine waters have been encountered, with pH ranging from 1.5 to -3.6 and total dissolved metal (mainly iron) and sulfate concentrations as high as 200 g L^{-1} and 760 g L^{-1}, respectively.[58] These are the most acidic waters known.

On reaching the surface, the acidic mine drainage water is mixed with air and oxygenated water, leading to rapid oxidation from the ferrous to the ferric form and the precipitation of the characteristic unsightly orange ochreous deposits of iron (hydr)oxides, *e.g.* ferrihydrite ($Fe(OH)_3$), goethite ($FeOOH$), observed along many stream and river beds.

$$4Fe^{2+} + O_{2(g)} + 4H^+ \rightleftharpoons 4Fe^{3+} + 2H_2O \qquad (3.84)$$

$$Fe^{3+} + 3H_2O \rightleftharpoons Fe(OH)_{3(s)} + 3H^+ \qquad (3.85)$$
$$\text{ferrihydrite}$$

$$Fe^{3+} + 2H_2O \rightleftharpoons FeOOH_{(s)} + 3H^+ \qquad (3.86)$$
$$\text{goethite}$$

The overall process may be represented as

$$2FeS_{2(s)} + 15/2O_{2(g)} + 7H_2O \rightleftharpoons 2Fe(OH)_{3(s)} + 4H_2SO_{4(1)} \ (3.87)$$

Microorganisms (*e.g. Thiobacillus thiooxidans, Thiobacillus ferrooxidans, Metallogenium*) are involved as catalysts in many of the oxidizing reactions, which would be extremely slow under the prevailing conditions of low pH (*e.g.* <4.5).[59] The combination of acid waters and coatings can have devastating effects upon aquatic biota, with depletion of free-swimming and bottom-dwelling organisms, the loss of spawning gravel for fish and direct fish mortalities. Typical approaches to treatment of acid mine drainage have included methods to remove the iron floc by oxidation and to adjust the pH through the use of limestone filter beds.

3.3.1.4 *Acid Mine Drainage and Release of Heavy Metals.*

The production of sulfuric acid from sulfide oxidation in mines can also lead to the leaching of metals other than iron, with the result that the emerging acidic waters may be laden with heavy metals. Thus in the former metalliferous mining area of south-west England, where there are now many abandoned mine workings, the closure and flooding of the famous Wheal Jane tin mine in 1992 led to a highly acidic cocktail of dissolved metals (Cu, Pb, Cd, Sn, As) entering the Carnon and Fal Rivers at 7–15 million litres per day and spreading throughout the surrounding estuaries and coastal waters.[60]

In the USA, the EPA has identified over 31,000 hazardous waste sites, with the largest complex of 'Superfund' sites to be remediated in western Montana, in the Clark Fork River Basin where there have been more than 125 years of copper and silver mining and smelting activities. Moore and Luoma[61] have characterized three types of contamination resulting from large-scale metal extraction: primary, consisting of wastes produced during mining, milling, and smelting and deposited near their source of origin; secondary, resulting from transport of contaminants away from these sites by rivers or through the atmosphere to soils, groundwaters, rivers, *etc.*; and tertiary, where contaminants may be remobilized many kilometres away from their point of origin. More generally, Furrer and co-workers,[62] in an examination of flocs from streams polluted from acid mine drainage in California and Germany, found that when aluminium-rich acid mine drainage mixes with near-neutral surface waters, fluffy aluminium oxyhydroxide flocs precipitate and move downstream as suspended solids, transporting adsorbed pollutants, *e.g.* heavy metal cations such as Pb^{2+}, Cu^{2+}, and Zn^{2+}. The flocs appear to be formed from aggregation of the aqueous polyoxocation, $AlO_4Al_{12}(OH)_{24}(H_2O)_{12}^{7+}$ [Al_{13}], which is toxic and possibly responsible for the declining fish populations in rivers polluted by acid mine drainage.

Some of the largest waste deposits occur in tailing ponds containing acid mine water. In the Clark Fork River Complex, it is estimated that ponds contain approximately 9000 t arsenic, 200 t cadmium, 90,000 t copper, 20,000 t lead, 200 t silver, and 50,000 t zinc. These metals enter streams and rivers as solutes and particulates and contaminate sediments in the river and reservoirs far downstream from the primary sources (Figure 4).[61] Downstream concentrations follow an exponential decline when viewed over several hundred kilometres. The sediment of Milltown Reservoir, more than 200 km from the mines and smelters at Butte and Anaconda, is highly contaminated with various metals and arsenic. It is believed that oxidation-reduction processes release arsenic from the reservoir sediments and cause contamination of an aquifer from which water drawn through wells, now closed, contains arsenic higher than the EPA drinking water standards.[61] Davis and Atkins,[63] however, have pointed out that, in the bottom sediments of the Clark Fork River channels, the fraction of fine-grained (<63 μm) clay/silt material, which contains high metal concentrations and to which benthic organisms are likely to be directly exposed, is small relative to sands and gravels. Furthermore, the metals occur predominantly in sulfides frequently armoured with an oxide rim and other sparingly soluble phases.

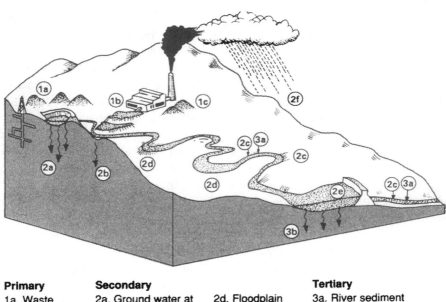

Primary	**Secondary**		**Tertiary**
1a. Waste rock	2a. Ground water at open pits	2d. Floodplain sediment/soil	3a. River sediment reworked from floodplain
1b. Tailings	2b. Ground water beneath ponds	2e. Reservoir sediment	3b. Groundwater from contaminated reservoir sediment
1c. Slag	2c. Sediment in river channels	2f. Soils from air pollution	

Figure 4   *Types of contamination resulting from large-scale metal extraction*
(Reprinted with permission from ref 61, © American Chemical Society, 1990)

When open-pit mining ended in the Clark Fork River Complex in 1982, pumping was discontinued, with the result that water began filling underground shafts and tunnels and the 390 m deep Berkeley pit. Contaminated acid water containing individual metal and sulfate concentrations thousands of times those in uncontaminated water could flow into an adjacent alluvial aquifer and eventually over the rim of the pit. Large-scale hard-rock mining in the western USA, especially Nevada, has greatly increased in recent years, with the result that deep 'pit lakes' are likely to form as open-pit metal mines intersecting groundwater are depleted and shut down.[64] Pit lakes in high sulfide rock will tend to have poor quality acidic water, although oxidized rock that contains significant carbonate will produce better quality, near-neutral pit lake water. Most of the larger existing pit lakes currently contain water that does not meet standards for drinking water, agricultural water quality, or aquatic life.[64] Factors such as the oxygen status of the lake, pH, the hydrogeologic flow system, composition of the wallrock, evapo-concentration, biological activity, and hydrothermal inputs are all important to the modelling of future water quality and impact.

Remediation approaches in circumstances like those described here are based on a variety of physical, chemical, and biological systems. These include the construction of ponds where sufficient organic matter is available to establish anaerobic conditions and immobilize at least some of the metals (*e.g.* Cu, Pb, Cd, Zn) as sulfides. Similar passive treatment of tailings-impacted groundwater has also employed precipitation of metal sulfides as the key clean-up step.[65] Active chemical treatment with lime, producing sludges, and biosorption by reeds, wetlands, *etc.*, are other methods which have been tried. Subsequent to the flooding of the Wheal Jane mine, an active treatment (chemical neutralization) plant and a passive treatment plant, the latter featuring three schemes differing only in the pre-treatment method (lime dosing, anoxic limestone drain, lime-free) used to modify the pH of the influent mine water, were set up. All three systems at the Wheal Jane Passive Treatment Plant included (i) constructed aerobic reed beds designed to remove iron and arsenic, (ii) an anaerobic cell to encourage reduction of sulfate and facilitate removal of zinc, copper, cadmium, and the remaining iron as metal sulfides, and (iii) aerobic rock filters designed to promote the growth of algae and facilitate the precipitation of manganese.[66] The study showed considerable success in the removal of key toxic metals and clearly demonstrated the potential for natural attenuation of acid mine drainage, particularly iron oxidation, by microbial populations.

3.3.2 Metals and Metalloids in Water

3.3.2.1 Arsenic in Groundwater. In what has been described as the largest arsenic poisoning epidemic in the world, hundreds of thousands of people in West Bengal (India) have been seriously affected by arsenic poisoning resulting from the consumption of water drawn from tube wells sunk some 20 to 150 m below the ground into aquifers.[67] In West Bengal, Bangladesh, and other parts of the world (*e.g.* Chile, Mexico, Argentina, Ghana, Mongolia, and Taiwan) where inorganic arsenic concentrations in drinking water are elevated (mg L^{-1} *vs.* WHO recommended limits of 10 µg L^{-1}), exposure has resulted in hyperpigmentation, hyperkeratosis, and cancer of the skin and various internal organs.[68-72]

Sulfide minerals are one of the most important natural sources of arsenic in groundwater. Oxidation of arsenopyrite (FeAsS), in analogous fashion to pyrite (FeS$_2$), may release high concentrations of arsenic into solution.[73,74]

$$4FeAsS_{(s)} + 13O_{2(g)} + 6H_2O \rightleftharpoons 4Fe^{2+} + 4AsO_4^{3-} + 4SO_4^{2-} + 12H^+$$
$$(3.88)$$

In Bengal and Bangladesh, it has been suggested that large-scale withdrawal of groundwater for irrigation produces seasonal fluctuation of the water table, which in turn results in intake of oxygen into the pore waters of sediments that are arsenic-rich in the form of arsenopyrite.[73,74] The exact speciation of soluble inorganic arsenic, *e.g.* as the undissociated forms or different oxyanions of the acids $H_3As^{III}O_3$ and $H_3As^VO_4$, will be dependent upon the prevailing pε and pH. Sorption of arsenic, especially pentavalent arsenate (*e.g.* AsO_4^{3-}, $HAsO_4^{2-}$, $H_2AsO_4^{-}$), onto ferric hydroxide (Fe(OH)$_3$) produced under oxidizing conditions may, however, restrict its mobility and availability, although fertilizer phosphate present in groundwater could perhaps displace the arsenate. Arsenite, especially as $H_3As^{III}O_3$ (pK_1 = 9.2), the predominant form under reducing conditions at pH < 9.2, is much less strongly sorbed.

As an alternative to the theory of arsenopyrite oxidation, however, especially where there is an absence of elevated concentrations of sulfate, it has been proposed that elevated arsenic concentrations in Bengal and Bangladesh groundwater could result from the release of adsorbed arsenate during the dissolution of hydrous iron oxides (similar to ferric hydroxide) under reducing conditions (see Section 3.3.3.2).[73-76] This phenomenon can be seen in those parts of Asia where sub-strata consist of recent alluvial sediments, rich in decaying organic matter, that were deposited and buried during the post-glacial period. Indeed, McArthur and co-workers[77] believe that the proximity of soluble organic matter to drive FeOOH reduction is the key and that this can only occur at the margins of buried peat basins. There is an alternative theory, propounded by Harvey and co-workers,[78] that the arsenic mobilization is associated with the inflow of younger carbon, probably untreated human waste in water that has been drawn into the aquifers of Bangladesh as a consequence of irrigation pumping.

Although trivalent inorganic arsenic, with its propensity for binding to the SH group of enzymes, is acknowledged to be more toxic to humans than pentavalent inorganic arsenic, it must be recognized that As^V can be converted to As^{III} in the human body as part of the reduction/biomethylation pathway of excretion

$$As^V \rightarrow As^{III} \rightarrow MMAA \rightarrow DMAA \qquad (3.89)$$

where MMAA ($CH_3AsO(OH)_2$, monomethylarsonic acid) and DMAA ($(CH_3)_2AsO(OH)$, dimethylarsinic acid) are less toxic metabolites, the latter predominating as the major form of arsenic excreted in urine.[68–72] There is evidence to suggest, however, that trivalent metabolic intermediates of MMAA and DMAA may be highly implicated in the initiation of cancer.[70–72] Methylation could also occur naturally by microbial action in freshwaters, although reported occurrences suggest that the effect is small.[79]

3.3.2.2 Lead in Drinking Water. The naturally soft, slightly acidic, plumbosolvent water of the Loch Katrine water supply for the Glasgow area was recognized many years ago to release lead from the lead pipes and tanks in the domestic plumbing of the Victorian and subsequent (even post-World War II) eras.[80]

$$O_{2(g)} + 4H^+ + 4e^- \rightleftharpoons 2H_2O \qquad E_H^0 = 1.229 \text{ V} \quad (3.90)$$

$$Pb^{2+} + 2e^- \rightleftharpoons Pb_{(s)} \qquad E_H^0 = -0.126 \text{ V} \quad (3.91)$$

$$O_{2(g)} + 4H^+ + 2Pb_{(s)} \rightleftharpoons 2Pb^{2+} + 2H_2O \qquad E_H^0 = 1.355 \text{ V} \quad (3.92)$$

Essentially, elemental lead becomes soluble in acidic conditions due to its oxidation by dioxygen. Furthermore, compounds such as carbonate and hydroxycarbonate compounds of lead, *i.e.* $PbCO_3$ and $Pb_3(CO_3)_2(OH)_2$, that may coat the pipes, will dissolve under acid conditions.

In view of the concern over detrimental effects of exposure to lead upon human health, in particular the possible impact upon intelligence and behaviour of young children, steps were taken in the mid-1970s to reduce the lead content of tap water in Glasgow and other susceptible areas, which often exceeded the WHO maximum guideline at the time of 100 μg L^{-1}. The method chosen was adjustment of pH to 8–9 by lime dosing. The effects of liming the Glasgow water supply were quite dramatic. Whereas pre-1978, when the pH was 6.3, only 50% of random daytime samples were <100 μg L^{-1} in lead, the figure increased to 80% during 1978–1980, when the adjusted pH was 7.8. After 1980, when the pH was increased further to 9.0, 95% of samples were <100 μg L^{-1}.[80] It appears that carbonate and hydroxycarbonate lead compounds present in the coatings on the pipes were stabilized. Significant reductions in blood lead levels of key exposed groups (*e.g.* pregnant women) were also observed.

Since 1989, as regulatory upper limits for lead in drinking water have fallen, *e.g.* to 50 μg L^{-1} (EC) and now to 10 μg L^{-1} (WHO), orthophosphate has been added to the water supply in Glasgow to precipitate insoluble lead compounds such as $Pb_3(PO_4)_2$ and $Pb_5(PO_4)_3OH$. This has resulted in a fall in the proportion of households with water lead >10 μg L^{-1} from 49% in 1981 to 17% in 1993.[81] Despite this improvement, an estimated 13% of infants were still exposed *via* bottle feeds to tap water lead concentrations in excess of 10 μg L^{-1} and it seems very unlikely that further treatment of the water supply will be able to guarantee water lead concentrations <10 μg L^{-1}.

Occasionally, elevated concentrations of lead in tap water arise from the illegal use of lead solder in capillary joints to join copper pipes.

$$Cu^{2+} + 2e^- \rightleftharpoons Cu_{(s)} \qquad\qquad E_H^0 = 0.337 \text{ V} \qquad (3.93)$$

$$Cu^{2+} + Pb_{(s)} \rightleftharpoons Cu_{(s)} + Pb^{2+} \qquad\qquad E_H^0 = 0.463 \text{ V} \qquad (3.94)$$

In Washington DC, USA, where the water disinfection programme was recently modified from chlorine to chloramines, an unexpected appearance of high levels of lead in the tap water has been partly attributed to similar galvanic corrosion when copper pipes are connected to lead-containing brass fittings in the presence of chloramines. With the shift to chloramines, the oxidizing potential of Washington DC's water was lowered and not only could the brass become highly anodic and the copper cathodic, but also the previously deposited PbO_2 scales on the inside of lead service pipes began to dissolve.[82]

3.3.2.3 Cadmium in Irrigation Water. In the 1950s in Japan, many people, especially menopausal women suffering from malnutrition, low vitamin D intake, and calcium deficiency, suffered a condition known as Itai-Itai (Ouch-Ouch) disease, with symptoms ranging from lumbago-type pains to multiple fractures of softened bones. The cause was irrigation water from a river (Jintsu) chronically contaminated with dissolved cadmium from a zinc mining and smelting operation. Contaminated rice from the irrigated paddy fields was eaten and the Ca^{2+} in bones replaced by Cd^{2+}, an ion of the same charge and size.

Although this is a particularly extreme example of acute effects resulting from high exposure to this non-essential element, concern has been expressed about chronic effects (*e.g.* kidney damage, hypertension) from possible enhanced exposure of humans through increased application of sewage sludge to agricultural land, in view of EC-enforced cessation of dumping at sea by 1999. Compared with other heavy metals, cadmium exhibits an especially high mean sludge concentration (20 mg kg^{-1}) relative to mean soil concentration (0.4 mg kg^{-1}).[83] In acidic soils the concentration of Cd^{2+} available for uptake by plants can be substantial,[84] as it adsorbs only weakly onto clays, whereas at pH >7 it readily precipitates, *e.g.* as $CdCO_3$ for which the solubility product, $K_{SP} = 1.8 \times 10^{-14}$, is indicative of the advantages of liming. Similarly, in drinking water, the presence of dissolved carbonate at a concentration of 5×10^{-4} mol L^{-1} can reduce the solubility of cadmium from 637 to 0.11 mg L^{-1}, in line with evidence that hard water, with a high calcium content, can protect against cadmium.[85]

3.3.2.4 Selenium in Irrigation Water. In 1983, high rates of embryonic deformity and death, attributed to selenium toxicosis, were found in wild aquatic birds at the Kesterson National Wildlife Refuge in the San Joaquin Valley of California.[86] Kesterson Reservoir was a regional evaporation pond facility where drainage waters, often containing high level of salts and trace elements (including selenium), from irrigated farmland had been collected since 1978.

With evapotranspiration, greatly in excess of precipitation, bringing soluble salts to the surface of farmland in the arid climate of the west-central San Joaquin Valley, and crop productivity, after irrigation, threatened by shallow saline groundwater near the root zone, grids of sub-surface tile drains were constructed to divert the saline waters to a collective drain (San Luis), which flowed into Kesterson Reservoir. It was the geologic setting of the San Joaquin Valley as well as the climate, however, which led not only to the soil salinization but also to the presence of selenium (in the form of SeO_4^{2-}) in Kesterson inflow waters at concentrations well in excess of the USEPA designation of 1000 µg L^{-1} for selenium as a toxic waste, never mind its 10 µg L^{-1} limit for drinking water and the $< 2.3 \, \mu g \, L^{-1}$ limit subsequently suggested for the protection of aquatic life.

To the west of the San Joaquin River (Figure 5), selenium in the soils is believed to be of natural origin. During the Jurassic and Cretaceous periods, there was deposition of marine sediments comprising sandstones, shales, and conglomerates, including seleno-sulfides of iron ($FeS_2 + FeSe_2$). Subsequent uplifting of these sediments produced the Coast Ranges and

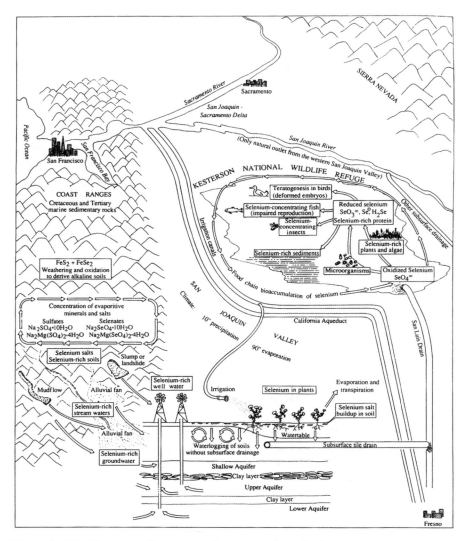

Figure 5   *The Kesterson effect: biogeochemical cycling of selenium in the Coast Ranges, San Joaquin Valley, and Kesterson National Wildlife Refuge of California* (Reproduced with permission from ref 86, © Springer Verlag, 1994)

subjected the sediments to weathering under oxidizing conditions. With resultant acid neutralized by the carbonate component of the sediments, the predominant form of selenium produced from oxidation of selenide, Se^{-II}, under alkaline conditions, is selenate, $Se^{VI}O_4^{2-}$, in preference to selenite, $Se^{IV}O_3^{2-}$, or biselenite, $HSe^{IV}O_3^-$.[87] Thus the soils on the Coast Ranges, which are alkaline, contain significant amounts of soluble mineral salts (*e.g.* $Na_2SO_4.10H_2O$, $Na_2Mg(SO_4)_2.4H_2O$), including selenates (*e.g.* $Na_2SeO_4.10H_2O$, $Na_2Mg(SeO_4)_2.4H_2O$).

The chemistry and mobility of selenite and selenate differ greatly in soils. Selenite is adsorbed by specific adsorption processes (*e.g.* on clays and hydrous metal oxides) and to a much greater extent than selenate, which is adsorbed, like sulfate, by comparatively weak non-specific processes. Thus, in aerated alkaline soils, such as those of semi-arid regions like the west-central San Joaquin Valley, mobile $Se^{VI}O_4^{2-}$ will be the dominant form and, through runoff, be capable of entering the groundwaters and sub-surface drainage waters, ultimately being removed to Kesterson Reservoir. There, it is taken up by the biota, with such devastating effects, and also converted to a range of selenium-containing species (*e.g.* $Se^{IV}O_3^{2-}$, Se^0, H_2Se, Se-rich protein), with much ultimately being deposited in the sediments. There, over geological time, natural diagenetic processes would presumably lead again to the formation of reduced seleno-sulfide species.

Considerable effort has been devoted to the clean-up of Kesterson Reservoir, which was drained in 1987, one year after the input of drainage water was stopped. Natural processes, stimulated by the addition of nutrients, of summertime microbial conversion and volatilization to the atmosphere of methylated forms such as dimethylselenide, $(CH_3)_2Se$, and downwards leaching of $Se^{VI}O_4^{2-}$ by percolating water in winter helped to dissipate 68–88% of total selenium from the topsoil (0–15 cm) over a period of eight years.[88] Phytoremediation techniques were also tested. Similarly, pilot bioreactors were set up to convert drainage water SeO_4^{2-} into commercially useful selenium-containing products. Although Kesterson Reservoir has been closed, selenium loading in the Central Valley of California, the San Joaquin River, the San Joaquin-Sacramento Delta, and San Francisco Bay still occurs and an emerging selenium contamination issue is developing in south-eastern Idaho, USA, in streams draining areas with phosphate mining activities.[89]

3.3.2.5 Chromium in Groundwater. Millions of tonnes of high-pH chromite ore processing residue (COPR) were deposited (*e.g.* as landfill material) in the past in urban areas such as Glasgow, Scotland,[90] and Hudson County, New Jersey, USA.[91] Even now the high-lime process responsible for the generation of this waste continues to be used in countries such as China, Russia, Kazakhstan, India, and Pakistan.[92] In Glasgow, the chromium content of this residue is high at 3–4% w/w but, more significantly, elevated concentrations as high as $\sim 1\%$ w/w of highly toxic and carcinogenic hexavalent chromium, which takes the form of chromate ($Cr^{VI}O_4^{2-}$) at the prevailing high pH of 9–12 characteristic of high-lime COPR, have been measured in the solid phase at the contaminated sites. Furthermore, on sites in the Glasgow area where COPR from a high-lime chemical works was last deposited in the 1960s, Cr^{VI} is still leaching out in such quantity that receiving groundwaters and streams in the area are seriously contaminated at concentrations of up to 100 mg L^{-1}, about 2000 times greater than Environmental Quality Standards (EQSs).[93]

There are many problems involved in the *in situ* application of remediation-by-reduction methods, *e.g.* employing Fe, Fe^{II}, or organic compounds, to reduce Cr^{VI} held in specific mineral phases within COPR to the much less harmful trivalent Cr^{III}. For example, the use of ferrous iron, Fe^{2+}, from ferrous sulfate, rather than reduce the $Cr^{VI}O_4^{2-}$ held within the COPR *via*

$$2CrO_4^{2-} + 6Fe^{2+} + 13H_2O \rightarrow Cr_2O_{3(s)} + 6Fe(OH)_{3(s)} + 8H^+ \quad (3.95)$$

actually increases the release of CrO_4^{2-} because of anion exchange of SO_4^{2-} from the ferrous sulfate for CrO_4^{2-} held in the layered double hydroxide mineral hydrocalumite within the COPR.[94] In principle, many of the problems involved in the *in situ* application of methods designed to reduce Cr^{VI} held within COPR to Cr^{III} can be avoided if the emphasis is switched to treatment of the Cr^{VI} released to groundwater and surface water from COPR. Where the objective is to treat groundwater either in a conventional pump and treat system or as it migrates through a permeable reactive barrier, there is much less impediment to chemical methods using Fe^{II} or Fe^0 or S^{2-}:[95–97]

$$2CrO_4^{2-} + 3H_2S_{(aq)} + 2H_2O \rightarrow 2Cr(OH)_{3(s)} + 3S_{(s)} + 4OH^- \quad (3.96)$$

More generally for Cr^{VI}-contaminated groundwater, which may also arise from other industrial activities such as tanning and electroplating, both *in situ*[98] and pump and treat methods could employ biological as well as chemical methods for the reduction of Cr^{VI} to Cr^{III}, for example in packed bed bioreactors.[99]

Another generic method for the reduction of Cr^{VI} to Cr^{III} is the addition of organic matter, for example as horse dung or molasses. Such treatment, however, whether for direct chemical reduction (*cf.* Equation

(3.111)) or, in conjunction with other nutrients, to stimulate bacterial reduction of Cr^{VI}, could provide a means of retention of Cr^{III} in solution in preference to either precipitation of $Cr(OH)_3$ or reoxidation to Cr^{VI} under the prevailing alkaline conditions of high-lime COPR sites. Farmer and co-workers[93] found that <3% of dissolved chromium was present as Cr^{VI} in groundwater at one Glasgow borehole site with a high dissolved organic carbon content of 300 mg L^{-1}. Here the chromium occurred predominantly as Cr^{III} in association with organic material of high molecular weight, with an implied reduction of Cr^{VI} by humic substances, perhaps *via* carboxylate groups, and the formation of Cr^{III}-humic complexes.

Although the drinking water supply of people in the Glasgow area is not threatened by the Cr^{VI}-contaminated groundwater, there are instances elsewhere of such problems. For example, a massive plume of Cr^{VI}-contaminated groundwater in the Mojave Desert, with concentrations as high as 12 mg L^{-1}, was reported to be within 40 m of the Colorado River, which supplies drinking water to California, USA.[100] This contamination arises from when an electric utility used Cr^{VI} to control corrosion and mould in water-cooling towers and then dumped untreated wastewater into percolation beds in the 1950s and 1960s. Current short-term emergency pumping, transport, and offsite treatment of the contaminated groundwater will have to be replaced by a long-term solution close to the river, possibly featuring *in situ* groundwater treatment or the installation of a permeable reactive barrier in which iron would reduce Cr^{VI} to Cr^{III}.

3.3.2.6 Aquatic Contamination by Gold Ore Extractants. The separate use of mercury and cyanide has led to the contamination of freshwater systems.

3.3.2.6.1 Mercury in the Amazon Basin. With the price of gold soaring in the late 1970s and early 1980s, there developed a modern-day gold rush in South America,[101] which reached its peak during 1990–1993.[102] In the relatively inaccessible Amazon area in Brazil, there were up to a million people operating on an informal basis. The technique which is used to extract the placer gold which has accumulated in many stream and alluvial deposits is based on amalgamation with mercury. Typically, high-pressure water hoses are used to dislodge alluvium, which is then taken into a large sluice by a motor suction pump. The sluices are lined with sacking into which mercury is added to amalgamate the gold particles, which lodge in the sacking. In other versions, no mercury is used in the sluices. Instead, material is collected from the sluice lining and crushed in an oil drum, where the mercury is then introduced, followed by manual panning.

The effects of all this activity can be seen in the increased turbidity, with clear rivers turning a muddy brown colour, and increased fish mortality. Furthermore, releases of mercury to the aquatic environment have resulted in elevated concentrations of mercury in sediments and fish. Commonly eaten fish species from the Madeira River were found to contain mean mercury levels of 0.7 mg kg^{-1} and up to 3.8 mg kg^{-1} in the Tapajos Valley.[103] Given the bioaccumulation and biomagnification of mercury in many fish species and the possible microbial methylation of mercury deposited in sediments to the even more toxic lipid-soluble methyl mercury, CH_3Hg^+, there is a real prospect of enhanced dietary uptake of the most toxic form of mercury by the local populations. This would be in addition to exposure from inhalation of mercury vapour during the handling of mercury and from subsequent post-amalgamation treatment steps, which usually involve refinement of the extracted gold by heating to drive the more volatile mercury off. It is thought that there is an annual discharge of about 100 tonnes of mercury into the Amazon ecosystem and perhaps the same amount released as mercury vapour. Thus there could be an increased risk of the neurological damage and foetal deformities commonly associated with mercury poisoning, as most notably exemplified in the fishing village population of Minimata, Japan, in the 1950s, when a chemical manufacturing plant discharged mercury salts and CH_3Hg^+ into the bay, to be taken up by fish and shellfish.

3.3.2.6.2 Cyanide. Another method of extracting gold that can lead to contamination of freshwater is cyanide heap leaching, which is used to extract tiny amounts of gold from huge volumes of low-grade deposits of gold-bearing rock. In this case, ore is dug from open-pit mines, crushed, and spread on asphalt or plastic pads. The heaps are then sprayed with cyanide, which dissolves the gold.

$$4Au_{(s)} + 8NaCN + 2H_2O + O_{2(g)} \rightarrow 4NaOH + 4NaAu(CN)_2 \quad (3.97)$$

The gold-bearing cyanide solutions run off on impermeable asphalt or plastic to collection vats and are then treated, for example with zinc, to extract gold.

$$2NaAu(CN)_2 + Zn_{(s)} \rightarrow Na_2Zn(CN)_4 + 2Au_{(s)} \quad (3.98)$$

One major environmental problem with this technique is that cyanide is highly toxic. Although attempts are made to loop the cyanide through a closed system to try to ensure that none is lost, cyanide does commonly escape to surface waters and groundwater. While cyanide quickly decomposes in oxygenated, acidic surface water, it can persist at toxic levels for much longer periods in groundwater and thus pose a longer-term threat.

3.3.3 Historical Pollution Records and Perturbatory Processes in Lakes

3.3.3.1 Records – Lead in Lake Sediments. Under ideal circumstances, freshwater lake sediments can preserve the record of temporal variations in anthropogenic input of contaminants *via* atmospheric deposition, catchment runoff, effluent inflow and dumping from industrial, transportation, mining, agricultural, and waste disposal sources.[104] Prerequisites are transfer of contaminants associated with settling inorganic particulates and/or biotic detritus from the water column to the sediments, no disturbance of sediments by physical mixing, slumping or bioturbation after deposition, no post-depositional degradation or mobility of the contaminants, and the establishment of a reliable time axis (*e.g. via* the use of the naturally-occurring radionuclide ^{210}Pb, half-life 22.35 years, and the nuclear fallout radionuclide ^{137}Cs, half-life 30.2 years).

The derived records of lead deposition to sediments in Loch Ness and the northern and southern basins of Loch Lomond, Scotland, over the past few hundred years since the onset of the Industrial Revolution are shown in Figure 6.[105] The noticeable change in the $^{206}Pb/^{207}Pb$ ratio of the lead deposited from the late 1920s onwards is attributed to car-exhaust emissions of lead of lower $^{206}Pb/^{207}Pb$ ratio as a result of the use of lead ores from Broken Hill, Australia, in the manufacture of alkyllead additives for the UK market. On the basis of the published literature, there appears to be reasonable confidence that, in the absence of post-depositional mixing, sediments do, in general, preserve a record of the deposition of lead to the sediments.[105–107] Many such records in Western Europe and North America now show concentrations of lead declining towards the sediment surface as a consequence of the withdrawal of leaded petrol and also a reduction in other emissions.[105,108] Longer-term palaeolimnological records, stretching back over several thousand years, demonstrate the influence of lead mining and smelting in Greek–Roman times and the later impact of the Mediaeval mining industry.[109]

3.3.3.2 Perturbatory Processes in Lake Sediments. The extent to which conditions for preservation of records are met depends upon the characteristics of the specific individual systems and contaminants under study. In reviewing the perturbation of historical pollution records in aquatic sediments, Farmer[110] has suggested that in the case

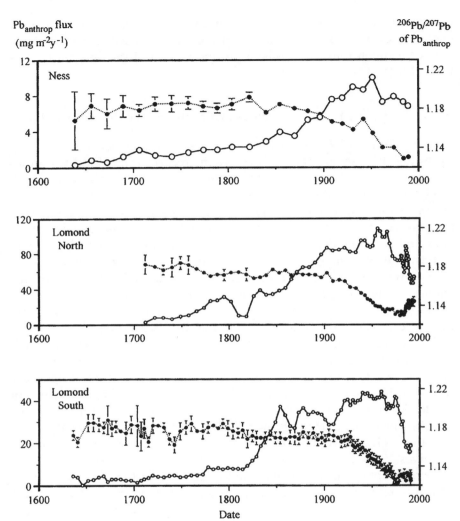

Figure 6 *Fluxes (○) and associated* $^{206}Pb/^{207}Pb$ *ratios (●) of anthropogenic lead deposited in the sediments of Loch Ness, northern Loch Lomond and southern Loch Lomond (Scotland), derived from analysis of sections of* ^{210}Pb-*dated sediment cores, vs. calendar date since the 17th century. Dates prior to approximately 1860 should be considered as extrapolations*
(Reprinted with permission from ref 105, © Elsevier, 2002)

of heavy metal behaviour in freshwater lakes there should ideally be investigation of

- different systems of varying status (*e.g.* oligotrophic, eutrophic, acid),

- seasonal influences,
- modes of element introduction and transport,
- element associations in the water column,
- depositional and redistributional processes and rates (*e.g. via* radionuclide and stable isotope studies) in the sediments,
- concentration profiles and speciation in the overlying water, solid sediment, and pore water,
- element/solid phase associations (*e.g. via* sequential chemical extraction procedures), and
- authigenic mineral identification in the sediments.

While opinions may differ as to whether particular sediment columns have been disturbed by mixing or the extent to which certain elements are vertically mobile, it is accepted that post-depositional redox-controlled cycling can affect the vertical profiles of manganese and iron.[111,112] The driving force is the microbiological decomposition of organic matter (represented as CH_2O) through bacterial utilization of O_2 and inorganic oxidizing agents in thermodynamically favoured sequence.[113]

$$\Delta G^0$$
$$(\text{kJ mol}^{-1}\ CH_2O)$$

$$CH_2O + O_{2(g)} \rightleftharpoons CO_{2(g)} + H_2O \qquad\qquad -475 \qquad (3.99)$$

$$5CH_2O + 4NO_3^- \rightleftharpoons 2N_{2(g)} + 4HCO_3^- + CO_{2(g)} + 3H_2O \qquad -448 \qquad (3.100)$$

$$CH_2O + 3CO_{2(g)} + H_2O + 2MnO_{2(s)} \rightleftharpoons 2Mn^{2+} + 4HCO_3^- \qquad -349 \qquad (3.101)$$

$$CH_2O + 7CO_{2(g)} + 4Fe(OH)_{3(s)} \rightleftharpoons 4Fe^{2+} + 8HCO_3^- + 3H_2O \qquad -114 \qquad (3.102)$$

$$2CH_2O + SO_4^{2-} \rightleftharpoons H_2S_{(g)} + 2HCO_3^- \qquad\qquad -77 \qquad (3.103)$$

$$2CH_2O \rightleftharpoons CH_{4(g)} + CO_{2(g)} \qquad\qquad -58 \qquad (3.104)$$

Upon dissolution of oxides and (hydr)oxides of manganese and iron under reducing conditions at depth, divalent cations of these elements can diffuse upwards through the pore waters to be oxidized and precipitated in near-surface oxic layers, leading to the characteristic near-surface enrichment of manganese and iron in the sediments of many well oxygenated lakes (see Section 3.2.4.3). Phosphorus and arsenic are also recycled but in their case it is the negatively charged anions that are released into solution from an association with solid phase ferric (hydr)oxides, which dissolve under sufficiently reducing conditions. Upward diffusion of reduced species leads ultimately to surface enrichment *via* co-precipitation or adsorption of phosphate and arsenate on iron (hydr)oxides in surface layers. As a result, arsenic concentrations as high as 675 mg kg^{-1} can be found near the surface of Loch Lomond sediments, compared with background values of 15–50 mg kg^{-1} at depth.[114] Enrichments of phosphorus and arsenic have similarly been found in association with authigenic iron oxides in sedimentary Fe/Mn layers in Lake Baikal where, however, enrichments of molybdenum were found associated with manganese oxides, perhaps through mixed oxide formation.[115]

For lakes which have undergone significant acidification, it has been suggested that heavy metals could be released from surface sections by pH-dependent dissolution, resulting in sub-surface maxima in sedimentary heavy metal concentrations. In two Canadian acid lakes, however, Carignan and Tessier[116] found that downward diffusive fluxes of dissolved zinc from overlying waters into anoxic pore waters were responsible for the pronounced sub-surface sediment maxima in solid phase zinc, presumably as the insoluble sulfide.

Traditionally, heavy metals such as lead, zinc, copper, and cadmium have been considered diagenetically (*i.e.* as a consequence of chemically and biologically induced changes in prevailing sedimentary conditions, *e.g.* pε, anionic composition, *etc.*) immobile and fixed in the sediment after deposition, partly as a result of the formation of comparatively insoluble sulfides under reducing conditions at depth. This may well be an over-simplified view, as interactions and formation of complexes between metals and dissolved organic matter (*e.g.* humic substances) may maintain dissolved metal concentrations at levels greater than those predicted by simple solubility product calculations (see Sections 3.2.3.5 and 3.2.4.1).[117] A combination of geochemical modelling, incorporating interactions of metals with mineral phases and naturally occurring organic matter,[34,118] and the development of increasingly sophisticated analytical technology, such as DET (diffusive equilibration in thin films)[119] and DGT (diffusive gradients in thin films) for the high-resolution determination on a sub-millimetre scale of metals in sedimentary pore waters,[120] offers considerable promise for greater understanding of redox-driven cycling of trace elements in lakes. Not only have fine structure and sharp-featured maxima been observed for manganese, iron, sulfide, and some trace metals in pore waters, but two-dimensional structure in concentration profiles has demonstrated the potential importance of sedimentary microniches for metal remobilization.[119,121] The increasing sophistication of the DGT technique is now also permitting comparisons of measured metal speciation in freshwaters with model (*e.g.* WHAM, ECOSAT) predictions.[122,123]

Finally, it should be noted that sediment cores continue to be used to investigate historical perspectives of environmental contamination by anthropogenic organic compounds, such as organochlorine pesticides,[124,125] polychlorinated biphenyls,[124,125] polychlorinated naphthalenes,[126] polycyclic aromatic hydrocarbons,[124,127,128] aliphatic hydrocarbons,[128,129] polychlorinated dibenzo-*p*-dioxins,[130] polychlorinated dibenzofurans,[130] and polybrominated diphenyl ethers[125,131] in environments as diverse as the Great Lakes[124,129–131] and Washington[128] in North America, Esthwaite Water[126] in England, remote mountain lakes[127] in the Pyrenees and Alps in Europe and Greenland.[125]

3.3.3.3 Onondaga Lake.

Onondaga Lake, near Syracuse, New York State, has been called 'one of the world's most polluted lakes' and provides a textbook case of the impact of industrial processes on the environment.[132] Rich natural resources, especially the local brine springs, made the Syracuse region an ideal location for chemical manufacturing. Unfortunately, the exploitation of these, despite the implementation of significant technological advances and, for the time, environmentally responsible operation of factories had a devastating impact upon the chemistry of the lake.

The Solvay Process, which was introduced for the manufacture of the industrially important chemical, 'soda ash', *i.e.* sodium carbonate (Na_2CO_3), along the west shore of Onondaga Lake in 1884, was cheaper and less polluting than the existing Leblanc Process. Essentially, it made use of two cheap and plentiful naturally occurring substances in the area – $NaCl$ from the deep brine springs and $CaCO_3$ from limestone outcroppings – in a simple reaction, which yielded two useful products.

$$CaCO_3 + 2NaCl \rightleftharpoons Na_2CO_3 + CaCl_2 \qquad (3.105)$$

In practice, this overall reaction required a series of individual reactions and chemical intermediates, one of the most important involving ammonia in a step to separate $NaHCO_3$ from NH_4Cl by fractional crystallization at 273 K.

$$NH_{3(g)} + CO_{2(g)} + NaCl_{(ag)} + H_2O_{(1)} \rightleftharpoons NaHCO_{3(s)} + NH_4Cl_{(aq)}$$
$$(3.106)$$

The $NaHCO_3$ was subsequently heated to yield Na_2CO_3 while NH_3, the most expensive compound used, was regenerated from the decomposition of NH_4Cl. So the very efficient Solvay Process, one of the first industrial processes to regenerate intermediates, maximized profits and, in theory, prevented environmental harm from unwelcome discharges.

Unfortunately, however, sales of $CaCl_2$ from the Onondaga Lake plant failed to match those of Na_2CO_3, with the result that excess $CaCl_2$ was allowed to be released into a tributary of the lake. In addition, substantial quantities of unmarketable salts (mainly $CaCl_2$) from the Solvay Process were dumped daily into the lake. Furthermore, waste slurries pumped into diked beds along the lake shoreline resulted in substantial leaching of Ca^{2+}, Na^+, and Cl^- ions into the lake by rainwater runoff. The Ca^{2+} ions reacted with CO_3^{2-} ions, resulting from

$$H_2O_{(1)} + CO_{2(g)} \rightleftharpoons H_2CO_{3(aq)} \rightleftharpoons 2H^+_{(aq)} + CO_3^{2-}{}_{(aq)} \quad (3.107)$$

to precipitate $CaCO_3$ *via*

$$Ca^{2+}_{(aq)} + CO_3^{2-}{}_{(aq)} \rightleftharpoons CaCO_{3(s)} \quad (3.108)$$

The effect of the deposition of $CaCO_3$, at up to 1.7×10^4 t per year, was to increase the sedimentation rate of the lake several-fold, with a large $CaCO_3$ delta where the major tributary, Ninemile Creek, flows into the lake, and create a layer of $CaCO_3$ about 1 m thick over the bottom sediments. As a result, Onondaga Lake became effectively a saturated solution of $CaCO_3$, with a pH of 7.6–8.2, and, most unusually for a lake in the north-eastern USA, immune to both acid precipitation and phosphate-induced eutrophication, the latter largely being avoided as a consequence of

$$3Ca^{2+}_{(aq)} + 2PO_4^{2-}{}_{(aq)} \rightleftharpoons Ca_3(PO_4)_{2(s)} \quad (3.109)$$

Since the closure of the Solvay Plant in 1980, the water quality has improved as salt concentrations have decreased from 3500 to 1000 mg L^{-1}.

The other major industrial process which has polluted Onondaga Lake has been the production of sodium hydroxide (NaOH) and chlorine, two of the world's most important chemicals, from sodium chloride *via* the electrolysis of NaCl solution and subsequent amalgamation of the sodium metal produced with the mercury cathode [*i.e.* Na(Hg)], prior to spraying into water to yield NaOH. With a single electrolysis cell containing up to 4 t Hg and a single plant containing dozens of cells, the escape of mercury (*e.g.* through leakage) at a rate of 5–10 kg per day from 1946–1970 into the lake resulted in mercury-contaminated sediments (>40 mg kg^{-1}) and fish, much of the mercury in the latter being in the form of the highly toxic CH_3Hg^+. Despite the closure of the chlor-alkali plant in the late 1980s, mercury is still flowing into the lake from tributaries, although concentrations of soluble mercury compounds in the lake are declining.[132] Kim and co-workers[133] have simulated mercury transport and speciation (Hg^{II}, Hg^0, CH_3Hg^+) in both the water column and the benthic sediment using a Water Quality Simulation Program. Model predictions generally agreed with measured values for the water column, in which advection, sorption, and settling were important mechanisms for mercury transport. Reduction, methylation, and demethylation affected mercury speciation in the water column and the benthic sediment. Natural attenuation showed no positive impact for remediation when compared with dredging and capping.

3.3.4   Nutrients in Water and Sediments

3.3.4.1   Phosphorus and Eutrophication.   Eutrophication can be considered as the excessive primary production of algae and higher plants through enrichment of waters by inorganic plant nutrients, usually nitrogen and phosphorus. The latter, in the form of phosphate, is normally the limiting nutrient because the amount of biologically available phosphorus is small in relation to the quantity required for algal growth.[134] Sources of nutrients can be discrete (*e.g.* specific sewage outfall) or diffuse (*e.g.* farmland fertilizers). Eutrophic lakes, highly productive and often turbid owing to the presence of algae, can be contrasted with oligotrophic lakes, which exhibit low productivity and are clear in summer. There have been many examples of unsightly algal blooms affecting freshwater bodies throughout the world, from Lake Erie in North America to the Norfolk Broads in East Anglia, England.[59] Public concern has increased along with the reported incidences of toxicity of the bloom-forming organisms, in particular the cyanobacteria (blue-green algae), which have been implicated in fish fatalities, for example in Loch Leven, Scotland.[135] The environmental damage costs of freshwater eutrophication in England and Wales have been estimated at £75–114.3 M per year.[136]

The chemical form of phosphorus in the water column available for uptake by biota is important. The biologically available phosphorus is usually taken to be 'soluble reactive phosphorus (orthophosphate)', *i.e.* which, upon acidification of a water sample, reacts with added molybdate to yield molybdophosphoric acid, which is then reduced with $SnCl_2$ to the intensely-coloured molybdenum blue complex and is determined spectrophotometrically ($\lambda_{max} = 882$ nm).[137] Reduction in inputs of phosphate, for example from point sources or by creating water meadows and buffer strips to contain diffuse runoff, has obviously been one of the major approaches to stemming eutrophication trends and encouraging the restoration of affected lakes. That this has not always been successful, however, can be attributed in many cases to the release/recycling of phosphorus previously deposited to and incorporated within the bottom sediments of the lake systems in question.[138]

The potential mobility and bioavailability of sedimentary phosphorus are to a large extent governed by the chemical associations and interactions of phosphorus with different sedimentary components.[139] Bearing in mind the removal or transport of phosphorus to the sediments, especially important phases are likely to be 'organic phosphorus' from deposited, dead, decaying biota, and 'sorbed orthophosphate' on inorganic particulates (*e.g.* iron (hydr)oxides). Many sequential extraction schemes have been developed to investigate phosphorus fractionation in lake sediments. Perhaps the most sophisticated is that of Psenner *et al.*[140] who identify, operationally define, and separate the following fractions: 'labile, loosely bound or adsorbed' (NH_4Cl-extractable); 'reductant-soluble, mainly from iron hydroxide surfaces' (buffered dithionite-extractable); 'adsorbed to metal oxides (*e.g.* Al_2O_3)' (NaOH-extractable), subsequently distinguishable from 'organic' (also NaOH-extractable); 'apatite-bound' (HCl-extractable); and 'residual' (persulfate-digestible). Pardo and co-workers[141] have reviewed such approaches in the light of the development by the European Commission of a harmonized protocol for the fractionation of sedimentary phosphorus.

Many such studies of sedimentary phosphorus profiles, also incorporating pore water measurement of soluble reactive phosphate, have demonstrated that redox-controlled dissolution of iron (hydr)oxides under reducing conditions at depth releases orthophosphate to solution. This then diffuses upwards (and downwards) from the pore water maximum to be re-adsorbed or co-precipitated with oxidized Fe^{III} in near-surface oxic sections. The downwards decrease in solid phase 'organic' phosphorus indicates increasing release of phosphorus from deposited organic matter with depth, some of which will become associated with hydrous iron and other metal oxides, added to the pool of mobile phosphorus in pore water or contribute to 'soluble unreactive phosphorus'. The characteristic reactions involving inorganic phosphorus in the sediments of Toolik Lake, Alaska, are shown in Figure 7.[142] If, at depth, the concentrations of Fe^{2+} and phosphate are high enough, authigenic vivianite ($Fe_3(PO_4)_2.8H_2O$) may precipitate out.

With redox control largely responsible for phosphorus mobility in sediments, what might the consequences of oxygen depletion in the hypolimnion be? If conditions in the surface sediments are not sufficiently oxidizing to precipitate iron (hydr)oxides and thereby adsorb the phosphate (*i.e.* the redox boundary for iron may be in the overlying

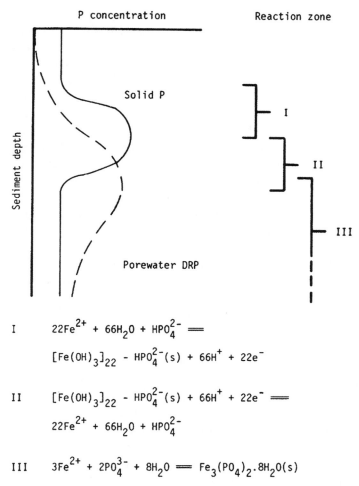

P concentration   Reaction zone

Solid P

Sediment depth

Porewater DRP

I   II   III

I $22Fe^{2+} + 66H_2O + HPO_4^{2-}$ ===

$[Fe(OH)_3]_{22} - HPO_4^{2-}(s) + 66H^+ + 22e^-$

II $[Fe(OH)_3]_{22} - HPO_4^{2-}(s) + 66H^+ + 22e^-$ ===

$22Fe^{2+} + 66H_2O + HPO_4^{2-}$

III $3Fe^{2+} + 2PO_4^{3-} + 8H_2O$ === $Fe_3(PO_4)_2 \cdot 8H_2O(s)$

Figure 7   *Characteristic reactions involving phosphorus in the sediments of Toolik Lake,*
Alaska. The primary processes controlling porewater phosphorus concentrations
are adsorption to and desorption from iron oxyhydroxides and the precipitation
of authigenic vivianite
(Reprinted with permission from ref 142, © E. Schweizerbart'sche Verlags-
buchhandlung, 1987)

water column), the phosphate from previously deposited sediments
would stream off into the water column and promote eutrophication.
This process is called internal loading of phosphorus.

Redox control may not be the only process affecting release of
phosphorus from sediments. During the enhanced photosynthesis of
algal blooms, the pH of lake water increases as CO_2 is used up and
HCO_3^- increases. Thus, in summer, both Lough Neagh[143] in N. Ireland
and Lake Glanningen[144] in Sweden have shown an increase in water

phosphate concentration. It seems likely that OH^- is exchanging with sorbed phosphate in alkaline lakes, thus releasing phosphorus from association with iron and aluminium (hydr)oxides (see Section 3.2.4.3). If the waters are calcium-rich, however, this could have the effect of precipitating phosphate as hydroxyapatite ($Ca_{10}(PO_4)_6(OH)_2$) under the high-pH conditions prevailing. There can be other factors, such as temperature, which promote phosphorus release. An increase in temperature can lead to increased bacterial activity, which increases oxygen consumption and decreases the redox potential. Turbulence may also be a factor. For example, wind-induced bottom currents in shallow lakes could destroy any pH gradients across a buffered sediment-water interface and re-suspend P-rich sediment in water of high pH.[145]

There have been many proposals for restoration of eutrophic lakes.[146] For example, 34 options have been put forward for Loch Leven,[147] Scotland, with the aim of reducing algal biomass and the incidence of the bloom-forming cyanobacteria in particular. These strategies fall into two categories: (i) those aimed at stemming the production of algae in the first place, including reduction in the supplies of light and nutrients and (ii) those aimed at reducing existing algal biomass, including physical methods such as increased flushing of the loch and harvesting of blooms, chemical treatment with algicides, and biological methods involving viruses, parasitic fungi and grazing protozoans, rotifers, and micro-crustaceans. Thus far, progress has largely been restricted to reducing external inputs of phosphorus from point sources to the loch, an approach which could perhaps be supplemented by the creation of water meadows and/or buffer strips to reduce inputs from diffuse runoff. Elsewhere, there was a surge in wetlands construction in the 1980s, using a mixture of plants to clean water contaminated with nitrates and phosphates.[148] One of the largest wetlands restoration projects is planned for the Florida Everglades, to the south and east of the rich farmlands around Lake Okeechobee, where runoff waters enriched in phosphorus from fertilizers have disrupted the flora and fauna. As water flows through cattails and sawgrass, phosphorus concentrations are expected to decline from 170 to 50 $\mu g\ L^{-1}$. An intriguing attempt to use ochre (*cf.* Section 3.3.1.3), discharged from abandoned mines, to strip phosphate from sewage and agricultural runoff and then use the phosphate-saturated ochre as a slow-release phosphate fertilizer to fertilize a range of grasses and cereal crops has been reported.[149] It should be noted that the addition of ferric sulfate or chloride to mop up phosphate in eutrophic lakes[59] may not work in the long term as there may be subsequent release of deposited phosphate from sediments under reducing conditions.

A novel use of phosphate to counter acidification has emerged in recent years.[150] Traditional remedial methods which have been adopted include the direct liming of lakes, *e.g.* in Sweden,[151] or of catchments, *e.g.* around Loch Fleet in SW Scotland.[152] Although helpful, such approaches based on neutralization are costly and usually need to be repeated at regular intervals. Furthermore, the resulting Ca-rich waters may turn out to support biota quite different from those found in natural softwater lakes. An alternative approach has been tried by Davison and co-workers on Seathwaite Tarn, an upland reservoir in the English Lake District.[131] Phosphate fertilizer was added to stimulate primary productivity and thereby increase the assimilation of nitrate. This generates base according to the equation

$$106CO_{2(g)}+138H_2O+16NO_3^- \rightleftharpoons (CH_2O)_{106} (NH_3)_{16}+16OH^-+138O_{2(g)}$$

$$(3.110)$$

As concentrations of nitrate are increasingly high in acid waters, the addition of modest amounts of phosphate may generate sufficient base to combat acidity without inducing excessive productivity. An increase in pH of 0.5 and a marked increase in biological productivity at all levels were observed over the three-year period of the experiment. In the longer term, additional quantities of base should be generated by the anoxic decomposition of organic material accumulating on the lake-bed (through the dissimilative reduction of inorganic oxidants such as nitrate, sulfate, or iron hydr(oxides) present in the sediments – Equations (3.100) to (3.103)). If oxygen is the electron acceptor there is no net gain of base (Equation (3.99)), there is no advantage in adding nitrate because 1 mol of nitrate is required to generate 1 mol of base, and the generated base (contributing to alkalinity) should not be confused with the temporary rise in pH associated with CO_2 consumption, which affects neither alkalinity nor acidity (see Section 3.2.4.1).

3.3.4.2 Nitrate in Groundwater. Principal sources of nitrate in water are runoff and drainage from land treated with agricultural fertilizers and also deposition from the atmosphere as a consequence of NO_x released from fossil fuel combustion. The nitrogen present in soil organic matter may also be released as nitrate through microbial action. Nitrate's role as a nutrient contributes significantly to blooms of algae, which upon death are decomposed first by aerobic bacteria, thereby depriving fish and other organisms of oxygen.[59,147]

There has long been concern expressed over the presence of nitrate in drinking water at concentrations exceeding the EC guideline of 50 mg L^{-1} because of the risk of methaemoglobinaemia (blue baby syndrome). Here,

NO_3^- is reduced in the baby's stomach to NO_2^-, which, on absorption, reacts with oxyhaemoglobin to form methaemoglobin. There have also been fears over the reaction of NO_2^- with secondary amines from the breakdown of meat or protein to produce carcinogenic *N*-nitroso compounds, but there is as yet no clear evidence of a link between stomach cancer and nitrate in water.[153] Nevertheless, there is growing concern over the contamination of groundwater by nitrate, for example in regions such as the Sierra Pelona Basin, California.[154] There the local groundwaters, the major source of drinking water from private water wells located near each private residence in the rural communities, often exceed the USEPA maximum contaminant level for drinking water of 10 mg L^{-1} (NO_3–N). Isotopic investigations, based upon $^{15}N/^{14}N$, have confirmed the predominance of anthropogenic, organic human, and/or animal waste and decay of irrigation-enhanced vegetation rather than natural nitrate sources. In south Dorset, UK, the cause of the increase in the nitrate concentrations of the Chalk groundwater from a baseline of 1.04 mg L^{-1} (NO_3-N) to 6.37 mg L^{-1} has been attributed to increased fertilizer use and increased livestock numbers, mostly during the final 25 years of the 20th century.[155] It is feared that nitrate concentrations in both surface and groundwaters of the Chalk in southern England may approach or exceed the EC limit of 11.3 mg L^{-1} (NO_3–N) for potable supplies in future years. Nitrate in manure and fertilizer is also now under suspicion as an endocrine disruptor (see Section. 3.3.5.2.3).[156]

3.3.5 Organic Matter and Organic Chemicals in Water

3.3.5.1 BOD and COD. The solubility of oxygen in water in equilibrium with the atmosphere at 25°C is 8.7 mg L^{-1}. Causes of oxygen depletion include decomposition of biomass (*e.g.* algal blooms) and the presence of oxidizable substances (*e.g.* sewage, agricultural runoff, factory effluents) in the water. The addition of oxidizable pollutants to streams produces a typical sag in the dissolved oxygen concentration. The degree of oxygen consumption by microbially mediated oxidation of organic matter in water is called the Biochemical (or Biological) Oxygen Demand (BOD) (*cf.* Equation (3.99)). Another index is the Chemical Oxygen Demand (COD), which is determined by using the powerful oxidizing agent, dichromate ($Cr_2O_7^{2-}$), to oxidize organic matter

$$2Cr_2O_7^{2-} + 3CH_2O + 16H^+ \rightleftharpoons 4Cr^{3+} + 3CO_{2(g)} + 11H_2O \quad (3.111)$$

followed by back-titration of excess added dichromate with Fe^{2+}

$$Cr_2O_7^{2-} + 6Fe^{2+} + 14H^+ \rightleftharpoons 2Cr^{3+} + 6Fe^{3+} + 7H_2O \quad (3.112)$$

As $Cr_2O_7^{2-}$ oxidizes substances not oxidized by O_2, the COD is usually greater than the BOD and to some extent overestimates the threat posed to oxygen content.

3.3.5.2 Synthetic Organic Chemicals. A large number of organic compounds are synthesized for agricultural use, mainly as pesticides, and for industrial use as solvents, cleaners, degreasers, petroleum products, plastics manufacture, *etc.*[157] Many organic micropollutants percolate into the soil and accumulate in aquifers or surface waters. They can contaminate drinking water sources *via* agricultural runoff to surface waters or percolation into groundwaters, industrial spillages to surface waters and groundwaters, runoff from roads and paved areas, industrial waste water effluents leaching from chemically treated surfaces, domestic sewage effluents, atmospheric fallout, and as leachate from industrial and domestic landfill sites.[157] In the last few years, emerging chemical contaminants such as pharmaceuticals and personal care products have also been detected in freshwaters.[158] In a widely publicized study, Kolpin and co-workers[159] investigated the occurrence of pharmaceuticals as well as other organic wastewater contaminants in a national reconnaissance study of 95 contaminants in 139 streams in 30 states of the USA in 1999–2000. As many as 38 of the 95 targeted compounds were found in a single water sample, although the average number of compounds in a given sample was seven and the concentrations of individual compounds were typically much less than 1 $\mu g\ L^{-1}$. Three classes of compounds (detergent metabolites, plasticizers, steroids) had the highest concentrations and the most frequently detected compounds were coprostanol (faecal steroid), cholesterol (plant and animal steroid), *N,N*-diethyltoluamide (insect repellent), tri(2-chloroethyl)phosphate (fire retardant) and 4-nonylphenol (non-ionic detergent metabolite) (Figure 8).[160] Although measured concentrations were generally low and rarely exceeded drinking water guidelines or aquatic-life criteria, little is known about the potential interactive effects that may occur from complex mixtures of organic wastewater contaminants in the environment.

3.3.5.2.1 Pesticides. Many pesticides in aquifers have resisted degradation and are more likely to persist there because of reduced microbial activity, absence of light, and lower temperatures.[161] Numerous aquifers, for example in eastern England, have been found to exceed the Maximum Admissible Concentrations (MAC) guidelines for total pesticides in drinking water (0.5 $\mu g\ L^{-1}$), due primarily to the presence of herbicides of the carboxy acid and basic triazine groups. Gray[157] has listed the 12 pesticides most often found in UK drinking waters in two

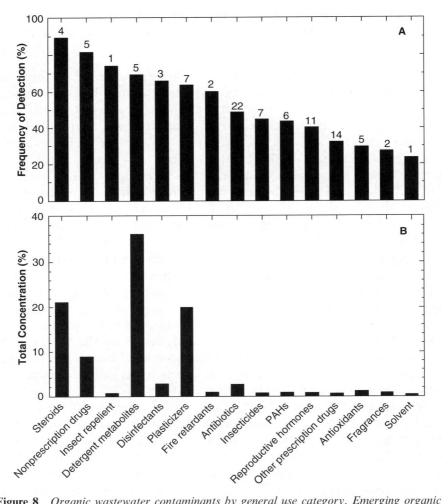

Figure 8 *Organic wastewater contaminants by general use category. Emerging organic contaminants in US streams, as reported by the US Geological Survey* [159,160]*, can be broken down into 15 categories. A shows the frequency of detection, and B shows the fraction of the total measured concentration (%). The number of compounds in each category is shown above the bars in A*
(Reprinted with permission from ref 159, © American Chemical Society, 2002)

groups, frequently occurring (atrazine, chlortoluon, isoproturon, MCPA, mecoprop, simazine) and commonly occurring (2,4-D, dicamba, dichlorprop, dimethoate, linuron, 2,4,5-T). A major seasonal source of insecticide in freshwaters is sheep dipping. With organophosphorus compounds, which themselves replaced the more persistent organochlorine insecticides such as lindane, now falling out of favour because of health risks to users, the use of synthetic pyrethroids as alternatives has resulted in the death of aquatic organisms in UK rivers. Pyrethroids, which do not persist in the environment and are largely non-toxic to mammals, are toxic to invertebrates and are estimated to be at least 100 times more toxic in the aquatic environment than organophosphorus pesticides. It appears to be the pouring of waste dip into holes in the ground that has caused the problem.[162] In order to protect the aquatic environment, the Environment Agency in England assesses water quality against EQSs. Defined as the concentration of a substance which must not be exceeded within the aquatic environment in order to protect it for its recognized uses, EQSs are specific to individual substances, including pesticides.

Atrazine, now banned in many European countries but which still accounts for 40% of all herbicide applied in the USA and is most often applied in the spring, is routinely present in streams, rivers, and reservoirs at levels of 1–10 μg L^{-1}, close to the USEPA standard of 3 μg L^{-1}.[163,164] Runoff peaks of 100–200 μg L^{-1}, however, have been recorded at the times when frogs are breeding and tadpoles are developing. Atrazine is therefore suspected of causing the gonadal abnormalities, such as retarded development and hermaphroditism, found in male wild leopard frogs in different regions of the USA.[163] (*cf.* Section 3.3.5.2.3)

3.3.5.2.2 Polychlorinated biphenyls (PCBs). The Great Lakes, the largest body of freshwater in the world, with long hydraulic residence times, long food chains, and multiple sources of PCBs, have been the focal point of PCB research in aquatic systems. The distribution of PCBs between the dissolved and particulate phases is dependent on the concentration of suspended particulate matter, dissolved and particulate organic carbon concentrations, and the extent to which the system is at equilibrium. In the 1990s, water concentrations of up to 0.6 ng L^{-1} were observed in the most contaminated lakes (Michigan, Erie, and Ontario), which may be compared with the USEPA Great Lakes Water Quality Guidance criteria of 0.017 ng L^{-1}. Sediment concentrations peaked in the 1970s, the period of maximum PCB production in the USA. Following a ban upon their North American production during the 1970s, there was a significant decline in the PCB concentrations of Great Lakes' fish from the mid-1970s to the mid-1980s but the rate of decrease has since slowed or stopped for all lakes.[165] In common with other organochlorine compounds, it is the stability, persistence, volatility, and lipophilicity of PCBs, which lead to considerable biomagnification along the food chain, often far from the place of release. To these traditional halogenated contaminants can now be added fire-retardant polybrominated diphenyl ethers (PBDEs),[166] and perfluorooctane sulfonates (PFOS),[167] which have been found in Great Lakes' water and biota.

3.3.5.2.3 Endocrine disruptors.   Concern has recently been expressed over the possible role of various synthetic organic chemicals (*e.g.* organochlorine pesticides, PCBs, phthalates, alkylphenolethoxylates (APEs), alkylphenols, *etc.*) as disruptors of endocrine systems of wildlife and perhaps even of humans.[168,169] Reproductive changes in male alligators from Lake Apopka, Florida, embryonic death, deformities, and abnormal nesting behaviour in fish-eating birds in the Great Lakes region and the occurrence of hermaphroditic fish near sewage outfalls on some British rivers have been attributed to postulated oestrogenic or anti-androgenic effects of some of these chemicals.[168,169] Most APEs, which are used as detergents, emulsifiers, wetting agents, and dispersing agents, enter the aquatic environment after disposal in wastewater. During biodegradation treatment of the latter, the APEs are transformed into more toxic, short-chain ethoxylates, alkylphenol carboxylic acids, and alkylphenols.[170] The threshold concentration of nonylphenol in water for production of the female egg yolk protein, vitellogenin, in male rainbow trout is ~ 10 µg L^{-1}.[171] It also appears that nonylphenol concentrations in many European rivers are up to 10 times higher than those found in USA rivers, which are typically < 1 µg L^{-1},[170,172] the EU limit. It has been suggested, however, that natural oestrogens (*e.g.* oestrone and 17β-oestradiol) may be responsible for the observed effects, perhaps after conversion of inactive excreted metabolites to active forms by bacterial enzyme action during sewage treatment.[173] The COMPREHEND (Community Programme of Research on Environmental Hormones and Disruptors) 1999–2001 survey found that in some domestic sewage effluents there were oestrogenic steroids present, almost certainly from human excretion, at concentrations sufficient to account for most of the oestrogenicity, causing widespread sexual disruption, in fish in rivers throughout England and Wales.[174] Maximum measured concentrations were 14 ng L^{-1} for oestradiol, 51 ng L^{-1} for its principal metabolite oestrone, 17 ng L^{-1} for another metabolite oestriol, and 2 ng L^{-1} for the synthetic steroid ethinylestradiol, the active ingredient in the birth control pill. As a result, the Environment Agency has set a threshold exposure limit for steroid estrogens of 1 ng L^{-1} for oestradiol, 3 ng L^{-1} for oestriol and 0.1 ng L^{-1} for ethinylestradiol in sewage effluent.[175] In industrial effluents, the oestrogenic steroids were not detected but nonylphenol (NP), NP mono- and diethoxylates, and bisphenol-A were found at concentrations up to 3, 7, and 1 µg L^{-1}, respectively.[174] The relatively high hydrophobicity of alkylphenols has led to their accumulation in sludge at activated sludge treatment works.[176]

3.3.5.2.4 Pharmaceuticals and personal care products (PPCPs). In Europe, ~ 3000 different substances are used in medicines such as painkillers, antibiotics, contraceptives, beta-blockers, lipid regulators, tranquilizers, and impotence drugs, and there are also thousands of personal care products (*e.g.* sunscreens, cosmetics, shampoos, *etc.*).[177] Municipal wastewater is one of the main entry routes of pharmaceuticals into the environment. Ternes[158] began monitoring for pharmaceuticals in Germany in the mid-1990s. Since then, in Switzerland, the detection of macrolide antibiotics, among the most important anti-bacterial agents used, in wastewater treatment plants and in the Glatt Valley Watershed has demonstrated that they are not eliminated completely in the treatment plants and that residual amounts reach receiving waters.[178] Also in Switzerland, chemicals from personal care products containing UV filters, such as benzophenone-3, 4-methylbenzylidene camphor (4-MBC), and ethyl-hexyl methoxy cinnamate (EHMC), have been found in untreated wastewater (maximum concentration 19 µg L^{-1}, EHMC), treated wastewater (max. 2.7 µg L^{-1}, 4-MBC) and, at much lower concentrations, lakes and rivers (<2–35 ng L^{-1}).[179] In the UK, Ashton and co-workers[180] investigated the environmental transport of 12 human pharmaceuticals to streams and found that while 10 were detected in sewage treatment works effluent samples (*e.g.* ibuprofen in 84% of samples, median concentration 3086 ng L^{-1}), there were fewer and at lower concentrations in streams (*e.g.* ibuprofen in 69% of samples, median concentration 826 ng L^{-1}).

In the USA, where similar studies have been carried out in treated waters and streams,[159,181–183] Stackelberg and co-workers[184] have demonstrated that many of the PPCPs can survive conventional drinking-water-treatment and occur in potable water supplies. Although concentrations in finished water were low (93% <0.5µg L^{-1}) and the maximum possible lifetime intake of, for example, carbamazepine would be only 13 mg, compared with a single therapeutic dose of ~ 100 mg, little is known about potential health effects associated with chronic ingestion *via* drinking water.

It is not yet known whether the relatively low environmental concentrations found for PPCPs produce adverse effects on aquatic and terrestrial biota or whether the toxicity of complex mixtures might be totally different from that of individual compounds.[177] Consideration should also be given to metabolites excreted by humans and to biological degradates at treatment plants and elsewhere in the environment.[184,185] Many pharmaceuticals are conjugated with glucuronic acid or sulfate prior to excretion, but these may be cleaved in wastewater treatment plants, releasing active pharmaceuticals.[177] Similarly, the disposal of contaminated sludge on land as a fertilizer should be viewed with concern.[177] The use of veterinary medicines may also cause environmental risks.[186]

3.3.5.2.5 Industrial solvents. Six are widely used in the UK – dichloromethane (DCM), trichloromethane, 1,1,1-trichloroethane (TCA), tetrachloromethane trichloroethene (TCE), tetrachloroethene or perchloroethylene (PCE).[157] Although there has been a steady decrease in use over the past 20 years, four (TCE, PCE, TCA, and DCM) are frequently found in drinking water. The major pollution threat is when solvents are discharged directly into or onto the ground, due to illegal disposal or accidental spillage. Since groundwater is not exposed directly to the atmosphere, the solvents are not able to escape by evaporation, so it is the aquifers that are most at risk from these chemicals. Pollution of groundwater by industrial solvents is a very widespread problem (*e.g.* Netherlands, Italy, USA, UK[187]), with aquifers underlying urbanized areas such as Milan, Birmingham, London, or New Jersey containing high concentrations of all solvents. The stability of TCE in particular and the inability of these solvents to readily evaporate mean that such contamination will last for many decades.[157]

A typical example of groundwater contaminated by synthetic organic chemicals from jet fuel and degreasing solvents (*e.g.* halogenated methanes, ethanes, and ethenes, including trichloroethylene, perchloroethylene, and vinyl chloride) can be found at Otis Air Force Base on Cape Cod, Massachusetts, USA, where the Cape's sole-source aquifer has been affected. The groundwater plumes, some in excess of 5 km in length, moving at ~ 0.5 m per day, contaminate 30 million litres of the Cape's drinking water every day. Reactive wall (Ni–Fe) technology is being employed to try to reduce concentrations of 5–150 µg L^{-1} TCE and tetrachloroethylene below the local drinking water limits of 0.5 µg L^{-1}.[188] The chlorinated ethenes are degraded in the presence of zerovalent iron by an abiotic reduction process.

3.3.5.2.6 MTBE. Methyl *tert*-butyl ether (MTBE), an oxygen-containing compound used as a fuel additive since the 1970s, moves quickly through soil, is highly soluble and does not biodegrade easily. It has been found in shallow groundwater in Denver, New England,[189] and elsewhere in the USA, most notably in Santa Monica, California, where a 1996 spill of MTBE-containing gasoline resulted in the contamination of wells at concentrations as high as 610 µg L^{-1}, well above the state's advisory limit of 35 µg L^{-1}.[190] In a preliminary assessment of the occurrence and possible sources of MTBE in groundwater in the United States in 1993/1994, as part of the US Geological Survey's National Water-Quality Assessment programme, Squillace and co-workers[191] found that, out of 60 volatile organic chemicals determined, MTBE was the second most frequently detected chemical in samples of shallow ambient groundwater from urban areas. MTBE was detected above 0.2 $\mu g\ L^{-1}$ in 17% of shallow groundwater samples from urban areas, 1.3% of shallow groundwater from agricultural areas, and 1.0% of deeper groundwater samples from major aquifers. Only 3% of the shallow wells sampled in urban areas had MTBE concentrations $>20\ \mu g\ L^{-1}$, the estimated lower limit of the USEPA draft drinking water health advisory. The 1999 Report of the Blue Ribbons Panel on oxygenates in gasoline found that 5–10% of community drinking water supplies in high-MTBE-use areas showed at least detectable concentrations of MTBE, with 1% $>20\ \mu g\ L^{-1}$.[192] As MTBE generally was not found, in shallow urban groundwater, along with benzene, toluene, ethylbenzene, or xylene, which are commonly associated with petrol spills, it was concluded that possible sources of MTBE in groundwater include leaking storage tanks and non-point sources such as recharge of precipitation and storm water runoff. In the summer of 1999, 13 municipal wells on the south shore of Lake Tahoe, located close to commercial areas with gasoline stations, were shut down as a result of actual or threatened MTBE contamination.[193] In Meyers, an unknown amount of gasoline leaked from a pipe into the soil for about two weeks before the release was repaired and five months later a groundwater sample collected at the gasoline station was 28,000 $\mu g\ L^{-1}$ in MTBE.[193] As MTBE is thought to be potentially carcinogenic to humans, its use poses an interesting dilemma for regulators, given that it helps to reduce carbon monoxide emissions from cars. Regardless of what happens to MTBE use in the future, it is clear that significant amounts of this compound are already present in the sub-surface.[192] In Europe, however, the problem seems much less severe.[194,195]

QUESTIONS AND PROBLEMS

(i) Calculate the ionic strength of a solution which is 0.01 mol L^{-1} with respect to potassium sulfate and 0.002 mol L^{-1} with respect to magnesium chloride. Use this value of ionic strength and the Güntelberg approximation to determine the mixed acidity constant, K', for methanoic acid ($K = 10^{-3.75}$).

(ii) Construct a plot to show how $-\log\{activity\}$ varies with pH for all species in one litre of groundwater containing 3×10^{-4} moles of 4-chlorophenol ($K' = 10^{-9.18}$). Annotate the plot to show the equilibrium pH of the groundwater.

(iii) Calculate the equilibrium pH of rainwater in equilibrium with $CO_{2(g)}$ at a partial pressure of 0.00035 atm (hint: the charge balance equation can be approximated by $\{H^+\} = \{HCO_3^-\}$).

	log K(288 K)
$CO_{2(g)} + H_2O \rightleftharpoons H_2CO_3^*$	-1.34
$H_2CO_3^* \rightleftharpoons H^+ + HCO_3^-$	-6.35
$HCO_3^- \rightleftharpoons H^+ + CO_3^{2-}$	-10.33

(iv) Construct a pε–pH diagram for selenium species present in natural waters. Assume that the total dissolved concentration of selenium is 10^{-6} mol L^{-1}. From the diagram, predict the major selenium species in

(i) oxic and acidic waters (pε ∼ 10, pH 4–7)
(ii) oxic and alkaline waters (pε ∼ 10, pH > 7)
(iii) anoxic groundwaters (pε ∼ −3, pH 4–8).

For each set of conditions, outline the implications for sorption onto solid phases.

	log K(298 K)
$SeO_4^{2-} + 4H^+ + 2e^- \rightleftharpoons H_2SeO_3 + H_2O$	41.40
$SeO_4^{2-} + 3H^+ + 2e^- \rightleftharpoons HSeO_3^- + H_2O$	38.76
$SeO_4^{2-} + 2H^+ + 2e^- \rightleftharpoons SeO_3^{2-} + H_2O$	30.26
$H_2SeO_3 \rightleftharpoons HSeO_3^- + H^+$	-2.75
$HSeO_3^- \rightleftharpoons SeO_3^{2-} + H^+$	-8.50
$H_2SeO_3 + 4H^+ + 4e^- \rightleftharpoons Se_{(s)} + 3H_2O$	48.58
$HSeO_3^- + 5H^+ + 4e^- \rightleftharpoons Se_{(s)} + 3H_2O$	51.33
$SeO_3^{2-} + 6H^+ + 4e^- \rightleftharpoons Se_{(s)} + 3H_2O$	59.83
$Se_{(s)} + 2H^+ + 2e^- \rightleftharpoons H_2Se_{(aq)}$	-3.52
$Se_{(s)} + H^+ + 2e^- \rightleftharpoons HSe^-$	-7.32
$H_2Se \rightleftharpoons HSe^- + H^+$	-3.80

(v) Draw, describe, and compare the concentration profiles of the chemical species of manganese, iron, arsenic, and phosphorus likely to be found in the solid and solution phases of unmixed sediments at the bottom of a seasonally anoxic, eutrophic freshwater lake during (i) summer and (ii) winter.

(vi) With respect to human health and ecological impact, discuss the short- and long-term consequences of the products of the synthetic organic chemical industry upon the quality of surface waters and groundwaters.

(vii) Explain the chemistry underlying (i) the major detrimental effects of metal mining upon rivers and groundwaters and (ii) associated preventive or remedial measures.

(viii) Critically discuss the effectiveness of measures to counter the effects of (i) acidification and (ii) eutrophication in freshwater lakes.

FURTHER READING

E.K. Berner and R.A. Berner, *Global Environment: Water, Air and Geochemical Cycles*, Prentice-Hall, Englewood Cliffs, NJ, 1996.

P.J. Boon and D.L. Howell (eds), *Freshwater Quality: Defining the Indefinable?*, The Stationery Office, Edinburgh, 1997.

R. Clarke and J. King, *The Atlas of Water*, Earthscan, London, 2004.

R.A. Downing, *Groundwater – Our Hidden Asset*, British Geological Survey, NERC, Keyworth, 1998.

J.I. Drever, *The Geochemistry of Natural Waters*, 3rd edn, Prentice-Hall, Englewood Cliffs, NJ, 1997.

M.E. Essington, *Soil and Water Chemistry: An Integrative Approach*, CRC Press, Boca Raton, FL, 2004.

A.J. Horne, *Limnology*, 2nd edn, McGraw-Hill, New York, 1994.

D. Langmuir, *Aqueous Environmental Geochemistry*, Prentice-Hall, Englewood Cliffs, NJ, 1997.

P.S. Maitland, P.J. Boon and D.S. McLusky (eds), *The Fresh Waters of Scotland*, Wiley, Chichester, 1994.

C.F. Mason, *Biology of Freshwater Pollution*, 3rd edn, Longman, Harlow, 1996.

J. Mather, D. Banks, S. Dumpleton and M. Fermor, *Groundwater Contaminants and their Migration* (eds), Geological Society, London, Special Publication No.128, 1998.

F.M. Morel and J.G. Hering, *Principles and Applications of Aquatic Chemistry*, Wiley, New York, 1993.

J. Pankow, *Aquatic Chemistry Concepts*, Lewis, Michigan, 1991.

REFERENCES

1. J. Rees, *Sci. Public Aff.*, 1997, (Winter), 20.
2. R. Clarke and J. King, *The Atlas of Water*, Earthscan, London, 2004.

3. E.K. Berner and R.A. Berner, *Global Environment: Water, Air and Geochemical Cycles*, Prentice-Hall, NJ, 1996.
4. P. Aldhous, *Nature*, 2003, **422**, 251.
5. M. Burke, *Environ. Sci. Technol.*, 1996, **30**, 162A.
6. M. Freemantle, *Chem. Eng. News*, 2004, **82**, 25.
7. R. Renner, *Environ. Sci. Technol.*, 1997, **31**, 466A.
8. J.I. Drever, *The Geochemistry of Natural Waters*, 3rd edn, Prentice-Hall, Englewood Cliffs, NJ, 1997.
9. R.A. Robinson and R.H. Stokes, *Electrolyte Solutions*, 2nd edn, Butterworths, London, 1970.
10. D.A. MacInnes, *J. Chem Soc.*, 1919, **41**, 1068.
11. D. Langmuir and J. Mahoney, Practical applications of groundwater geochemistry, *Proceedings of the 1st Canadian/American Conference on Hydrogeology*, B. Hitchon and E. I. Wallick (eds), National Water Well Association, Worthington, OH, 1985, 69.
12. D. Langmuir, *Aqueous Environmental Geochemistry*, Prentice-Hall, Englewood Cliffs, NJ, 1997.
13. D.R. Turner, M. Whitfield and A.G. Dickson, *Geochim. Cosmochim. Acta*, 1981, **45**, 855.
14. R. Bates, B.R. Staples and R.A. Robinson, *Anal. Chem.*, 1970, **42**, 867.
15. W. Stumm and J.J. Morgan, *Aquatic Chemistry*, 3rd edn, Wiley, New York, 1996.
16. R.M. Pytkowicz, *Equilibria, Nonequilibria and Natural Waters*, Wiley, New York, 1983.
17. F. Millero, *Mar. Chem.*, 2000, **70**, 5.
18. A.H. Truesdell and B.F. Jones, *U.S. Geol. Survey J. Research*, 1974, **2**, 233.
19. D.L. Parkhurst, Chemical modelling of aqueous systems II, D.C. Melchior and R.L. Bassett (eds), in *Am. Chem. Soc. Symp. Ser.*, Washington, DC, 1990, **416**, 30.
20. C.W. Davies, *Ion Association*, Butterworth, Washington DC, 1962.
21. F.M.M. Morel and J.G. Hering, *Principles and Applications of Aquatic Chemistry*, Wiley, New York, 1993.
22. J.D. Allison, D.S. Brown and K.J. Novo-Gradac, MINTEQA2, A Geochemical Assessment Data Base and Test Cases for Environmental Systems: Version 3.0 User's Manual, Report EPA/600/3-91/-21, US EPA, Athens, GA, 1991.
23. E.A. Guggenheim, *Philos. Mag.*, 1935, **19**, 588.
24. G. Scatchard, *Equilibrium in Solutions*, Harvard University Press, Cambridge, MA, 1976.

25. I. Grenthe and H. Wanner, *Guidelines for the Extrapolation to Zero Ionic Strength*, Report NEA-TDB-2.1, F-91191, OECD, Nuclear Energy Agency Data Bank, Giv-sur Yvette, France, 1989.

26. D.K. Nordstrom and J.L. Munoz, *Geochemical Thermodynamics*, Benjamin/Cummings, Menlo Park, CA, 1985.

27. J.N. Brønsted, *J. Am. Chem. Soc.*, 1922, **44**, 877.

28. G. Scatchard, *J. Am. Chem. Soc.*, 1961, **83**, 2636.

29. K.S. Pitzer, Thermodynamic modelling of geological materials: Minerals, fluids and melts, in *Reviews in Mineralogy*, vol 17, I.S.E. Carmichael and H. P. Eugster (eds), Mineralogical Society of America, 1987, 97.

30. J.W. Ball and D.K. Nordstrom, Users Manual for WATEQ 4F, with Revised Thermodynamic Data Base and Test Cases for Calculating Speciation of Major, Trace and Redox Elements in Natural Waters, US Geological Survey Open File Report, Menlo Park, CA, 1991, 91.

31. F.J. Stevenson, *Humus Chemistry: Genesis, Composition, Reactions*, Wiley, New York, 1994.

32. D.J. Wesolowski and D.A. Palmer, *Geochim. Cosmochim. Acta*, 1994, **58**, 2947.

33. L. Sigg and H.B. Xue, *Chemistry of Aquatic Systems: Local and Global Perspectives*, G. Bidoglio and W. Stumm (eds), Kluwer Academic, Dordrecht, 1994.

34. D.G. Kinniburgh and D.M. Cooper, *Environ. Sci. Technol.*, 2004, **38**, 3641.

35. D.L. Parkhurst, Users Guide to PHREEQ C – A Computer Programme for Speciation, Reaction-path, Advective Transport, and Inverse Geochemical Calculations, US Geological Survey Water Resources Inv. Report, 1995, 95.

36. J.C.L. Meeussen, *Environ. Sci. Technol.*, 2003, **37**, 1175.

37. E. Tipping, *Comput. Geosci.*, 1994, **20**, 973.

38. D.G. Kinniburgh, W.H. Van Reimsdijk, L.K. Koopal, M. Borkovec, M.F. Benedetti and M.J. Avena, *Environ. Sci. Technol.*, 1995, **29**, 446.

39. T. Hiemstra, W.H. Van Reimsdijk and G.H. Bolt, *J. Colloid Interface Sci.*, 1989, **133**, 91.

40. J.S. Geelhoed, J.C.L. Meeussen, S. Hillier, D.G. Lumsdon, R.P. Thomas, J.G. Farmer and E. Paterson, *Geochim. Cosmochim. Acta*, 2002, **66**, 3927.

41. R.W. Battarbee, R.J. Flower, A.C. Stevenson and B. Rippey, *Nature*, 1985, **314**, 350.

42. V.J. Jones, A.C. Stevenson and R.W. Battarbee, *Nature*, 1986, **322**, 157.
43. C.T. Driscoll and W.D. Schecher, *Environ. Geochem. Health*, 1990, **12**, 28.
44. N.J. Bunce, *Environmental Chemistry*, 2nd edn, Wuerz, Winnipeg, 1994.
45. S.M. Palmer and C.T. Driscoll, *Nature*, 2002, **417**, 242.
46. C.T. Driscoll, K.M. Driscoll, K.M. Roy and M.J. Mitchell, *Environ. Sci. Technol.*, 2003, **37**, 2036.
47. J.S. Kahl, J.L. Stoddard, R. Haeuber, S.G. Paulsen, R. Birnbaum, F.A. Deviney, J.R. Webb, D.R. Dewalle, W. Sharpe, C.T. Driscoll, A.T. Herlihy, J.H. Kellogg, P.S. Murdoch, K. Roy, K.E. Webster and N.S. Urquhart, *Environ. Sci. Technol.*, 2004, **38**, 484A.
48. D.M. Cooper and A. Jenkins, *Sci. Total Environ.*, 2003, **313**, 91.
49. R.F. Wright, T. Larssen, L. Camarero, B.J. Cosby, R.C. Ferrier, R. Helliwell, M. Forstus, A. Jenkins, J. Kopčček, V. Majer, F. Moldan, M. Posch, M. Rogora and W. Schöpp, *Environ. Sci. Technol.*, 2005, **39**, 64A.
50. R.C. Massey and D. Taylor (Eds), *Aluminium in Food and the Environment*, Royal Society of Chemistry, Cambridge, 1989.
51. N.D. Priest, *J. Environ. Monit.*, 2004, **6**, 375.
52. P. Altmann, J. Cunningham, U. Dhanesha, M. Ballard, J. Thompson and F. Marsh, *Brit. Med. J.*, 1999, **319**, 807.
53. J.D. Birchall, *Chem. Br.*, 1990, **26**, 141.
54. C. Exley, *J. Inorg. Biochem.*, 2003, **97**, 1.
55. F. Case, *Chem. World*, 2005, **2**, 28.
56. R.J. Pentreath, in *Issues in Environmental Science and Technology*, R.E. Hester and R.M. Harrison (eds), vol. 1, Royal Society of Chemistry, Cambridge, 1994, 121.
57. J.E. Andrews, P. Brimblecombe, T.D. Jickells and P.S. Liss, *An Introduction to Environmental Chemistry*, Blackwell Science, Oxford, 1996.
58. D.K. Nordstrom, C.N. Alpers, C.J. Ptacek and D.W. Blowes, *Environ. Sci. Technol.*, 2000, **34**, 254.
59. C.F. Mason, *Biology of Freshwater Pollution*, 3rd edn, Longman, Harlow, 1996.
60. L.E. Hunt and A.G. Howard, *Mar. Pollut. Bull.*, 1994, **28**, 33.
61. J.N. Moore and S.N. Luoma, *Environ. Sci. Technol.*, 1990, **24**, 1279.
62. G. Furrer, B.L. Phillips, K.U. Ulrich, R. Pothig and W.H. Casey, *Science*, 2002, **297**, 2245.
63. A. Davis and D. Atkins, *Environ. Sci. Technol.*, 2001, **35**, 3501.

64. G.C. Miller, W.B. Lyons and A. Davis, *Environ. Sci. Technol.*, 1996, **30**, 118A.
65. R.J. Allan, *Proceedings of the 10th International Conference on Heavy Metals in the Environment*, vol. 1, CEP, Edinburgh, 1995, 293.
66. P.G. Whitehead, G. Hall, C. Neal and H. Prior, *Sci. Total Environ.*, 2005, **338**, 41.
67. T.R. Chowdhury, B. Kr. Mandal, G. Samanta, G. Kr. Basu, P.P. Chowdhury, C.R. Chanda, N. Kr. Karan, D. Lodh, R. Kr. Dhar, D. Das, K.C. Saha and D. Chakraborti, in *Arsenic: Exposure and Health Effects*, C.O. Abernathy, R.L. Calderon and W.R. Chappell (eds), Chapman & Hall, London, 1997, 93.
68. W.R. Chappell, C.O. Abernathy and C.R. Cothern (eds), *Arsenic: Exposure and Health*, Science and Technology Letters, Northwood, 1994.
69. C.O. Abernathy, R.L. Calderon and W.R. Chappell (eds), *Arsenic: Exposure and Health Effects*, Chapman & Hall, London, 1997.
70. W.R. Chappell, C.O. Abernathy and R.L. Calderon (eds), *Arsenic: Exposure and Health Effects*, Elsevier, Oxford, 1999.
71. W.R. Chappell, C.O. Abernathy and R.L. Calderon (Eds), *Arsenic: Exposure and Health Effects IV*, Elsevier, Oxford, 2001.
72. W.R. Chappell, C.O. Abernathy, R.L. Calderon and D.J. Thomas (eds), *Arsenic: Exposure and Health Effects V*, Elsevier, Oxford, 2003.
73. P.L. Smedley and D.G. Kinniburgh, *Appl. Geochem.*, 2002, **17**, 517.
74. P.L. Smedley and D.G. Kinniburgh, in *Essentials of Medical Geology*, O. Selinus, B.J. Alloway, J.A. Centeno, R.B. Finkelman, R. Fuge, U. Lindh and P. Smedley (eds), Elsevier, Burlington, 2005, 263.
75. R. Nickson, J. McArthur, W. Burgess, K.M. Ahmed, P. Ravenscroft and M. Rahman, *Nature*, 1998, **395**, 338.
76. R.T. Nickson, J.M. McArthur, P. Ravenscroft, W.S. Burgess and K.M. Ahmed, *Appl. Geochem.*, 2000, **15**, 403.
77. J.M. McArthur, D.M. Banerjee, K.A. Hudson-Edwards, R. Mishra, R. Purohit, P. Ravenscroft, A. Cronin, R.J. Howarth, A. Chatterjee, T. Talukder, D. Lowry, S. Houghton and D.K. Chadha, *Appl. Geochem.*, 2004, **19**, 1255.
78. C.F. Harvey, C.H. Swartz, A.B.M. Badruzzaman, N. Keon-Blute, W. Yu, M.A. Ali, J. Jay, R. Beckie, V. Niedan, D. Brabander, P.M. Oates, K.N. Ashfaque, S. Islam and H.F. Hemond, *Science*, 2002, **298**, 1602.

79. W.R. Cullen and K.J. Reimer, *Chem. Rev.*, 1989, **89**, 713.
80. M.R. Moore, W.N. Richards and J.G. Sherlock, *Environ. Res.*, 1985, **38**, 67.
81. G.C.M. Watt, A. Britton, W.H. Gilmour, M.R. Moore, G.D. Murray, S.J. Robertson and J. Womersley, *Brit. Med. J.*, 1996, **313**, 979.
82. R. Renner, *Environ. Sci. Technol.*, 2004, **38**, 224A.
83. P. O'Neill, *Environmental Chemistry*, 2nd edn, Chapman & Hall, London, 1993.
84. J.E. Fergusson, *The Heavy Elements: Chemistry, Environmental Impact and Health Effects*, Pergamon Press, Oxford, 1990.
85. J.E. Fergusson, *Inorganic Chemistry and the Earth*, Pergamon Press, Oxford, 1982.
86. T.S. Presser, *Environ. Management*, 1994, **18**, 437.
87. R.H. Neal, in *Heavy Metals in Soils*, 2nd edn, B.J. Alloway (ed), Blackie, Glasgow, 1995, 260.
88. M. Flury, W.T. Frankenberger Jr. and W.A. Jury, *Sci. Total Environ.*, 1997, **198**, 259.
89. S.J. Hamilton, *Sci. Total Environ.*, 2004, **326**, 1.
90. J.G. Farmer, M.C. Graham, R.P. Thomas, C. Licona-Manzur, E. Paterson, C.D. Campbell, J.S. Geelhoed, D.G. Lumsdon, J.C.L. Meeussen, M.J. Roe, A. Conner, A.E. Fallick and R.J.F. Bewley, *Environ. Geochem. Health*, 1999, **21**, 331.
91. B.R. James, *Environ. Sci. Technol.*, 1996, **30**, 248A.
92. R.G. Darrie, *Environ. Geochem. Health*, 2001, **23**, 187.
93. J.G. Farmer, R.P. Thomas, M.C. Graham, J.S. Geelhoed, D.G. Lumsdon and E. Paterson, *J. Environ. Monit.*, 2002, **4**, 235.
94. J.S. Geelhoed, J.C.L. Meeussen, M.J. Roe, S. Hillier, R.P. Thomas, J.G. Farmer and E. Paterson, *Environ. Sci. Technol.*, 2003, **37**, 3206.
95. R.J.F. Bewley, R. Jeffries, S. Watson and D. Granger, *Environ. Geochem. Health*, 2001, **23**, 267.
96. D.W. Blowes, C.J. Ptacek and J.L. Jambor, *Environ. Sci. Technol.*, 1997, **31**, 3348.
97. S.M. Ponder, J.G. Darab and T.E. Mallouk, *Environ. Sci. Technol.*, 2000, **34**, 2564.
98. J. Fruchter, *Environ. Sci. Technol.*, 2002, **36**, 464A.
99. E.M.N. Chirwa and Y.-T. Wang, *Environ. Sci. Technol.*, 1997, **31**, 1446.
100. R. Renner, *Environ. Sci. Technol.*, 2004, **38**, 178A.
101. D. Cleary and I. Thornton, in *Issues in Environmental Science and Technology*, R.E. Hester and R.M. Harrison (eds), vol 1, Royal Society of Chemistry, Cambridge, 1994, 17.

102. M.H.D. Pestana and M.L.L. Formoso, *Sci. Total Environ.*, 2003, **307**, 125.

103. L.D. Lacerda, O. Malm, J.R.D. Guimardes, W. Salomons and R.-D. Wilken, in *Biogeodynamics of Pollutants in Soils and Sediments*, W. Salomons and W.M. Stigliani (eds), Springer, Berlin, 1995, 213.

104. D.H.M. Alderton, *Historical Monitoring*, MARC Report No. 31, 1985, 1.

105. L.J. Eades, J.G. Farmer, A.B. MacKenzie, A. Kirika and A.E. Bailey-Watts, *Sci. Total Environ.*, 2002, **292**, 55.

106. J.R. Graney, A.N. Halliday, G.J. Keeler, J.O. Nriagu, J.A. Robbins and S.A. Norton, *Geochim. Cosmochim. Acta*, 1995, **59**, 1715.

107. H.C. Moor, T. Schaler and M. Sturm, *Environ. Sci. Technol.*, 1996, **30**, 2928.

108. E. Callender and P.C. Van Metre, *Environ. Sci. Technol.*, 1997, **31**, 424A.

109. M.L. Brännvall, R. Bindler, O. Emteryd and I. Renberg, *J. Palaeolimnol.*, 2001, **25**, 421.

110. J.G. Farmer, *Environ. Geochem. Health*, 1991, **13**, 76.

111. W. Davison, *Earth Sci. Rev.*, 1993, **34**, 119.

112. C.L. Bryant, J.G. Farmer, A.B. MacKenzie, A.E. Bailey-Watts and A. Kirika, *Limnol. Oceanogr.*, 1997, **42**, 918.

113. R.A. Berner, *Early Diagenesis*, Princeton University Press, Princeton, NJ, 1980.

114. J.G. Farmer and M.A. Lovell, *Geochim. Cosmochim. Acta*, 1986, **50**, 2059.

115. B. Müller, L. Granina, T. Schaller, A. Ulrich and B. Wehrli, *Environ. Sci. Technol.*, 2002, **36**, 411.

116. R. Carignan and A. Tessier, *Science*, 1985, **228**, 1524.

117. H. Elderfield and A. Hepworth, *Mar. Pollut. Bull.*, 1975, **6**, 85.

118. E. Tipping, *Computer Geosci.*, 1994, **20**, 973.

119. S.M. Shuttleworth, W. Davison and J. Hamilton-Taylor, *Environ. Sci. Technol.*, 1999, **33**, 4169.

120. H. Zhang, W. Davison, S. Miller and W. Tych, *Geochim. Cosmochim. Acta*, 1995, **59**, 4181.

121. P.R. Teasdale, S. Hayward and W. Davison, *Anal. Chem.*, 1999, **71**, 2186.

122. H. Zhang, *Environ. Sci. Technol.*, 2004, **38**, 1421.

123. E.R. Unsworth, H. Zhang and W. Davison, *Environ. Sci. Technol.*, 2005, **39**, 624.

124. A.R. Schneider, H.M. Stapleton, J. Cornwell and J.E. Baker, *Environ. Sci. Technol.*, 2001, **19**, 3809.

125. C. Malmquist, R. Bindler, I. Renberg, B. van Bavel, E. Karlsson, N.J. Anderson and M. Tysklind, *Environ. Sci. Technol.*, 2003, **37**, 4319.

126. B. Gevao, T. Harner and K.C. Jones, *Environ. Sci. Technol.*, 2000, **34**, 33.

127. P. Fernández, R.M. Vilanova, C. Martínez, P. Appleby and J.O. Grimalt, *Environ. Sci. Technol.*, 2000, **34**, 1906.

128. S.G. Wakeham, J. Forrest, C.A. Masiello, Y. Gélinas, C.R. Alexander and P.R. Leavitt, *Environ. Sci. Technol.*, 2004 **38**, 431.

129. P.V. Doskey, *Environ. Sci. Technol.*, 2001, **35**, 247.

130. R.F. Pearson, D.L. Swackhamer, S.J. Eisenreich and D.T. Long, *Environ. Sci. Technol.*, 1997, **31**, 2903.

131. W. Song, J.C. Ford, A. Li, W.J. Mills, D.R. Buckley and K.J. Rockne, *Environ. Sci. Technol.*, 2004, **38**, 3286.

132. A.T. Schwartz, D.M. Bunce, R.G. Silberman, C.L. Stanitski, W.J. Stratton and A.P. Zipp, in *Chemistry in Context*, C. Wheatley (ed), W.C. Brown Publishers, Dubuque, IA, 1997, 219.

133. D. Kim, Q. Wang, G.A. Sorial, D.D. Dionysiou and D. Timber-lake, *Sci. Total Environ.*, 2004, **327**, 1.

134. D.M. Harper, *Eutrophication of Freshwaters: Principles, Problems and Restoration*, Chapman & Hall, London, 1992.

135. S.G. Bell and G.A. Codd, in *Issues in Environmental Science and Technology*, R.E. Hester and R.M. Harrison (eds), vol. 5, Royal Society of Chemistry, Cambridge, 1996, 109.

136. J.N. Pretty, C.F. Mason, D.B. Nedwell, R.E. Hine, S. Leaf and R. Dils, *Environ. Sci. Technol.*, 2003, **37**, 201.

137. J. Murphy and J.R. Riley, *Analyt. Chim. Acta*, 1962, **27**, 31.

138. M.W. Marsden, *Freshwater Biol.*, 1989, **21**, 139.

139. J.G. Farmer, A.E. Bailey-Watts, A. Kirika and C. Scott, *Aquat. Conserv.*, 1994, **4**, 45.

140. R. Psenner, B. Bostrom, M. Dinka, K. Pettersson, R. Pucsko and M. Sager, *Arch. Hydrobiol. Ergebn. Limnol.*, 1988, **30**, 98.

141. P. Pardo, G. Rauret and J.F. López-Sánchez, *J. Environ. Monit.*, 2003, **5**, 312.

142. J.C. Cornwell, *Arch. Hydrobiol.*, 1987, **109**, 161.

143. B. Rippey, in *Interactions between Sediments and Freshwater*, H.L. Golterman (ed), Dr. W. Junk B.V. Publishers, The Hague, 1977, 349.

144. S.-O. Ryding and C. Forsberg, in *Interactions between Sediments and Freshwater*, H.L. Golterman (ed), Dr. W. Junk B.V. Publishers, The Hague, 1977, 227.

145. L. Hakanson and M. Jansson, *Principles of Lake Sedimentology*, Springer, Berlin, 1983.
146. A.J.D. Ferguson, M.J. Pearson and C.S. Reynolds, in *Issues in Environmental Science and Technology*, R.E. Hester and R.M. Harrison (eds), vol 5, Royal Society of Chemistry, Cambridge, 1996, 27.
147. A.E. Bailey-Watts, I.D.M. Gunn and A. Kirika, *Loch Leven Past and Current Water Quality and Options for Change*, Report to the Forth River Purification Board, Institute of Freshwater Ecology, Edinburgh, 1993.
148. P. Young, *Environ. Sci. Technol.*, 1996, **30**, 292A.
149. S. Hadlington, *Chem. World*, 2005, **2**, 12.
150. W. Davison, D.G. George and N.J.A. Edwards, *Nature*, 1995, **377**, 504.
151. P. Nyberg and E. Thornelof, *Water, Air Soil Pollut.*, 1988 **41**, 3.
152. G. Howells and T.R.K. Dalziel (Eds), *Restoring Acid Waters Loch Fleet 1984–1990*, Elsevier, London, 1992.
153. T.M. Addiscott, in *Issues in Environmental Science and Technology*, R.E. Hester and R.M. Harrison (eds), vol 5, Royal Society of Chemistry, Cambridge, 1996, 1.
154. A.E. Williams, L.J. Lund, J.A. Johnson and Z.J. Kabala, *Environ. Sci. Technol.*, 1998, **32**, 32.
155. K.J. Limbrick, *Sci. Total Environ.*, 2003, **314–316**, 89.
156. J. Pelley, *Environ. Sci. Technol.*, 2003, **37**, 162A.
157. N.F. Gray, *Drinking Water Quality*, Wiley, Chichester, 1994.
158. T.A. Ternes, *Water. Res.*, 1998, **32**, 3245.
159. D.W. Kolpin, E.T. Furlong, M.T. Meyer, E.M. Thurman, S.D. Zaugg, L.B. Barber and H.T. Buxton, *Environ. Sci. Technol.*, 2002, **36**, 1202.
160. B.E. Erickson, *Environ. Sci. Technol.*, 2002, **36**, 140A.
161. K.R. Eke, A.D. Barnden and D.J. Tester, in *Issues in Environmental Science and Technology*, R.E. Hester and R.M. Harrison (eds), vol. 5, Royal Society of Chemistry, Cambridge, 1996, 43.
162. F. Pearce, *New Sci.*, 1997, **11**(January), 4.
163. T. Hayes, K. Haston, M. Tsui, A. Hoang, C. Haeffele and A. Vonk, *Nature*, 2002, **419**, 895.
164. J. Withgott, *Science*, 2002, **296**, 447.
165. D. Swackhamer, in *Issues in Environmental Science and Technology*, R.E. Hester and R.M. Harrison (eds), vol. 6, Royal Society of Chemistry, Cambridge, 1996, 137.
166. A. Schaefer, *Environ. Sci. Technol.*, 2001, **35**, 139A.

167. B. Boulanger, J. Vargo, J.L. Schnoor and K.C. Hornbuckle, *Environ. Sci. Technol.*, 2004, **38**, 4064.

168. T. Colborn, J.P. Meyers and D. Dumanoski, *Our Stolen Future*, Little, Brown and Company, Boston, 1996.

169. R.E. Hester and R.M. Harrison (eds), *Issues in Environmental Science and Technology*, vol 12, Royal Society of Cambridge, Cambridge, 1999.

170. R. Renner, *Environ. Sci. Technol.*, 1997, **31**, 316A.

171. S. Jobling, D. Sheahan, J.A. Osborne, P. Matthiessen and J.P. Sumpter, *Environ. Toxicol. Chem.*, 1996, **15**, 194.

172. C.P. Rice, I. Schmitz-Afonso, J.E. Loyo-Rosales, E. Link, R. Thoma, L. Fay, D. Altfater and M.J. Camp, *Environ. Sci. Technol.*, 2003, **37**, 3747.

173. J. Kaiser, *Science*, 1996, **274**, 1837.

174. A.D. Pickering and J.P. Sumpter, *Environ. Sci. Technol.*, 2003, **37**, 331A.

175. M. Burke, *Environ. Sci. Technol.*, 2004, **38**, 362A.

176. A.C. Johnston and J.P. Sumpter, *Environ. Sci. Technol.*, 2001, **35**, 4697.

177. T.A. Ternes, A. Joss and H. Siegrist, *Environ. Sci. Technol.*, 2004, **38**, 392A.

178. C.S. McArdell, E. Molnar, M.J.-F. Suter and W. Giger, *Environ. Sci. Technol.*, 2003, **37**, 5479.

179. M.E. Balmer, H.-R. Buser, M.D. Müller and T. Poiger, *Environ. Sci. Technol.*, 2005, **39**, 953.

180. D. Ashton, M. Hilton and K.V. Thomas, *Sci. Total Environ.*, 2004, **333**, 167.

181. D.W. Kolpin, M. Skopec, M.T. Meyer, E.T. Furlong and S.D. Zaugg, *Sci. Total Environ.*, 2004, **328**, 119.

182. G.R. Boyd, H. Reemtsma, D.A. Grimm and S. Mitra, *Sci. Total Environ.*, 2003, **311**, 135.

183. G.R. Boyd, J.M. Palmeri, S. Zhang and D.A. Grimm, *Sci. Total Environ.*, 2004, **333**, 137.

184. P.E. Stackelberg, E.T. Furlong, M.T. Meyer, S.D. Zaugg, A.K. Henderson and D.B. Reissman, *Sci. Total Environ.*, 2004, **329**, 99.

185. A.B.A. Boxall, C.J. Sinclair, K. Fenner, D. Kolpin and S.J. Maund, *Environ. Sci. Technol.*, 2004, **38**, 368A.

186. A.B.A. Boxall, D.W. Kolpin, B. Halling-Sørensen and J. Tolls, *Environ. Sci. Technol.*, 2003, **37**, 286A.

187. M.O. Rivett, D.N. Lerner and J.W. Lloyd, *Water Environ. Management*, 1990, **4**, 242.

188. E.L. Appleton, *Environ. Sci. Technol.*, 1996, **30**, 536A.

189. J.D. Ayotte, D.M. Argue and F.J. McGarry, *Environ. Sci. Technol.*, 2005, **39**, 9.
190. M. Cooney, *Environ. Sci. Technol.*, 1997, **31**, 269A.
191. P.J. Squillace, J.S. Zogorski, W.G. Wilber and C.V. Price, *Environ. Sci. Technol.*, 1996, **30**, 1721.
192. R. Johnson, J. Pankow, D. Bender, C. Price and J. Zogorski, *Environ. Sci. Technol.*, 2000, **34**, 210A.
193. L.S. Dernbach, *Environ. Sci. Technol.*, 2000, **34**, 516A.
194. C. Achten, A. Kolb, W. Püttmann, P. Seel and R. Gihr, *Environ. Sci. Technol.*, 2002, **36**, 3652.
195. C. Achten, A. Kolb and W. Püttmann, *Environ. Sci. Technol.*, 2002, **36**, 3662.

CHAPTER 4

Chemistry of the Oceans

STEPHEN J. DE MORA

International Atomic Energy Agency, Marine Environment Laboratory, 4 quai Antoine 1er, MC 98000, Principality of Monaco

4.1 INTRODUCTION

The World Ocean is a complex mixture containing all the elements, albeit in dilute amounts in some cases. Seawater contains dissolved gases and, apart from some exceptional environments, is consequently both well oxygenated and buffered at a pH of about 8. There are electrolytic salts, the ionic strength of seawater being approximately 0.7, and many organic compounds in solution. At the same time, there is a wide range of inorganic and organic particles in suspension. However, these comfortable distinctions become quite confused in seawater because some molecules present in true solution are sufficiently large to be retained by a filter. Moreover, surface adsorption allows particles to scavenge-dissolved elements and accumulate coatings of organic material from solution. Some elements, particularly those with biochemical functions, may be rapidly removed from solution. Concurrently, reactions involving geological time scales are proceeding slowly. Yet despite this apparent complexity, many aspects of the composition of seawater and chemical oceanography can now be explained with recourse to the fundamental principles of chemistry. This chapter serves to bridge the gap between those with environmental expertise and those with a traditional chemical background.

4.1.1 The Ocean as a Biogeochemical Environment

A traditional approach utilised in geochemistry, and now also in environmental chemistry, is to consider the system under investigation a reservoir. For a given component, the reservoir has sources (inputs) and sinks (outputs). The system is said to be at equilibrium, or operating under steady-state conditions, when a mass balance between inputs and outputs is achieved. An imbalance could signify that an important source or sink has been ignored. Alternatively, the system may be perturbed, possibly mediated through anthropogenic activities, and therefore be changing towards a new equilibrium state.

Processes within the reservoir that affect the temporal and spatial distribution of a given component are transportation and transformations. Both physics and biology within the system play a role. Clearly, transport effects are dominated by the hydrodynamic regime. Although transformations involve both chemical (dissolution, redox reactions, speciation changes) and geological (sedimentation) processes, biological activity generally controls nutrient and trace metal distributions. Furthermore, the biota influences concentrations of O_2 and CO_2, which in turn determine the pH and pε (*i.e.*, the redox potential), respectively. For these reasons, some fundamental aspects of descriptive physical and biological oceanography are included in this chapter.

In terms of biogeochemical cycling, the ocean constitutes a large reservoir. The surface area is 361.11×10^6 km^2, encompassing nearly 71% of the earth's surface. The average depth is 3.7 km, but depths in the submarine trenches can exceed 10 km. The ocean contains about 97% of the water in the global hydrological cycle. A schematic representation of the oceanic reservoir is presented in Figure 1. The material within it can be operationally defined, usually based on filtration, as dissolved or particulate. For simplicity here, the ocean is divided into two layers, with distinct surface and deepwaters. The boundary regions are also distinguished, as the composition in these regions can be quite different to that of bulk seawater. Furthermore, interactions within these environments can alter the mass transfers across the boundary. The rationale for such features will be presented in subsequent sections.

Material supplied to the ocean originates from the atmosphere, rivers, glaciers and hydrothermal waters. The relative importance of these pathways depends upon the component considered and geographic location. River runoff commonly constitutes the most important source. Transported material may be either dissolved or particulate, but discharges are into surface waters and confined to coastal regions. Hydrothermal waters are released from vents on the seafloor. Such hydrothermal waters are formed when seawater circulates into the fissured rock matrix, and under conditions of elevated temperature and pressure, compositional changes in the aqueous phase occur due to seawater – rock interactions. This is an important source of some elements, such as Li, Rb and Mn. The atmosphere supplies particulate material globally to the surface of the ocean. In recent years, this has been the most prominent pathway to the World

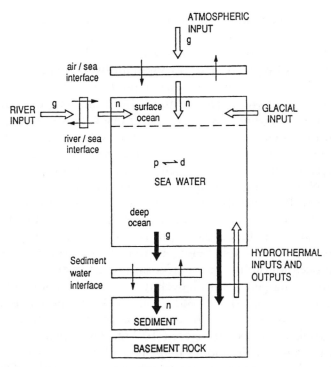

Figure 1 *A schematic representation of the ocean reservoir. The source and sink fluxes are designated as g and n, referring to gross and net fluxes, thereby indicating that interactions within the boundary regions can modify the mass transfer. Within seawater, the p ⇌ d term signifies that substances can undergo particulate–dissolved interactions. However, it must be appreciated that several transportation and transformation processes might be operative* (Adapted from Chester, 1990.[1])

Ocean for Pb, identified by its isotopic signature as originating from petrol additives. Wind-borne transport is greatest in low latitudes and the Sahara Desert is known to act as an important source of dust. Also, the airborne flux of nutrients, notably nitrogenous compounds, has become increasingly recognised as important both in some coastal waters, such as Chesapeake Bay, and large seas, including the Mediterranean Sea. In contrast, glacial activity makes little impact on the World Ocean. Glacier-derived material tends to be comprised of physically weathered rock residue, which is relatively insoluble. In addition, the input is largely confined to Polar Regions, with Antarctica responsible for approximately 90% of the material.

Although volatilisation and subsequent evasion to the atmosphere can be important for elements, such as Se and Hg that undergo biomethylation, sedimentation acts as the major removal mechanism. The mechanism is a geological/physical process, essentially the deposition of suspended particles in response to changes in the hydrodynamic regime. Such removal is especially important in coastal environments. Marine organisms play an important role in the open sea through the sinking of shells of microorganisms and faecal pellets. The chemical precipitation of salts from seawater occurs only under special conditions. $CaCO_3$ can precipitate in warm tropical lagoons and Mn^{2+} can be oxidised to the relatively insoluble MnO_2, a process of importance near deep-sea hydrothermal vents.

Some definitions help in the interpretation of chemical phenomena in the ocean. Conservative behaviour signifies that the concentration of a constituent or absolute magnitude of a property varies only due to mixing processes. Components or parameters that behave in this manner can be used as conservative indices of mixing. Examples are salinity and potential temperature, the definitions for which are presented in subsequent sections. In contrast, non-conservative behaviour indicates that the concentration of a constituent may vary as a result of biological or chemical processes. Examples of parameters that behave non-conservatively are dissolved oxygen and pH.

Residence time, τ, is defined as

$$\tau = \frac{A}{(\mathrm{d}A/\mathrm{d}T)} \qquad (4.1)$$

where A is the total amount of constituent A in the reservoir and $\mathrm{d}A/\mathrm{d}T$ can be either in rate of supply or the rate of removal of A. This represents the average lifetime of the component in the system and is, in effect, a reciprocal rate constant (his concept is discussed further in Chapter 7).

The photic zone refers to the upper surface of the ocean in which photosynthesis can occur. This is typically taken to be the depth at which sunlight radiation has declined to 1% of the magnitude at the surface. This might typically be >100 m for visible light or photosynthetically active radiation (PAR), but generally <20 m for UV wavelengths, of recent interest due to the enhanced input invoked by stratospheric ozone depletion.

4.1.2 Properties of Water and Seawater

Water is a unique substance, with unusual attributes because of its structure. The molecule consists of a central oxygen atom with two attached hydrogen atoms forming a bond angle of about 105°. As oxygen is more electronegative than hydrogen, it attracts the shared electrons to a greater extent. In addition, the oxygen atom has a pair of lone orbitals. The overall effect produces a molecule with a very strong dipole moment, having distinct negative (O) and positive (H) ends. While there are several important consequences, two will be considered here. Firstly, the positive H atoms of one molecule are attracted towards the negative O atom in adjacent molecules giving rise to hydrogen bonding. This has important implications with respect to a number of physical properties, especially those relating to thermal characteristics. Secondly, the large dipole moment ensures that water is a very polar solvent.

Considering firstly the physical properties, water has much higher freezing and melting points than would be expected for a molecule of molecular weight 18. Water has high latent heats of evaporation and fusion. Therefore, considerable energy is required to cause phase changes, the energy being utilised in hydrogen-bond rupturing. Moreover, it has a high specific heat and is a good conductor of heat. Consequently, heat transfer in water by advection and conduction gives rise to uniform temperatures. The density of pure water exhibits anomalous behaviour. In ice, O atoms have 4 H atoms orientated about them in a tetrahedral configuration. These units are packed together with a hexagonal symmetry. At the freezing point, 0 °C, ice is less dense than water. Heating breaks some hydrogen bonds and the molecules can achieve slightly closer packing causing the density to increase. Thus, the maximum density occurs at 4 °C because thermal expansion at higher temperatures compensates for this compression effect. As will be discussed later, seawater differs in this respect. Thus, fresh ice floats on water, which in part explains how rivers and lakes can freeze over, while remaining liquid at depth. With respect to other properties, water has a high surface tension that is manifest in stable droplet formation and has a relatively low molecular viscosity and therefore is quite a mobile fluid.

Water is an excellent solvent. It is extremely polar and can dissolve a wider range of solutes and in greater amounts than any other substance. Water has a very high dielectric constant, a measure of the solvent's ability to keep apart oppositely charged ions. The solvating characteristics of individual ions influence their behaviour in solution, *i.e.*, in terms of hydration, hydrolysis and precipitation. Although water exhibits amphoteric behaviour, electrolytic dissociation is quite small. Furthermore, dissociation gives equal ion concentrations of both H_3O^+ and OH^- and so pure water is neutral. The amphoteric behaviour enhances dissolution of introduced particulate matter through surface hydrolysis reactions.

While the concept will be considered in detail below, the term salinity (S‰) is introduced here as a measure of the salt content of seawater, a typical value for oceanic waters being 35 g kg^{-1}. In an oceanographic context, the most important consequence of the addition of salt to water is the effect on density. However, many of the characteristics outlined above are also altered. The addition of electrolytes can cause a small increase in the surface tension. This effect is not commonly observed in seawater due to the presence of surfactants, which decrease the surface tension and so facilitate foam formation. As illustrated in Figure 2, the presence of salt does depress the temperature of maximum density and the freezing point of the solution relative to pure water. Thus, seawater with a typical salt content of 35 g kg^{-1} freezes at approximately $-1.9\,^{\circ}$C and the resulting ice is denser than the solution. However, more often than the formation of sea ice itself, the freezing process tends to produce fresh ice overlying a more concentrated brine solution. Salts can be precipitated at much lower temperatures, *i.e.*, mirabilite ($Na_2SO_4.2H_2O$) at $-8.2\,^{\circ}$C and halite (NaCl) at $-23\,^{\circ}$C. Some brine inclusions and salt crystals can become incorporated into the ice.

From an oceanographic perspective, the fundamental properties of seawater are temperature, salinity and pressure (*i.e.*, depth dependent). Together, these parameters control the density of the water, which in turn determines the buoyancy of the water and pressure gradients. Small density differences integrated over oceanic scales cause considerable pressure gradients and result in currents.

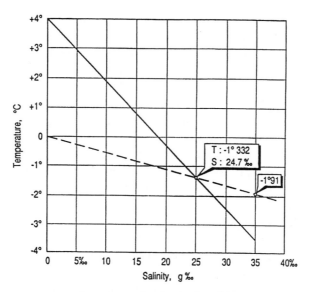

Figure 2 *The temperature of maximum density (—) and freezing point (- -) of seawater as a function of dissolved salt content*
(Adapted from Tchernia, 1980.[2])

Surface water temperatures are extremely variable, obviously influenced by location and season. The minimum temperature found in polar latitudes approaches the freezing point of nearly −2 °C. Equatorial oceanic waters can reach 30 °C. Temperature variations with depth are far from consistent. In a region where mixing prevails, as observed especially in the surface waters, a layer forms with a relatively uniform temperature. The zone immediately beneath normally exhibits a sharp change in temperature, known as the thermocline. The thermocline in the ocean extends down to about 1000 m within equatorial and temperate latitudes. It acts as an important boundary in the ocean, separating the surface and deep layers and limiting mixing between these two reservoirs.

Below the thermocline, the temperature changes only little with depth. The temperature is a non-conservative property of seawater because adiabatic compression causes a slight increase in the *in situ* temperature measured at depth. For instance in the Mindanao Trench in the Pacific Ocean, the temperature at 8500 and 10,000 m is 2.23 and 2.48 °C, respectively. The term potential temperature is defined to be the temperature that the water parcel would have if raised adiabatically to the ocean surface. For the examples above, the potential temperatures are 1.22 and 1.16 °C, respectively. Potential temperature of seawater is a conservative index.

Salinity in the surface waters in the open ocean range between 33 and 37 (Figure 3), the main control being the balance between evaporation and precipitation. The highest salinities occur in regional seas, where the evaporation rate is extremely high, namely the Mediterranean Sea (38–39) and the Red Sea (40–41). Within the World Ocean, the salinity is greatest in latitudes of about 20° where the evaporation exceeds precipitation. Lower salinities occur poleward as evaporation diminishes and near the equator where precipitation is very high. Local effects can be important, as evident in the vicinity of large riverine discharges that dilute the salinity. Salinity variations with depth are related to the origin of the deepwaters and so will be considered in the section on oceanic circulation. A zone in which the salinity exhibits a marked gradient is known as a halocline.

Whereas the density of pure water is 1.000 g ml^{-1}, the density of seawater (S‰ = 35) is about 1.03 g ml^{-1}. The term 'sigma-tee', σ_t, is used to denote the density (actually the specific gravity and hence it is a dimensionless number) of water at atmospheric pressure based on temperature and salinity *in situ*. Density increases, and so the buoyancy decreases, with an increase in σ_t. It is defined as

$$\sigma_t = (\text{specific gravity}_{S‰,T} - 1) \times 1000$$

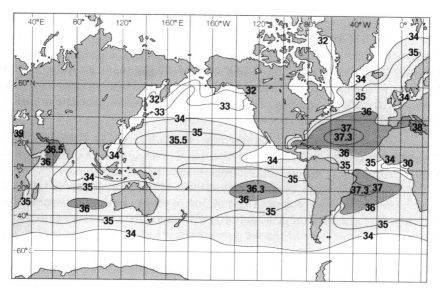

Figure 3 *The distribution of mean annual salinity in the surface waters of the ocean* (Adapted from The Open University, 1989.[3])

In a plot of temperature against salinity (a T–S diagram), constant σ_t appear as curved lines, which denote waters of constant pressure and are known as isopycnals. A zone in which the pressure changes greatly is known as a pycnocline. Within the water column, a pycnocline, therefore, separates water with distinctive temperature and salinity characteristics, usually indicative of different origins. A T–S diagram can also be used to estimate the properties resulting from the mixing of two water masses. As noted above, the temperature is not a conservative property, and therefore σ_t is also non-conservative. To circumvent the associated difficulties of interpretation, an analogous term known as the potential density, σ_θ, is defined on the basis of potential temperature instead of *in situ* temperature. The σ_θ is therefore a conservative index.

4.1.3 Salinity Concepts

Salinity is a measure of the salt content of seawater. Developments in analytical chemistry have led to an historical evolution of the salinity concept. Intrinsically, it would seem to be a relatively straightforward task to measure. This is true for imprecise determinations that can be quickly performed using a hand-held refractometer. The salinity affects seawater density and thus, the impetus for high precision in salinity measurements came from physical oceanographers.

The first techniques utilised for the determination of salinity, involving the gravimetric analysis of salt left after evaporating seawater to dryness, were fraught with difficulties. Variable amounts of water of crystallisation might be retained. Some salts, such as $MgCl_2$, can decompose leaving residues of uncertain composition. Other constituents, especially organic material, might be volatilised or oxidised. Overall, such methods led to considerable inconsistencies and inaccuracies.

The second set of procedures for salinity measurement made use of the observation from the *Challenger* expedition of 1872–1876 that sea-salt composition was apparently invariant. Hence, the total salt content could be calculated from any individual constituent, such as Cl^- that could be readily determined by titration with Ag^+. At the turn of the century, Knudsen defined salinity to be the weight in grams of dissolved inorganic matter contained in 1 kg of seawater, after bromides and iodides were replaced by an equivalent amount of chloride and carbonate was converted to oxide. Clearly from the adopted definition, the analytical technique was not specific to Cl^- and so the term chlorinity was introduced. Chlorinity (Cl‰) is the chloride concentration in seawater, expressed as g kg^{-1}, as measured by Ag^+ titration (*i.e.*, ignoring other halide contributions by assuming Cl^- to be the only reactant). The relationship of interest was that between S‰ and Cl‰, given as
$$S‰ = 1.805Cl‰ + 0.030$$

As a calibrant solution for the $AgNO_3$ titrant, Standard Seawater was prepared that had certified values for both chlorinity and salinity. Unfortunately, the above salinity–chlorinity relationship was derived from only nine seawater samples that were somewhat atypical. It has since been redefined using a much larger set of samples representative of oceanic waters to become
$$S‰ = 1.80655Cl‰$$

The third category of salinity methodologies was based on conductometry, as the conductivity of a solution is proportional to the total salt content. Standard Seawater, now also certified with respect to conductivity, provides the appropriate calibrant solution. The conductivity of a sample is measured relative to the standard and converted to salinity in practical salinity units (psu). Note that although psu has replaced the outmoded ‰, usually units are ignored altogether in modern usage. These techniques continue to be the most widely used methods because conductivity measurements can provide salinity values with a precision of ±0.001 psu. Highly precise determinations require temperature control of samples and standards to within ±0.001 °C. Application of a non-specific technique like conductometry relies upon the assumption that the sea-salt matrix is invariant, both spatially and temporally. Thus, the technique cannot be reliably employed in marine boundary environments where the seawater composition differs to the bulk characteristics.

There are two types of conductometric procedures commonly used. Firstly, a Wheatstone Bridge circuit can be set up, whereby the ratio of the resistance of unknown seawater to standard seawater balances the ratio of a fixed resistor to a variable resistor. The system uses alternating current to minimise electrode fouling. Alternatively, the conductivity can be measured by magnetic induction, in which case the sensor consists of a plastic tube containing sample seawater that links two transformers. An oscillator establishes a current in one transformer that induces current flow within the tube, the magnitude of which depends upon the salinity of the sample. This in turn induces a current in the second transformer, which can then be measured. This design has been exploited for *in situ* conductivity measurements.

4.1.4 Oceanic Circulation

The distribution of chemical components within the ocean is determined by both transportation and transformation processes. A brief outline of oceanic circulation is necessary to ascertain the relative influences. Two main flow systems must be considered. Surface circulation is established by tides and the prevailing wind patterns and deep circulation is determined by gravitational forces. Both are modified by Coriolis force, the acceleration due to the earth's rotation. It acts to deflect moving fluids (*i.e.*, both air and water) to the right in the northern hemisphere and to the left in the southern hemisphere. The magnitude of the effect is a function of latitude, being nil at the equator and increasing poleward.

Surface oceanic circulation is depicted in Figure 4. For the most part, the circulation patterns describe gyres constrained by the continental boundaries. The prevailing winds acting under the influence of Coriolis force result in clockwise and counter-clockwise flow in the northern and southern hemispheres, respectively. The flow fields are non-uniform, exhibiting faster currents along the western margins. These are manifest, for example, as the Gulf Stream, Kuroshio Current and Brazil Current. Circulation within the Indian Ocean is exceptional in that there are distinct seasonal variations in accord with the monsoons. The absence of other continents within the immediate boundary region of Antarctica gives rise to a circumpolar current within the Southern Ocean.

The surface circulation is restricted to the upper layer influenced by the wind, typically about 100 m. However, underlying water can be transported up into this zone when horizontal advection is insufficient to

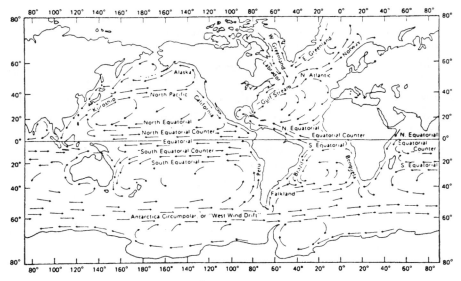

Figure 4   *The surface circulation in the ocean*
(Adapted from Stowe, 1979.[4])

maintain the superimposed flow fields. This process is called upwelling and is of considerable importance in that biochemical respiration of organic material at depth ensures that the ascending water is nutrient-rich. Upwelling occurs notably in the eastern oceanic boundaries where longshore winds result in the offshore transport of the surface water. Examples are found off the coasts of Peru and West Africa. Similar processes cause upwelling in the Arabian Sea, but this is seasonal due to the monsoon effect. A divergence is a zone in which the flow fields separate. In such a case, upwelling may result as observed in the equatorial Pacific. It should be noted that a region in which the streamlines come together is known as a convergence, and the water sinks in this zone.

The density of the water controls the deepwater circulation. If the density of a water body increases, it has a tendency to sink. Subsequently, it will spread out over a horizon of uniform σ_θ. As the density can be raised due to either an increase in the salinity or a decrease in the temperature, the deepwater circulatory system is also known as thermohaline circulation. As shown in Figure 5 of the ocean conveyor belt, the densest oceanic waters are formed in Polar Regions due to the relatively low temperatures and the salinity increase that results from ice formation. Antarctic Bottom Water (ABW) is generated in the Weddell Sea and flows northward into the South Atlantic. North Atlantic Deep Water (NADW)

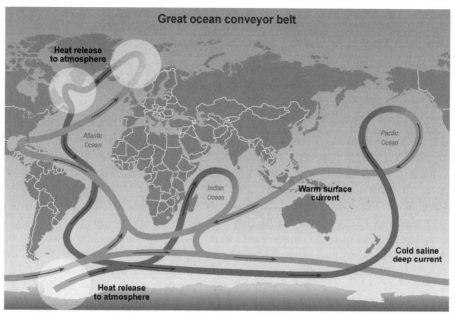

Figure 5 *A schematic diagram of the thermohaline circulation of the world ocean, also know as the great ocean conveyor belt, highlighting polar regions of deepwater formation, deepwater circulation eastward from the poles and the returning westward surface water flow*
(Adapted from IPCC, 2001.[5])

is formed in the Norwegian Sea and off the southern coast of Greenland. The flow of the NADW can be traced southwards through the Atlantic Ocean to Antarctica, where it is diverted eastward into the Southern Indian Ocean and South Pacific. There it heads northwards and either enters the North Pacific or becomes mixed upward into the surface layer in the equatorial region. The transit time is on the order of 1000 years. As noted previously, the thermocline acts as an effective barrier against mixing of dissolved components in the ocean. Consequently, this deepwater formation process in high latitudes is important because it facilitates the relatively rapid transport of material from the surface of the ocean down to great depths. The deep advection of atmospherically derived CO_2 is a pertinent example. The formation of NADW could diminish with climate change because surface waters would be warmer and less dense due to enhanced melting of ice in the Arctic region.

Diverse processes can form intermediate waters within the water column. In the southern South Atlantic, the NADW overrides the denser ABW. Antarctic Intermediate Water results from water sinking along the Antarctic Convergence ($\sim 50°$S). Relatively warm, saline water exits the Mediterranean Sea at depth and can be identified as a distinctive layer within the North and South Atlantic.

4.2 SEAWATER COMPOSITION AND CHEMISTRY

4.2.1 Major Constituents

The major constituents in seawater are conventionally taken to be those elements present in typical oceanic water of salinity 35 that have a concentration greater than 1 mg kg^{-1}, excluding Si, which is an important nutrient in the marine environment. The concentrations and main species of these elements are presented in Table 1. One of the most significant observations from the *Challenger* expedition of 1872–1876 was that these major components existed in constant relative amounts. As already explained, this feature was exploited for salinity determinations. Inter-element ratios are generally constant, and often expressed as a ratio to Cl‰ as shown in Table 1. This implies conservative behaviour, with concentrations depending solely upon mixing processes, and indeed, salinity itself is a conservative index.

Because of this behaviour, individual seawater constituents can be utilised for source apportionment studies in non-marine environments. For instance, an enrichment factor (*EF*) for a substance X is defined as

$$EF_X = \frac{(X/Na^+)_{\text{sample}}}{(X/Na^+)_{\text{seawater}}} \tag{4.2}$$

An EF of 1 indicates that the substance exists in comparable relative amounts in the sample and in seawater, thereby giving a good indication

Table 1 *Chemical species and concentrations of the major elements in seawater*

Element	Chemical species	Concentration (mol L^{-1})	For S = 35 (g kg^{-1})	Ratio to chlorinity (Cl = 19.374‰)
Na	Na$^+$	4.79×10^{-1}	10.77	5.56×10^{-1}
Mg	Mg^{2+}	5.44×10^{-2}	1.29	6.66×10^{-2}
Ca	Ca^{2+}	1.05×10^{-2}	0.4123	2.13×10^{-2}
K	K$^+$	1.05×10^{-2}	0.3991	2.06×10^{-2}
Sr	Sr^{2+}	9.51×10^{-5}	0.00814	4.20×10^{-4}
Cl	Cl$^-$	5.59×10^{-1}	19.353	9.99×10^{-1}
S	SO$_4^{2-}$, NaSO$_4^-$	2.89×10^{-2}	0.905	4.67×10^{-2}
C (inorganic)	HCO$_3^-$, CO$_3^{2-}$	2.35×10^{-3}	0.276	1.42×10^{-2}
Br	Br$^-$	8.62×10^{-4}	0.673	3.47×10^{-3}
B	B(OH)$_3$, B(OH)$_4^-$	4.21×10^{-4}	0.0445	2.30×10^{-3}
F	F$^-$, MgF$^+$	7.51×10^{-5}	0.00139	7.17×10^{-5}

Source: based on Dyrssen and Wedborg, 1974.[6]

of a marine origin. If $EF_X > 1$, then it is enriched with respect to seawater. Conversely, depletion is signified when values $EF_X < 1$. Another example of the application of inter-element ratios can be found in examining the geochemical cycle of sulfur. Concentrations of SO_4^{2-} and Na^+ in ice cores and marine aerosols exhibit a SO_4^{2-}/Na^+ greater than that observed in seawater. This excess can be readily calculated and is known as non-sea salt sulfate (NSSS). Contributions to NSSS include SO_2 derived from both volcanic and anthropogenic sources, together with dimethylsulfide (DMS) of marine biogenic origin.

Not all the major constituents consistently exhibit conservative behaviour in the ocean. The most notable departures occur in deepwaters, where Ca^{2+} and HCO_3^- exhibit anomalously high concentrations due to the dissolution of calcite. The concept of relative constant composition does not apply in a number of atypical environments associated with boundary regions. Inter-element ratios for major constituents can be quite different in estuaries and near hydrothermal vents. Obviously, these are not solutions of sea salt with the implication that accuracy of salinity measurements by chemical and conductometric means is limited.

The residence times for some elements are presented in Table 2. The major constituents normally have long residence times. The residence time is a crude measure of a constituent's reactivity in the reservoir. The aqueous behaviour and rank ordering can be appreciated simply in terms the ionic potential given by the ratio of electronic charge to ionic radius (Z/r). Elements with $Z/r < 3$ are strongly cationic. The positive charge density is relatively diffuse, but sufficient to attract an envelope of water molecules to form a hydrated cation. As the ionic potential increases, the force of attraction towards the water similarly rises to the extent that one oxygen – hydrogen bond in the molecule breaks. This causes the solution pH to fall and metal hydroxides to form. Neutral hydroxides tend to be relatively insoluble and so precipitate. However, in the more extreme case for which $Z/r > 12$, the attraction towards the oxygen is so great that both bonds in the associated water molecules are broken. The reaction product is an oxyanion, usually quite soluble because of the associated anionic charge. Thus in seawater, those elements (Al, Fe) having a tendency to form insoluble hydroxides have short residence times. This is also true for elements that exist preferentially as neutral oxides (Mn, Ti). Hydrated cations (Na^+, Ca^{2+}) and strongly anionic species (Cl^-, Br^-, $UO_2(CO_3)_2^{4-}$) have long residence times. This treatment is, of course, somewhat of an over-simplification ignoring the rather significant role that biological organisms play in nutrient and trace element chemistry.

Table 2   *The residence time and speciation of some elements in the ocean*

Element	Principal species	Concentration $(mol\ L^{-1})$	Residence time $(year)$
Li	Li^+	2.6×10^{-5}	2.3×10^6
B	$B(OH)_3$, $B(OH)_4^-$	4.1×10^{-4}	1.3×10^7
F	F^-, MgF^+	6.8×10^{-5}	5.2×10^5
Na	Na^+	4.68×10^{-1}	6.8×10^7
Mg	Mg^{2+}	5.32×10^{-2}	1.2×10^7
Al	$Al(OH)_4^-$, $Al(OH)_3$	7.4×10^{-8}	1.0×10^2
Si	$Si(OH)_4$	7.1×10^{-5}	1.8×10^4
P	HPO_4^{2-}, PO_4^{3-}, $MgHPO_4$	2×10^{-6}	1.8×10^5
Cl	Cl^-	5.46×10^{-1}	1×10^8
K	K^+	1.02×10^{-2}	7×10^6
Ca	Ca^{2+}	1.02×10^{-2}	1×10^6
Sc	$Sc(OH)_3$	1.3×10^{-11}	4×10^4
Ti	$Ti(OH)_4$	2×10^{-8}	1.3×10^4
V	$H_2VO_4^-$, HVO_4^{2-}, $NaHVO_2^-$	5×10^{-8}	8×10^4
Cr	CrO_4^{2-}, $NaCrO_4^-$	5.7×10^{-9}	6×10^3
Mn	Mn^{2+}, $MnCl^+$	3.6×10^{-9}	1×10^4
Fe	$Fe(OH)_3$	3.5×10^{-8}	2×10^2
Co	Co^{2+}, $CoCO_3$, $CoCl^+$	8×10^{-10}	3×10^4
Ni	Ni^{2+}, $NiCO_3$, $NiCl^+$	2.8×10^{-8}	9×10^4
Cu	$CuCO_3$, $CuOH^+$, Cu^{2+}	8×10^{-9}	2×10^4
Zn	$ZnOH^+$, Zn^{2+}, $ZnCO_3$	7.6×10^{-8}	2×10^4
Br	Br^-	8.4×10^{-4}	1×10^8
Sr	Sr^{2+}	9.1×10^{-5}	4×10^6
Ba	Ba^{2+}	1.5×10^{-7}	4×10^4
La	La^{3+}, $LaCO_3^+$, $LaCl^{2+}$	2×10^{-11}	6×10^2
Hg	$HgCl_4^{2-}$	1.5×10^{-10}	8×10^4
Pb	$PbCO_3$, $Pb(CO_3)_2^{2-}$, $PbCl^+$	2×10^{-10}	4×10^2
Th	$Th(OH)_4$	4×10^{-11}	2×10^2
U	$UO_2(CO_3)_2^{4-}$	1.4×10^{-8}	3×10^6

Source: based on Brewer, 1975[7] and Bruland, 1983.[8]

4.2.2   Dissolved Gases

4.2.2.1   Gas Solubility and Air–Sea Exchange Processes.   The ocean contains a vast array of dissolved gases. Some of the gases, such as Ar and chlorofluorocarbons (CFCs), behave conservatively and can be utilised as tracers for water mass movements and ventilation rates. Equilibrium processes at the air–sea interface generally lead to saturation, and the concentration then remains unchanged once the water sinks. Thus, the gas concentration is characteristic of the lost contact with the atmosphere. There are several important non-conservative gases, which exhibit wide variations in concentration due to biological activity. O_2 determines the redox potential in seawater and CO_2 buffers the ocean at pH 8. Ocean–atmosphere exchange processes for gases such as CO_2 and DMS may play an important role in climate change.

Both temperature and salinity affect the solubility of gases in water. Empirical relationships can be found elsewhere.[9,10] The trends are such that gas solubility increases with a decrease in temperature or an increase in salinity. The changes in solubility are non-linear and differ dramatically for various gases. Whereas Figure 6 depicts the solubility of several gases as a function of temperature, Figure 7 shows the influence of both salinity and temperature on the solubility of molecular nitrogen.

At the ocean–atmosphere interface, the exchange of gases occurs to achieve equilibrium between the two systems, and consequently gases become saturated in the water. However, supersaturation can be achieved by several mechanisms. Firstly, bubbles that form from white cap activity can be entrained and are dissolved at depth. The slight but significantly greater pressure at depth relative to the surface favours gas dissolution, and establishes a higher equilibrium concentration. Secondly as evident from Figure 6, if two water masses that have been

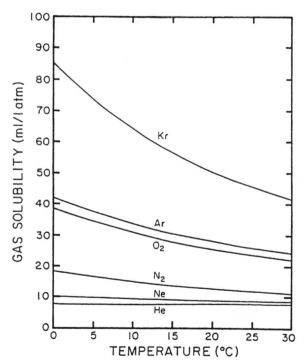

Figure 6 *The solubility of various gases in seawater as a function of temperature* (Adapted from Broecker and Peng, 1982.[11])

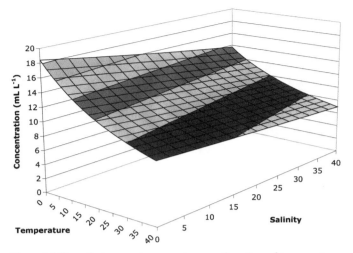

Figure 7   *The solubility of nitrogen in seawater as a function of temperature and salinity*

equilibrated at different temperatures were mixed, then the resulting water body would be supersaturated. Thirdly, gases like oxygen that are produced *in situ* by biological activity may become supersaturated, particularly when evasion to the atmosphere is hindered.

The gas solubility for a water body in equilibrium with the overlying air mass can be expressed in several ways. It is convenient to consider Henry's Law that states

$$H = c_a c_w^{-1}$$

where H is the Henry's law constant and c_a and c_w refer to the concentration of a gas in air and water, respectively. As discussed by Liss (1983),[12] air–sea exchange occurs when a concentration gradient exists (*i.e.*, $\Delta C = c_a H^{-1} - c_w$) and the magnitude of the consequential flux, F, is given as

$$F = K \, \Delta C$$

where the proportionality constant, K, has dimensions of velocity and so is frequently referred to as the transfer velocity. The concept is elaborated upon in Chapter 7.

Air–sea exchange processes are consequently dependent upon the concentration gradient and the transfer velocity. The transfer velocity is not a constant, but rather depends upon several physical parameters, such as temperature, wind speed and wave state. The exchange can also be attenuated by the presence of a surface film or slick. Alternatively, the exchange can be facilitated by bubble formation. The concentration

Table 3 *The net global fluxes of some trace gases across the air/sea interface*

Gas	Global air–sea direction[a]	Flux magnitude (g yr^{-1})
CH_4	+	10^{12}–10^{13}
CO_2	−	6×10^{15}
N_2O	+	6×10^{12}
{ CCl_4	−	10^{10}
{ CCl_4	=	~ 0
{ CCl_3F	−	5×10^9
{ CCl_3F	=	~ 0
CH_3I	+	3–13×10^{11}
CO	+	$100 \pm 90 \times 10^{12}$
H_2	+	$4 \pm 2 \times 10^{12}$
Hg	+	$\sim 2 \times 10^9$

^{a} + sea to air; − air to sea;=no net flux.
Source: from Chester, 1990.[1]

gradient determines the direction of the flux, into or out of the ocean. Net global fluxes for some gases are presented in Table 3. The atmosphere serves as the source of material for conservative gases, especially those of anthropogenic origin, but several gases produced *in situ* by biological activity evade from the ocean.

4.2.2.2 Oxygen. Oxygen is a non-conservative gas and a typical oceanic profile is shown in Figure 8. The concentration varies throughout the water column, its distribution being greatly influenced by biological activity. The generalised chemical equation for carbon fixation is often given as

$$nCO_2 + nH_2O \rightleftharpoons (CH_2O)_n + nO_2$$

During photosynthesis this reaction proceeds to the right, thereby producing organic material, as designated by $(CH_2O)_n$, and O_2. The surface waters become equilibrated with respect to atmospheric O_2, but they can get supersaturated during periods of intense photosynthetic activity. Respiration occurs as the above reaction proceeds to the left and O_2 is consumed. Photosynthesis is obviously restricted to the photic zone in the upper ocean and ordinarily exceeds respiration. However, the relative importance of the two processes changes with depth. The oxygen compensation depth is the horizon in the water column at which the rate of O_2 production by photosynthesis equals the rate of respiratory O_2 oxidation.

Below the photic zone, O_2 is utilised in chemical and biochemical oxidation reactions. As evident in Figure 8, the concentration diminishes

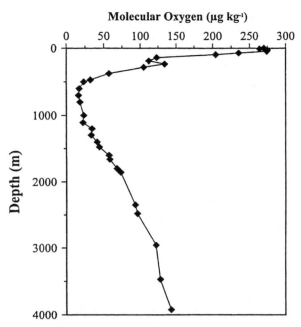

Figure 8 *A profile of molecular oxygen in the North Pacific Ocean* (Data from Bruland, 1980.[13])

with depth to develop an oxygen minimum zone. Thereafter, the O_2 concentration in deeper waters begins to increase because these waters originated from Polar Regions. They were cold and in equilibrium with atmospheric gases at the time of sinking, but subsequently lost little of the dissolved O_2 because the flux of organic material to deepwaters is relatively small.

The dissolved O_2 content of seawater has a significant control on the redox potential, often designated in environmental chemistry by pε (see also Chapter 3). This is defined with reference to electron activity in an analogous fashion to pH and thus

$$p\varepsilon = -\log\{\,e^-\,\}$$

The relationship between pε and the more familiar electrode potential E or E^H is

$$p\varepsilon = \frac{F}{2.303RT}E \tag{4.3}$$

and for the standard state

$$p\varepsilon^{\circ} = \frac{F}{2.303RT} E^{\circ} \qquad (4.4)$$

where F is Faraday's constant, R the universal gas constant and T the absolute temperature in Kelvin. Whereas a high value of $p\varepsilon$ indicates oxidising conditions, a low value signifies reducing conditions. Oxygen plays a role *via* the reaction

$$O_2 + 4H^+ + 4e^- \rightleftharpoons 2H_2O$$

At 20 °C, $K = 10^{83.1}$ and so water of pH=8.1 in equilibrium with atmospheric O_2 ($pO_2 = 0.21$ atm) has $p\varepsilon = 12.5$. This conforms to surface conditions, but the $p\varepsilon$ decreases as the O_2 content diminishes with depth. The oxygen minimum is particularly well developed beneath the highly productive surface waters of the eastern tropical Pacific Ocean, where there is a large flux of organic material to depth and subsequently considerable oxidation. The O_2 becomes sufficiently depleted (*i.e.*, hypoxia) that the resulting low redox conditions causes NO_3^- to be reduced to NO_2^-. Aeolian transport of nitrate to Chesapeake Bay can lead to low O_2 conditions. Similarly, intermittent hypoxia develops in parts of the Gulf of Mexico due to the riverine transport of nutrients derived from agricultural uses in the Mississippi catchment.

When circulation is restricted vertically due to thermal or saline stratification and horizontally by topographic boundaries, the water becomes stagnant and the oxygen may be completely utilised producing anoxic conditions. Such regions represent atypical marine environments where reducing conditions prevail. Well-known examples include the Black Sea, which is permanently anoxic below 200 m, and the Cariaco Trench, a depression in the Venezuelan continental shelf. Some fjords, such as Saanich Inlet in western Canada and Dramsfjord of Norway, may be intermittently anoxic. Periodic flushing of these inlets by dense, oxygenated waters displaces deep anoxic water to the surface causing massive fish mortality.

O_2 can be used as a tracer to help identify the origin of water masses. The warm, saline intrusion into the Atlantic Ocean from the Mediterranean Sea is relatively O_2 deficient. Alternatively, the waters downwelling from Polar Regions have elevated O_2 concentrations.

4.2.2.3 Carbon Dioxide and Alkalinity. Marine chemists sometimes adopt activity conventions quite different to those traditionally used in chemistry. It is useful to preface a discussion about the carbon dioxide–calcium carbonate system in the oceans with a brief outline of pH scales.

Although originally introduced in terms of ion concentration, today the definition of pH is based on hydrogen ion activity and is

$$pH = -\log a_H$$

where a_H refers to the relative hydrogen ion activity (*i.e.*, dimensionless, as is pH). Defined using concentration scales, the pH can be

$$pH = -\log\left(c_H \gamma_H / c^{\circ}\right)$$

or

$$pH = -\log\left(m_H \gamma_H / m^{\circ}\right)$$

where c_H and m_H represent molar and molal concentrations, c° and m° are the respective standard state conditions (1 mol L^{-1} and 1 mol kg^{-1}), and γ_H is the appropriate activity coefficient. Obviously γ_H differs in these two expressions because $c^{\circ} \neq m^{\circ}$. However, different activity scales may also be used. In the *infinite dilution activity scale*, $\gamma_H \Rightarrow 1$ as the concentration of hydrogen ions and all other ions approach 0. For analyses, pH meters are calibrated using dilute buffers prepared in pure water. Alternatively, in the *constant ionic medium activity scale*, $\gamma_H \Rightarrow 1$ as the concentration of hydrogen ions approaches 0, while all other components are maintained at some constant level. Calibrant buffers are prepared in solutions of constant ionic composition, and in marine chemistry, this is often a solution of synthetic seawater. While these two methodologies are equally justifiable from a thermodynamic point of view, it is important to appreciate that pH scales so defined are quite different. As a further consequence, the absolute values for dissociation constants also differ.

The biogeochemical cycle of inorganic carbon in the ocean is extremely complicated. It involves the transfer of gaseous carbon dioxide from the atmosphere into solution. Not only is this a reactive gas that readily undergoes hydration in the ocean, but also it is fixed as organic material by marine phytoplankton. Inorganic carbon can be regenerated either by photochemical oxidation in the photic zone or *via* respiratory oxidation of organic material at depth. Surface waters are supersaturated with respect to aragonite and calcite, different solid phases of $CaCO_3$, but precipitation is limited to coastal lagoons such as found in the Bahamas. However, several marine organisms utilise calcium carbonate to form shells. Sinking shells can remove inorganic carbon from surface waters that is then regenerated following dissolution in the under-saturated waters found at depth. Nonetheless, calcitic oozes of biogenic origin constitute a major component in marine sediments.

Finally, the inorganic carbon equilibrium is responsible for buffering seawater at a pH near 8 on timescales of centuries to millennia.

There are several equilibria to be considered (see also Chapter 3). Firstly, CO_2 is exchanged across the air–sea interface

$$CO_{2g} \rightleftharpoons CO_{2aq}$$

The equilibrium process obeys Henry's Law, but the dissolved CO_2 reacts rapidly with water to become hydrated as

$$CO_{2aq} + H_2O \rightleftharpoons H_2CO_3$$

Relative to the exchange process, the hydration reaction forming carbonic acid occurs quite quickly. This means that the concentration of dissolved CO_2 is extremely low. The two processes can be considered together as

$$CO_{2g} + H_2O \rightleftharpoons H_2CO_3$$

The equilibrium constant is then

$$K_{CO_2} = \frac{\{H_2CO_3\}}{p_{CO_2}\{H_2O\}} \tag{4.5}$$

where p_{CO_2} is the partial pressure of CO_2 in the marine troposphere. Carbonic acid undergoes dissociation

$$H_2CO_3 + H_2O \rightleftharpoons H_3O^+ + HCO_3^-$$

$$HCO_3^- + H_2O \rightleftharpoons H_3O^+ + HCO_3^{2-}$$

for which the first and second dissociation constants (using $\{H^+\}$ rather than $\{H_3O^+\}$) are

$$K_1 = \frac{\{H^+\}\{HCO_3^-\}}{\{H_2CO_3\}} \tag{4.6}$$

$$K_2 = \frac{\{H^+\}\{CO_3^{2-}\}}{\{HCO_3^-\}} \tag{4.7}$$

The hydrogen ion activity can be established with a pH meter. However, as discussed above, this measurement must be operationally defined. On the other hand, the individual ion activities of bicarbonate and carbonate ions cannot be measured. Instead, ion concentrations are determined, as outlined below, by titration. Accordingly, the equilibrium constants are redefined in terms of concentrations. These are then known as apparent, rather than true, equilibrium constants and distinguished using a prime notation. It must be appreciated that apparent equilibrium constants are not invariant, but rather are affected by temperature, pressure, salinity, and, as outlined previously, the pH scale adopted. The apparent dissociation constants are

$$K'_1 = \frac{\{H^+\}[HCO_3^-]}{[H_2CO_3]} \tag{4.8}$$

$$K'_2 = \frac{\{H^+\}[CO_3^{2-}]}{[HCO_3^-]} \tag{4.9}$$

It should be noted that whereas ion activities are denoted by curly brackets {}, concentrations are designated by square brackets []. Analogous to the pH, pK conventionally refers to $-\log K$. Numerical values for the constants pK_{CO_2}, pK'_1 and, pK'_2 based on a constant ionic medium scale (*i.e.*, seawater with chlorinity=19‰) are given in Table 4. This provides sufficient information to calculate the speciation of carbonic acid in seawater at a given temperature as a function of pH. This is shown for carbonic acid at 20 °C in both pure water and seawater equilibrated with atmospheric carbon dioxide in Figure 9. While there are several confounding features, the pH of seawater can be considered to be buffered by the bicarbonate/carbonate pair. The pH is generally about 8, but is sensitive to the concentration ratio $[HCO_3^-]{:}[CO_3^{2-}]$ as evident from rearranging the expression for K'_2 to become

$$\{H^+\} = K'_2 \frac{[HCO_3^-]}{[CO_3^{2-}]} \tag{4.10}$$

To understand the response of the oceanic CO_2 system to *in situ* biological activity or enhanced CO_2 concentrations in the atmosphere,

Table 4 *Equilibrium constants for the carbonate system*

$T\ (°C)$	pK_{CO_2}	pK'_1	pK'_2
0	1.19	6.15	9.40
5	1.27	6.11	9.34
10	1.34	6.08	9.28
15	1.41	6.05	9.23
20	1.47	6.02	9.17
25	1.53	6.00	9.10
30	1.58	5.98	9.02

Source: adapted from Stumm and Morgan, 1996.[14]

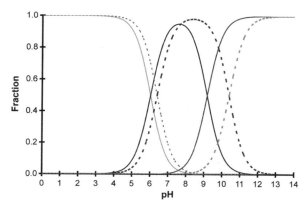

Figure 9 *The distribution of carbonic acid species at 20 °C as a function of pH in seawater of salinity 35 (solid lines) and pure water (dotted lines)*

it is necessary to consider in more detail the factors influencing the inorganic carbon cycle. Two useful parameters can be introduced. Firstly, the total concentration of inorganic carbon, $\sum CO_2$, in seawater is

$$\sum CO_2 = [CO_2] + [H_2CO_3] + [HCO_3^-] + [CO_3^{2-}]$$

The first term is negligible and as evident in Figure 9, the major species at pH 8 are HCO_3^- and CO_3^{2-}.

Alkalinity is defined as a measure of the proton deficit in solution and should not to be confused with basicity. Alkalinity is operationally defined by titration with a strong acid to the carbonic acid end point. This is known as the titration alkalinity (TA). Seawater contains weak acids other than bicarbonate and carbonate that are titrated, and therefore TA is given as

$$TA = [HCO_3^-] + 2[CO_3^{2-}] + [B(OH)_4^-] + [OH^-] - [H^+]$$

This is a simplified version of the more general expression in Chapter 3. The influence of $[OH^-]$ and $[H^+]$ on the TA is small and can often be ignored. The borate contributes about 3% of the TA, and if not determined independently, can be estimated from the apparent boric acid dissociation constants and the salinity, relying upon the relative constancy of composition of sea salt. This correction would give the carbonate alkalinity (CA)

$$CA = [HCO_3^-] + 2[CO_3^{2-}]$$

Considering the dissociation constants above, this can be alternatively expressed as

$$CA = \frac{K'_1 [H_2CO_3]}{\{H^+\}} + \frac{2 K'_1 K'_2 [H_2CO_3]}{\{H^+\}^2} \qquad (4.11)$$

or

$$CA = \frac{K'_1 \, K_{CO_2} \, pCO_2}{\{H^+\}} + \frac{2 \, K'_1 \, K'_2 \, K_{CO_2} \, pCO_2}{\{H^+\}^2} \qquad (4.12)$$

This equation can be rearranged to give the following quadratic expression that can be solved for the pH

$$CA\{H^+\}^2 - K'_1 \, K_{CO_2} \, pCO_2\{H^+\} - 2 \, K'_1 \, K'_2 \, K_{CO_2} \, pCO_2 = 0$$

Worked Example 1

For seawater (35 psu, 15 °C) with an alkalinity of 2.30 meq l^{-1} and in equilibrium with atmospheric $CO_2 = 3.65 \times 10^{-4}$ atm, calculate (i) the pH and (ii) the speciation of carbonic acid.

(i) pH calculation

Data, including constants from Table 4 are $K_{CO2} = 10^{-1.41}$, $K'_1 = 10^{-6.05}$, $K'_2 = 10^{-9.23}$, CA=2.30×10^{-3}, pCO_2=3.65×10^{-4}

The pH is obtained from the calculation of $\{H^+\}$ using the above quadratic equation. Thus

$$\{H^+\} = \frac{-b + \sqrt{b^2 - 4ac}}{2a}$$

where

$$a = CA$$
$$= 2.30 \times 10^{-3}$$
$$b = - K'_1 \, K_{CO2} \, pCO_2$$
$$= - 10^{-6.5} \cdot 10^{-1.41} \, 10^{-3.46}$$
$$= - 1.20 \times 10^{-11}$$
$$c = - 2 \, K'_1 \, K'_2 \, K_{CO2} \, pCO_2$$
$$= - 2 10^{-6.05} \cdot 10^{-9.23} \cdot 10^{-1.41} \, 10^{-3.46}$$
$$= - 1.42 \times 10^{-20}$$

Giving $\{H^+\}$=6.22×10^{-9} and pH=8.21.

(ii) Carbonic acid speciation calculations

Knowing the pH, each of the three major species (H_2CO_3, HCO_3^-, CO_3^{2-}) can be calculated as a fraction (or percentage) of the $\sum CO_2$.

Note that the negligible contribution due to dissolved CO_2 is ignored. The necessary expressions are derived from the definitions of the K'_1 and K'_2 given previously.

$$[HCO_3^-]\frac{K'_1}{\{H^+\}}[H_2CO_3] \tag{4.13}$$

$$[CO_3^{2-}] = \frac{K'_2}{\{H^+\}}[HCO_3^-] = \frac{K'_1 K'_2}{\{H^+\}^2}[H_2CO_3]$$

Substituting into the expression for the summation of all carbonic species gives

$$\sum CO_2 = [H_2CO_3] + [HCO_3^-] + [CO_3^{2-}] = [H_2CO_3] + K'_1\{H^+\}^{-1}[H_2CO_3] + K'_1 K'_2\{H^+\}^{-2}[H_2CO_3] = [H_2CO_3](1 + K'_1\{H^+\}^{-1} + K'_1 K'_2\{H^+\}^{-2})$$

Thereafter, the fractional contribution of each species to the total can be calculated using

1. $\frac{[H_2CO_3]}{\sum [CO2]} =$

   $(1 + K'_1\{H^+\}^{-1} + K'_1 K'_2\{H^+\}^{-2})^{-1}$

2. $\frac{[HCO_3^-]}{\sum [CO_2]} = (K'_1\{H^+\})$

   $\left(1 + K'_1\{H^+\}^{-1} + K'_1 K'_2\{H^+\}^{+2}\right)^{-1}$

3. $\frac{[CO_3^{2-}]}{\sum [CO2]} = (K'_1 K'_2\{H^+\}^{-2})$

   $(1 + K'_1\{H^+\}^{-1} + K'_1 K'_2\{H^+\}^{-2})^{-1}$

Substituting the values for K'_1 and K'_2 and using the previously calculated pH of 8.21, the fractional contribution of each species is 0.006, 0.907 and 0.087 for H_2CO_3, HCO_3^-, and CO_3^{2-}, respectively.

Consider now the effect of altering the pCO_2 in the water. The alkalinity should not change in response to variations in CO_2 alone because the hydration and dissociation reactions give rise to equivalent amounts of H^+ and anions. CO_2 can be lost by evasion to the atmosphere (a process usually confined to equatorial regions) or by photosynthesis. This causes the $\sum CO_2$ to diminish and the pH to rise, an effect that can be quite dramatic in tidal rock pools in which pH may then rise to 9. Conversely, an increase in pCO_2, either by invasion from the atmosphere or release following respiration, prompts an increase in

$\sum CO_2$ and a fall in pH. Thus, the depth profiles of pH would mimic that of O_2 but the $\sum CO_2$ would exhibit a maximum at the oxygen minimum.

There are further confounding influences, in particular concerning the solid phases of $CaCO_{3(s)}$. Many organisms make calcareous shells (testes) of $CaCO_{3(s)}$ in the form of either aragonite or calcite. The shells sink and dissolve when the organism dies. The solubility is governed by

$$Ca^{2+} + CO_3^{2-} \rightleftharpoons CaCO_{3(s)}$$

Surface waters are supersaturated with respect to $CaCO_{3(s)}$, but precipitation rarely occurs, possibly due to an inhibitory effect by Mg^{2+}_{aq} forming ion pairs with $CO_3^{2-}_{aq}$. The solubility of $CaCO_{3(s)}$ increases with depth, due to both a pressure effect and the decrease in pH following respiratory release of CO_2, causing the shells to dissolve. This behaviour not only increases the alkalinity, but also accounts for the non-conservative nature of Ca^{2+} and inorganic carbon in deepwaters. The depth at which appreciable dissolution begins is known as the lysocline. At a greater depth, designated as the carbonate compensation depth (CCD), no calcareous material is preserved in the sediments. The depths of the lysocline and CCD are influenced by the flux of organic material and shells, and tend to be deeper under high productivity zones.

The CO_2 and $CaCO_{3(s)}$ systems are coupled in that the pH buffering in the ocean is due to the reaction

$$CO_2 + H_2O + CaCO_{3(s)} \rightleftharpoons Ca^{2+} + 2HCO_3^-$$

In addition to the effects noted previously, an input of CO_2 promotes the dissolution of $CaCO_{3(s)}$. The reaction does not proceed to the right without constraint, but rather meets a resistance given by the Revelle factor, R

$$R = \frac{dpCO_2/pCO_2}{d\sum CO_2/\sum CO_2} \tag{4.14}$$

This value is approximately 10, indicating that the ocean is relatively well buffered against changes in $\sum CO_2$ in response to variations in atmospheric pCO_2. Although the ocean does respond to an increase in the atmospheric burden of CO_2, the timescales involved are quite considerable. The surface layer can become equilibrated on the order of decades, but as the thermocline inhibits exchange into deepwaters, the equilibration of the ocean as a whole with the atmosphere proceeds on the order of centuries. The ventilation of deepwater by downwelling water masses in polar latitudes only partly accelerates the overall process.

4.2.2.4 Dimethylsulfide and Climatic Implications. The Gaia hypothesis of Lovelock[15] states that the biosphere regulates the global environment for self-interest. This pre-supposes that controls, perhaps poorly understood or unknown, serve to maintain the present *status quo*. Charlson *et al.*[16] have made use of this hypothesis to suggest that biogenic production of DMS and the consequent formation of atmospheric cloud condensation nuclei (CCN, *i.e.*, small particles onto which water can condense) acts as a feedback mechanism to counteract the global warming resulting from elevated greenhouse gas concentrations in the atmosphere. The cycle is illustrated in Figure 10. Global warming, with concurrent warming of the ocean surface, leads to enhanced phytoplankton productivity. This promotes the production and evasion to the atmosphere of DMS. The DMS undergoes oxidation to form CCN that promote cloud formation and increase the planetary *albedo* (*i.e.*, reflectivity with respect to sunlight) thereby causing a cooling effect. From a biogeochemical perspective, the two key features are the controls on the biogenic production of DMS and the formation of CCN following aerial oxidation of DMS. These will be considered below in more detail. With respect to the physics, the most important aspects of the proposed climate control mechanism are that the enhancement of the *albedo* is due to an increase in the number and type of CCN, and that this CCN production occurs in the marine boundary layer. The *albedo* of calm seawater is very low ($\sim 2\%$) in comparison to vegetated regions (10–25%), deserts ($\sim 35\%$), and snow-covered surfaces ($\sim 90\%$).

That biological processes within the oceans act as a major source of reduced sulfur gases is well established,[17] and of particular importance is the generation of DMS. Surface concentrations, approximately in the range of 0.7–17.8 nmol L^{-1}, exhibit large temporal and geographic variations. Oceanic distributions indicate that DMS is produced within the photic zone, that is consistent with a phytoplankton source, but DMS concentrations are poorly correlated with normal indicators of primary productivity. While *Phaeocystis* and *Coccolithoporidae* have been identified as important DMS producers, there is uncertainty as to the full potential for biological DMS formation. With respect to climate modification, questions remain as to the biological response to global warming. For the model of Charlson *et al.*[16] to hold, organisms might either increase DMS formation or biological succession could change in such a way as to favour DMS producers. Thus, marine biogenic source

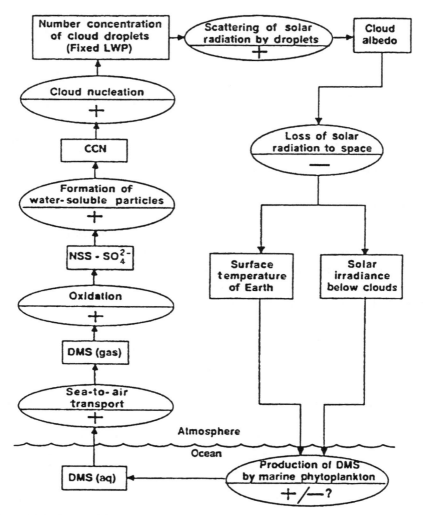

Figure 10 *The possible climatic influence of DMS of marine biogenic origin*
(Adapted from Charlson *et al.*, 1987.[16])

strengths and the controlling factors remain important unresolved issues in sulfur biogeochemistry.

DMS concentrations in the remote marine troposphere vary in the range of 0.03–32 nmol m^{-3}. Not surprisingly and as with the seawater concentrations, considerable temporal and geographic disparities occur. Furthermore, atmospheric DMS concentrations exhibit diurnal variations, with a night time maximum and an afternoon minimum consistent with a photochemical sink. Whereas oxidation involves HO free radicals during the day, a reaction with NO_3 may be important at night. Relatively low levels are associated with air masses derived from continental areas, owing to the enhanced concentrations of oxidants. While oceanic venting rates are dependent upon a number of meteorological and oceanographic conditions, there is no question that the marine photic zone acts as the major source of DMS to the overlying troposphere.

The oxidation of DMS in the atmosphere (see also Chapter 2) could yield several products, namely dimethylsulfoxide (DMSO), methanesulfonate (MSA) or sulfate. Insofar as aerosol formation is concerned, the two key products are MSA and SO_4^{2-}. Atmospheric particles in the sub-micron size range exert a significant influence on the earth's climate. The effect can be manifested *via* three mechanisms. Firstly, the particles themselves may enhance backscatter of solar radiation. Secondly, they act as CCN promoting cloud formation and so increasing the earth's *albedo*. Thirdly, such clouds affect the hydrological cycle. The evidence for such a biofeedback mechanism limiting global warming remains circumstantial.

4.2.3 Nutrients

Although several elements are necessary to sustain life, traditionally in oceanography the term "nutrients" has referred to nitrogen (notably, but not exclusively, nitrate), phosphorus (usually as phosphate), and silicon (as silicate). The rationale for this classification was that analytical techniques had long been available that allowed the precise determination of these constituents despite their relatively low concentrations. They were observed to behave in a consistent and explicable manner, but quite differently to the major constituents in seawater.

The distributions of these three nutrients are determined by biological activity. Nitrate and phosphate become incorporated into the soft parts of organisms. As evident in the modified carbon fixation equation of Redfield[18] given below

$$106CO_2 + 16NO_3^- + HPO_4^{3-} + 122H_2O$$
$$+ 18H^+ \rightleftharpoons C_{106}H_{263}O_{110}N_{16}P + 138O_2$$

The uptake of these nutrients into tissues occurs in constant relative amounts. The ratio (*i.e.*, Redfield ratio) for C/N/P is 106:16:1. Silicate is utilised by some organisms, particularly diatoms (phytoplankton) and radiolaria (zooplankton), to form siliceous skeletons. Such skeletons consist of an amorphous, hydrated silicate, $SiO_2 \cdot nH_2O$, often called opaline sili*ca*.

Photosynthetic carbon fixation depends on the availability of the nutrients, and of course, it ceases when one of the nutrient supplies is exhausted. The limiting nutrient in the ocean is usually nitrogen, but there are conditions under which Fe may limit biological productivity. Biological productivity is greatest in regions with the greatest supply of nutrients, such as coastal regions and near zones of upwelling. Moreover, biological activity is seasonal at most latitudes, being influenced by PAR coupled with the depletion of nutrients from surface waters that become thermally stratified due to summer warming.

The biogeochemical cycling of nitrogen in the marine environment is quite complex, largely owing to the diversity of forms in which it is present, coupled with the ease with which organisms can either assimilate or transform the various nitrogenous species. Nitrogen exhibits the range of oxidation states, from $N(V)$ in nitrate to $N(-III)$ in NH_3 and numerous organic compounds. Also, N occurs as dissolved gaseous molecules, including N_2 and N_2O; dissolved inorganic nitrogen (DIN) such as NO_3^- and NO_2^-; dissolved organic nitrogen (DON) present in organic compounds like amino acids; and particulate organic nitrogen (PON) in organisms and detritus. Few marine organisms are capable of assimilating N_2 directly, (known as nitrogen fixation) despite its relative abundance in seawater. The nitrogenous species more commonly used are nitrate, nitrite, urea and ammonium.

Bacteria play an important role in the redox chemistry of nitrogen species in seawater. Starting with PON, the first step is remineralisation in which PON is converted to DON. The breakdown of some of the DON to DIN follows with the first product being NH_3; the process is relatively rapid and known as ammonification. NH_3 is protonated to a limited extent in seawater, giving rise to NH_4^+ ions. Nitrification is the stepwise oxidation of NH_4^+ to NO_2^- and eventually to NO_3^-. Denitrification, the reduction of nitrogen species to N_2, can occur under conditions of hypoxia or anoxia. In such cases, bacteria respire organic material using NO_3^- and NO_2^- as electron acceptors.

A depth profile of nitrate, phosphate and silicate in the North Pacific Ocean is presented in Figure 11. Nutrients behave much like $\sum CO_2$ and are removed in the surface layer, especially in the photic zone. Thus, concentrations can become quite low, and indeed sufficiently low to limit further photosynthetic carbon fixation. The organisms sink following death. The highest concentrations occur where respiration and bacterial decomposition of the falling organic material are greatest, that is at the oxygen minimum. The nutrients, including silica, are consequently

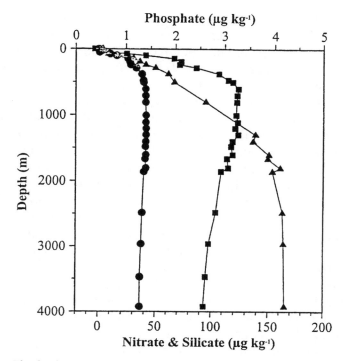

Figure 11 *The depth distribution of nitrate (●), phosphate (■), and silicate (▲) in the North Pacific Ocean*
(Data from Bruland, 1980.[13])

regenerated and their concentrations in deepwaters are much greater than those observed in the surface waters, thereby accounting for the fertilising effect of upwelling. It should be noted that the siliceous remains behave differently than the calcareous shells discussed previously. The oceans everywhere are under-saturated with respect to silica. Its solubility exhibits no pronounced variation with depth and there is no horizon analogous to the CCD (see Section 4.2.2.3). Silica is preserved to any great extent only in deep-sea sediments associated with the highly productive upwelling zones in the ocean.

4.2.4 Trace Elements

Trace elements in seawater are taken to be those that are present in quantities less than 1 mg L^{-1}, excluding the nutrient constituents. The distribution and behaviour of minor elements have been reviewed in the light of data that conform to an oceanographically consistent manner.[1,8]

Analytical difficulties are readily comprehensible when it is appreciated that the concentration for some of these elements can be extremely low, *i.e.*, a few pg L^{-1} for platinum group metals.[19] Some trace elements, such as Cs$^+$, behave conservatively and therefore absolute concentrations depend upon salinity. More often, the elements are non-conservative and their distributions in both surface waters and the water column vary greatly, reflecting the differing source strengths and removal processes in operation. Generalisations regarding residence times cannot be made in many cases because biologically active elements are removed from seawater relatively rapidly. Nevertheless, conservative constituents and platinum group metals have rather long residence times on the order of 10^5 years.

Considering firstly the distribution in surface waters, several elements exhibit high concentrations in coastal waters in comparison to levels in the centres of oceanic gyres. Typically, this distribution arises because the elements originate predominantly from riverine inputs or through diffusion from coastal sediments. However, as they are effectively removed from the surface waters in the coastal regions, little material is advected horizontally to the open sea. Examples of elements that behave in this way are Cd, Cu and Ni. In contrast, the concentration of Pb, including ^{210}Pb, is greater in the gyres. This results from a strong widespread aeolian input coupled with less effective removal from surface waters in the gyres.

Clearly, the removal mechanisms have an appreciable effect on dissolved elemental abundances. The two major processes in operation are uptake by biota and scavenging by suspended particulate material. In the first instance, the constituent mimics the behaviour of nutrients. This is evident in the metal/nutrient correlation for Cd/P and Zn/Si (Figure 12).

No consistent pattern for depth profiles of trace elements exists. Conservative elements trend with salinity variations provided they have no significant submarine sources. Non-conservative elements may exhibit peak concentrations at different depths in oxygenated waters as

    (i) surface enrichment,
    (ii) maximum at the O$_2$ minimum,
    (iii) mid-depth maximum not associated with the O$_2$ minimum, and
    (iv) bottom enrichment.

The criteria for an element, such as Pb, to exhibit a maximum concentration in surface waters are that the only significant input must be at the surface (aeolian supply) and it must be effectively removed from the water column. Constituents such as As, Ba, Cd, Ni and Zn exhibit nutrient type behaviour. Those elements (Cd) associated with the soft parts of the organism are strongly correlated with phosphate and are

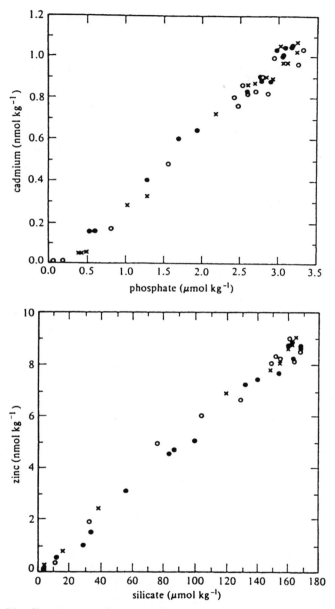

Figure 12 *Metal/nutrient correlations in the North Pacific for Cd/P and Zn/Si* (Adapted from Bruland, 1980.[13])

regenerated at the O_2 minimum. Elements (*i.e.*, Zn) associated with the skeletal material may exhibit a smoothly increasing trend with depth. The third case pertains to elements, notably Mn, that have a substantial input from hydrothermal waters. These are released into oceanic waters from spreading ridges. Ocean topography is such once these waters are advected away from such regions towards the abyssal plains, they are then found at some intermediate depth. Bottom enrichment is observed for elements (Mn) that are remobilised from marine sediments. The behaviour of Al combine features outlined above, resulting in a mid-depth minimum concentration. Surface enrichment evident in mid (41° N) but not high ($\sim 60°$ N) latitudes in the North Atlantic results from the solubilisation of aeolian material. Removal occurs *via* scavenging and incorporation into siliceous skeletal material. Subsequent regeneration by shell dissolution increases deepwater Al levels.

4.2.5 Physico-Chemical Speciation

Physico-chemical speciation refers to the various physical and chemical forms in which an element may exist in the system. In oceanic waters, it is difficult to determine chemical species directly. Whereas some individual species can be analysed, others can only be inferred from thermodynamic equilibrium models as exemplified by the speciation of carbonic acid in Figure 9. Often an element is fractionated into various forms that behave similarly under a given physical (*e.g.*, filtration) or chemical (*e.g.*, ion exchange) operation. The resulting partition of the element is highly dependent upon the procedure utilised, and so known as *operationally defined*. In the following discussion, speciation will be exemplified with respect to size distribution, complexation characteristics, redox behaviour and methylation reactions.

Physico-chemical speciation determines the environmental mobility of an element, especially with respect to partitioning between the water and sediment reservoirs. The influence can be manifested through various mechanisms as summarised in Figure 13. Settling velocities, and by implication the residence time, are controlled by the size of the particle. Thus, dissolved to particulate interactions involving adsorption, precipitation or biological uptake can readily remove a constituent from the water column. The redox state can have a comparable influence. Mn and Fe are reductively remobilised in sediments. Following their release in reduced forms from either hydrothermal sources or interstitial waters, they rapidly undergo oxidation to form colloids, which then are quickly removed from the water column. Speciation also determines the bioavailability of an element for marine organisms. It is generally accepted

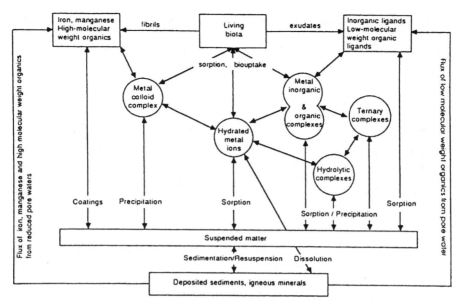

Figure 13 *Speciation of metal ions in seawater and the main controlling mechanisms* (Adapted from Öhman and Sjöberg, 1988.[20])

that the uptake of trace elements is limited to free ions and some types of lipid–soluble organic complexes. This is important in that some elements may be essential, but toxic at elevated concentrations. *Via* complexation reactions, organisms may be able to modify the free ion concentration. Thus, siderophores can sequester Fe, often in limited supply. Alternatively, metallothionein detoxify contaminants, notably Cd and Hg, through chelation.

An element can exist in natural waters in a range of forms that exhibit a size distribution as indicated in Table 5. Entities in true solution include ions, ion pairs, complexes and a wide range of organic molecules that can span several size categories. At the smallest extreme are ions, which exist in solution with a co-ordinated sphere of water molecules as discussed previously. Na^+, K^+ and Cl^- exist predominantly as free hydrated ions. Collisions of oppositely charged ions occur due to electrostatic attraction and can produce an ion pair. An ion pair is the transient coupling of a cation and anion during which each retains its own co-ordinated water envelope. While impossible to measure directly, concentrations can be calculated with knowledge of the ion activities and the stability constant. For the formation of the ion pair $NaSO_4^-$ *via*

$$Na^+ + SO_4^{2-} \rightleftharpoons NaSO_4^-$$

Table 5   *The size distribution of trace metal species in natural waters*

Size range	Metal species	Examples	Phase state
< 1 nm	Free metal ions	Mc^{2+}, Cd^{2+}	Soluble
1–10 nm	Inorganic ion pairs	$NiCl^+$	Soluble
	Inorganic complexes	$HgCl_4^{2-}$	
	Low molecular mass organic complexes	Zn-fulvates	
10–100 nm	High molecular mass organic complexes	Pb-humates	Colloidal
100–1000 nm	Metal species adsorbed onto inorganic colloids	$Co–MnO_2$	Particulate
	Metals associated with detritus	$Pb–Fe(OH)_3$	
> 1000 nm	Metals absorbed into living cells	Cu–clays	Particulate
	Metals adsorbed onto or incorporated into mineral solids and precipitates	$PbCO_{3(s)}$	

Source: adapted from de Mora and Harrison, 1984.[21]

the stability constant is defined as

$$K = \frac{\{NaSO_4^-\}}{\{Na^+\}\{SO_4^{2-}\}} \tag{4.15}$$

Ion pair formation is important for Ca^{2+}, Mg^{2+}, SO_4^{2-} and HCO_3^-.

If the attraction is sufficiently great, a dehydration reaction can occur leading to covalent bonding. A complex consists of a central metal ion sharing a pair of electrons donated by another constituent, termed a ligand, acting as a Lewis base. The metal ion and ligand share a single water envelope. Ligands can be neutral (*e.g.*, H_2O) or anionic (*e.g.*, Cl^-, HCO_3^-) species. A metal ion can co-ordinate with one or more ligands, which need not be the same chemical entity. Alternatively, the cation can share more than one electron pair with a given ligand thereby forming a ring structure. This type of complex, known as a chelate, exhibits enhanced stability largely due to the entropy effect of releasing large numbers of molecules from the co-ordinated water envelopes.

Complex formation is an equilibrium process. Ignoring charges for the general case of a metal, M and ligand, L, complex formation occurs as

$$M + L \rightleftharpoons ML$$

for which the formation constant, K_1, is given by

$$K_1 = \frac{\{ML\}}{\{M\}\{L\}} \tag{4.16}$$

A second ligand may then be co-ordinated as

$$M + L \rightleftharpoons ML_2$$

$$K_2 = \frac{\{ML_2\}}{\{ML\}\{L\}} \qquad (4.17)$$

The equilibrium constant for ML_2 can be expressed solely in terms of the activities of the M and L

$$\beta_2 = \frac{\{ML_2\}}{\{M\}\{L\}^2} \qquad (4.18)$$

where β_2 is the product of $K_1 K_2$ and is known as the stability constant. The case can be extended to include n ligands as

$$\beta_n = \frac{\{ML_n\}}{\{M\}\{L\}^n} \qquad (4.19)$$

Worked Example 2

Assuming all γ's $= 1$, calculate the speciation of mercury in typical seawater (35 psu at 25 °C) given the following values for stepwise stability constants for successive chlorocomplexes ($K_1 = 10^{6.74}$, $K_2 = 10^{6.48}$, $K_3 = 10^{0.85}$, $K_4 = 10^{1.00}$). Note that [Cl$^-$] is 0.559 mmol L^{-1} and that it is not necessary to know the mercury concentration in seawater. The total mercury concentration is given as the sum of all contributing species. Thus

$$[Hg]_T = [Hg^{2+}] + [HgCl^+] + [HgCl_2] + [HgCl_3^-] + [HgCl_4^{2-}]$$

From the definition of the stability constants, we know that
$[HgCl^+] = K_1[Hg^{2+}][Cl^-]$ and $[HgCl_2] = \beta_2[Hg^{2+}][Cl^-]^2$, *etc.* Thus,
$$[Hg]_T = [Hg^{2+}](1 + K_1[Cl^-] + \beta_2[Cl^-]^2 + \beta_3[Cl^-]^3 + \beta_4[Cl^-]^4).$$

Now let $D = (1 + K_1[Cl^-] + \beta_2[Cl^-]^2 + \beta_3[Cl^-]^3 + \beta_4[Cl^-]^4)^{-1}$, and note that this is a constant for a stipulated chloride concentration. Thus, at the seawater chloride concentration ([Cl$^-$] $= 10^{-0.25}$) this gives
$D = (1 + 10^{6.74} \, 10^{-0.25} + 10^{6.74} 10^{6.48} 10^{-0.25} + 10^{6.74} 0^{6.48} 10^{0.85} 10^{-0.25} + 10^{6.74} 10^{6.48} 10^{0.85} 10^{1.00} \, 10^{-0.25})^{-1}$
$= 7.27 \times 10^{-15}$

The fractional (or percentage) contribution of each species can be determined using

Fractional contribution	Expression	Value for $[Cl^-] = 0.559 \, \text{mmol L}^{-1}$
$[Hg^{2+}] : [Hg^{2+}]_T$	D	7.3×10^{-15}
$[HgCl^+] : [Hg^{2+}]_T$	$D \cdot K1[Cl^-]$	2.2×10^{-8}
$[HgCl_2] : [Hg^{2+}]_T$	$D \cdot \beta_2[Cl^-]^2$	$3.7 \times 10^{-2}(4\%)$
$[HgCl_3] : [Hg^{2+}]_T$	$D \cdot \beta_3[Cl^-]^3$	$1.5 \times 10^{-1}(15\%)$
$[HgCl_4^{2-}] : [Hg^{2+}]_T$	$D \cdot \beta_4[Cl^-]^4$	$8.2 \times 10^{-1}(82\%)$

Thus, at the high chloride concentration found in seawater, the mercury speciation is dominated by the tetra- and tri-chloro species. Note (1) If the total mercury content for a given seawater sample were known, the concentration of each species is readily calculated. That is

$$[Hg^{2+}] = f_{HG}{}^{2+} [Hg^{2+}]T;$$

(2) This is a simplified example that has not considered other species (bromides, hydroxides, *etc.*) that might be important in seawater.

The speciation of constituents in solution can be calculated if the individual ion activities and stability constants are known. This information is relatively well known with respect to the major constituents in seawater, but not for all trace elements. Some important confounding variables create considerable difficulties in speciation modelling. Firstly, it is assumed that equilibrium is achieved, meaning that neither biological interference nor kinetic hindrance prevents this state. Secondly, seawater contains appreciable amounts (at least in relation to the trace metals) of organic matter. However, the composition of the organic matrix, the number of available binding sites, and the appropriate stability constants are poorly known. Nevertheless, speciation models can include estimates of these parameters. Organic material can form chelates with relatively high-stability constants and dramatically decrease the free ion activity of both necessary and toxic trace elements. Organisms may make use of such chemistry, producing compounds either to sequester metals in limited supply (*e.g.*, siderophores to complex Fe) or to detoxify contaminants (*e.g.*, metallothionein to chelate Cd, Hg, *etc.*). Thirdly, surface adsorption of dissolved species onto colloids or suspended particles may remove them from solution. As with organic matter, an exact understanding of the complexation characteristics of the suspended particles is not available, but approximations can also be incorporated into speciation models.

Elements may be present in a variety of phases other than in true solution. Colloidal formation is particularly important for elements such as Fe and Mn, which produce amorphous oxyhydroxides with very great complexation characteristics. Adsorption processes cannot be ignored in biogeochemical cycling. Particles tend to have a much shorter residence time in the water column than do dissolved constituents. Scavenging of trace components by falling particles accelerates deposition to the sediment sink.

Several elements in seawater may undergo alkylation *via* either chemical or biological mechanisms.[22] Type I mechanisms involve methyl radical or carbonium ion transfer and no formal change in the oxidation state of the acceptor element. The incoming methyl group may be derived for example from methylcobalamin coenzyme, *S*-adenosylmethionine, betaine or iodomethane. Elements involved in Type I mechanisms include Pb, Tl, Se and Hg. Other reaction sequences involve the oxidation of the methylated element. The methyl source can be a carbanion from methylcobalamin coenzyme. Oxidative addition from iodomethane and enzymatic reactions has also been suggested. Some elements that can undergo such methylation processes are As, Sb, Ge, Sn and S. Methylation can enhance the toxicity of some elements, especially for Pb and Hg. The environmental mobility can also be affected. Methylation in the surface waters can enhance volatility and so favour evasion from the sea, as observed for S, Se and Hg. Methylation within the sediments may facilitate transfer back into overlying waters.

Elements may exhibit multiple oxidation states in seawater. Redox processes can be modelled in an analogous manner to the ion pairing and complexation outlined previously. The information is often presented graphically in the form of a predominance area diagram, which is a plot of pε *vs.* pH showing the major species present for the designated conditions. Although a single oxidation state might be anticipated from equilibrium considerations, there are several ways in which multiple oxidation states might arise. Biological activity can produce non-equilibrium species, as evident in the alkylated metals discussed above. Whereas Mn(IV) and Cu(II) might be expected by thermodynamic reasoning, photochemical processes in the surface waters can lead to the formation of significant amounts of Mn(II) and Cu(I). Fe(III) is the favoured redox state of Fe in seawater, but it is relatively insoluble and exists predominantly in a colloidal phase. Photochemical reduction to Fe(II), which only slowly oxidises to Fe(III), might act as a very important mechanism rendering Fe bioavailable to marine organisms.

Goldberg[19] has presented impressive information (given that seawater concentrations are as low as 1.5 pg L^{-1} for Ir and 2 pg L^{-1} for Ru) on the speciation, including redox state, of platinum group metals as a means of interpreting distributions in seawater and marine sediments. Pt and Pd are stabilised in seawater as tetrachloro-divalent anions. Their relative abundance of 5 Pt:1 Pd agreeing with a factor of five difference in β_4. Pt is enriched in ferromanganese nodules following oxidation to a stable (IV) state, behaviour not observed for Pd. Rh exists predominantly in the heptavalent state, but accumulates in reducing sediments as lower valence sulfides. Au and Ag are present predominantly in solution as monovalent forms. Ag, but not Au, accumulates in anoxic coastal sediments.

4.3 SUSPENDED PARTICLES AND MARINE SEDIMENTS

4.3.1 Description of Sediments and Sedimentary Components

The sediments represent the major sink for material in the oceans. The main pathway to the sediments is the deposition of suspended particles. Such particles may be only in transit through the ocean from a continental origin or be formed *in situ* by chemical and biological processes. Sinking particles can scavenge material from solution. Accordingly, this section introduces the components found in marine sediments, but emphasises processes that occur within the water column that lead to the formation and alteration of the deposited material.

Marine sediments cover the ocean floor to a thickness averaging 500 m. The deposition rates vary with topography. The rate may be several millimetres per year in nearshore shelf regions, but is only from 0.2 to 7.5 mm per 1000 years on the abyssal plains. Oceanic crustal material is formed along spreading ridges and moves outwards eventually to be lost in subduction zones, the major trenches in the ocean. Because of this continual movement, the sediments on the seafloor are no older than Jurassic in age, about 166 million years.

The formation of marine sediments depends upon chemical, biological, geological and physical influences. There are four distinct processes that can be readily identified. Firstly, the source of the material obviously is important. This is usually the basis for classifying sediment components and will be considered below in more detail. Secondly, the material and its distribution on the ocean floor are influenced by its transportation history, both to and within the ocean. Thirdly, there is the deposition process that must include particle formation and alteration in the water column. Finally, the sediments may be altered after deposition, a process known as diagenesis. Of particular importance are reactions leading to changes in the redox state of the sediments.

The components in marine sediments are classified according to origin. Examples are given in Table 6. Lithogenous (or terrigenous) material comes from the continents as a result of weathering processes.

Table 6 *The four categories of marine sedimentary components with examples of mineral phases*

Classification	Mineral example	Chemical formula
Lithogenous	Quartz	SiO_2
	Microcline	$KAlSi_3O_8$
	Kaolinite	$Al_4Si_4O_{10}(OH)_8$
	Montmorillonite	$Al_4Si_8O_{20}(OH)_4 \cdot nH_2O$
	Illite	$K_2Al_4(Si, Al)_8O_{20}(OH)_4$
	Chlorite	$(Mg,Fe^{2+})_{10}Al_2(Si,Al)_8O_{20}(OH,F)_{16}$
Hydrogenous	Fe–Mn mineral	$FeO(OH) - MnO_2$
	Carbonate fluoroapatite	$Ca_5(PO_4)_{3-x}(CO_3)_xF_{1+x}$
	Barite	$BaSO_4$
	Pyrite	FeS_2
	Aragonite	$CaCO_3$
	Dolomite	$CaMg(CO_3)_2$
Biogenous	Calcite	$CaCO_3$
	Aragonite	$CaCO_3$
	Opaline silica	$SiO_2 \cdot nH_2O$
	Apatite	$Ca_5(F,Cl)(PO_4)_3$
	Barite	$BaSO_4$
	Organic matter	
Cosmogenous	Cosmic spherules	
	Meteoric dusts	

Source: adapted from Harrison and de Mora, 1996.[23]

The relative contribution of lithogenous material to the sediments will depend on the proximity to the continent and the source strength of material derived elsewhere. The most important components in the lithogenous fraction are quartz and the clay minerals (kaolinite, illite, montmorillonite and chlorite). The distribution of the clay minerals varies considerably. Illite and montmorillonite tend to be ubiquitous in terrestrial material, but the latter has a secondary origin associated with submarine volcanic activity. Kaolinite typifies intense weathering observed in tropical and desert conditions. Therefore, it is relatively enriched in equatorial regions. On the other hand, chlorite is indicative of the high-latitude regimes where little chemical weathering occurs. The lithogenous components tend to be inert in the water column and represent detrital deposition. Nonetheless, the particle surfaces can act as important sites for adsorption of organic material and trace elements.

Hydrogenous components, also known as chemogenous or halmeic material, are those produced abiotically within the water column. This may comprise primary material formed directly from seawater upon exceeding a given solubility product, termed authigenic precipitation. The best-known examples of authigenic material are the various types of ferromanganese nodules found throughout the oceans. Alternatively, secondary material may be formed as components of continental or volcanic origin become altered by low-temperature reactions in seawater, a mechanism known as halmyrolysis. Halmyrolysis reactions can occur in the estuarine environment, being essentially an extension of chemical weathering of lithogenous components. Such processes continue at the sediment–water interface. Accordingly, there are considerable overlaps between the terms weathering, halmyrolysis and diagenesis. Owing to the importance that surface chemistry has on the final composition, authigenic precipitation and halmyrolysis are considered further in Section 4.3.2.

Biogenous (or biotic) material is produced by the fixation of mineral phases by marine organisms. The most important phases are calcite and opaline silica, although aragonite and magnesian calcite are also deposited. As indicated in Table 7, several plants and animals are involved, but the planktonic organisms are the most important with respect to the World Ocean. The source strength depends upon the biological species composition and productivity of the overlying oceanic waters. For instance, siliceous oozes are found in polar latitudes (diatoms) and along the equator (radiolaria). The relative contribution of biogenous material to the sediments depends upon its dilution by material from other sources and the extent to which the material can be dissolved in seawater. As noted previously, both calcareous and siliceous skeletons are subject to considerable dissolution in the water column and at the sediment–water interface.

There are two sources that give rise to minor components in the marine sediments. Cosmogenous material is that derived from an extra-terrestrial source. Such material tends to comprise small (*i.e.*, <0.5 mm) black

Table 7 *Quantitatively important plants and animals that secrete calcite, aragonite, Mg-calcite, and opaline silica*

Mineral	Plants	Animals
Calcite	Coccolithophorids[a]	Foraminifera[a]
		Molluscs
		Bryozoans
Aragonite	Green algae	Molluscs
		Corals
		Pteropods[a]
		Bryozoans
Mg-calcite	Coralline (red) algae	Benthic foraminifera
		Echinoderms
		Serpulids (tubes)
Opaline silica	Diatoms[a]	Radiolaria[a]
		Sponges

[a] Planktonic organisms
Source: adapted from Berner and Berner, 1987.[24]

micrometeorites or cosmic spherules. The composition is either magnetite or a silicate matrix including magnetite. They are ubiquitous but scarce, with relative contributions to the sediments decreasing with an increase in sedimentation rate. Finally, there are anthropogenic components, notably heavy metals and Sn, which can have a significant influence on sediments in coastal environments.

As noted in Section 4.1.1, the principal modes of transport of particulate material to the ocean are by rivers or *via* the atmosphere. Within the oceans, distribution is further affected by ice rafting, turbidity currents, organisms and oceanic currents. Turbidity currents refer to the turbid and turbulent flow of sediment-laden waters along the seafloor caused by sediment slumping. They are especially important in submarine canyons and can transport copious amounts of material, including coarse-grained sediments, to the deep sea. Ice rafting can also transport substances to the deep sea. Although, ice rafting is presently confined to the polar latitudes (40°N and 55°S), there have been considerable variations in ice limits within the geologic record. Organisms are notable not only for biogenous sedimentation, but also they can influence fine-grained lithogenous material that becomes incorporated into faecal pellets consequently accelerating the settling rate. Ocean currents are important for the distribution of material with a long residence time. Major surface currents are zonal and tend to reinforce the pattern of aeolian supply. On the other hand, deepwater currents are of little consequence as velocities are slow relative to the settling rates.

4.3.2 Surface Chemistry of Particles

4.3.2.1 Surface Charge.   Parcles in seawater tend to exhibit a negative surface charge. There are several mechanisms by which this might arise. Firstly, the negative charge can result from crystal defects (*i.e.*, vacant cation positions) or cation substitution. Clay minerals are layered structures of octahedral AlO_6 and tetrahedral SiO_4. Either substitution of Mg(II) and Fe(II) for the Al(III) in octahedral sites or replacement of Si(IV) in tetrahedral location by Al(III) can cause a net negative charge. Secondly, a surface charge can result from the differential dissolution of an electrolytic salt such as barite ($BaSO_4$). A charge will develop whenever the rate of dissolution of cations and anions differs. Thirdly, organic material can be negatively charged due to the dissociation of acidic functional groups.

Adsorption processes can also lead to the development of a negatively charged particle surface. One example is the specific adsorption of anionic organic compounds onto the surfaces of particles. Another mechanism relates to the acid–base behaviour of oxides in suspension. Metal oxides (most commonly Fe, Mn) and clay minerals have frayed edges on account of broken metal–oxygen bonds. The surfaces can be hydrolysed and exhibit amphoteric behaviour

$$-X - O^-_{(s)} + H^+_{aq} \rightleftharpoons -X - OH_{(s)}$$

$$-X - O^-_{(s)} + H^+_{aq} \rightleftharpoons -X - OH^+_{2(s)}$$

The hydroxide surface exhibits a different charge depending upon the pH. Cations other than H^+ can act as the potential determining ion. The point of zero charge (PZC) is the negative log of the activity at which the surface exhibits no net surface charge. At the PZC

$$[-X - O^-_{(s)}] = [-X - OH^+_{2(s)}]$$

The PZC for some mineral solids found in natural waters are shown in Table 8. Clearly, the extent to which such surfaces can adsorb metal cations will be dependent upon the pH of the solution. At the pH typical of seawater, most of the surfaces indicated in Table 8 would be negatively charged and would readily adsorb metal cations.

4.3.2.2 Adsorption Processes. Physical or non-specific adsorption involves relatively weak attractive forces, such as electrostatic attraction and van der Waals forces. Adsorbed species retain their co-ordinated

Table 8 *The PZC for some mineral phases*

Mineral	pH_{PZC}
α-Al_2O_3	9.1
α-$Al_2(OH)_3$	5.0
γ-AlOOH	8.2
CuO	9.5
Fe_3O_4	6.5
α-FeOOH	7.8
γ-Fe_2O_3	6.7
'Fe(OH)$_3$' (amorphous)	8.5
MgO	12.4
δ-MnO_2	2.8
β-MnO_2	7.2
SiO_2	2.0
$ZrSiO_4$	5
Feldspars	2–2.4
Kaolinite	4.6
Montmorillonite	2.5
Albite	2.0
Chrysotile	>12

Source: adapted from Stumm and Morgan, 1996.[14]

sphere of water and hence, cannot approach the surface closer than the radius of the hydrated ion. Adsorption is favoured by ions having a high charge density, i.e., trivalent ions in preference to univalent ones. Additionally, an entropy effect promotes the physical adsorption of polymeric species, such as Al and Fe oxides, because a large number of water molecules and monomeric species is displaced.

Chemisorption or specific adsorption involves greater forces of attraction than physical adsorption. As hydrogen bonding or π–orbital interactions are utilised, the adsorbed species lose their hydrated spheres and can approach the surface as close as the ionic radius. Whereas multilayer adsorption is possible in physical adsorption, chemisorption is necessarily limited to monolayer coverage.

As outlined previously, hydrated oxide surfaces have sites that are either negatively charged or readily deprotonated. The oxygen atoms tend to be available for bond formation, a favourable process for transition metals. Several mechanisms are possible. An incoming metal ion, M^{z+}, may eliminate an H^+ ion as

$$-X - O - H + M^{z+} \rightleftharpoons -X - O - M^{(z-1)+} + H^+$$

Alternatively, two or more H^+ ions may be displaced, thereby forming a chelate as

$$
\begin{array}{ccc}
-X-O-H & & -X-O \diagdown \\
& + M^{z+} \rightleftharpoons & \qquad M^{(z-2)+} + 2H^+ \\
-X-O-H & & -X-O \diagup
\end{array}
$$

A metal complex, ML_n^{z+}, may be co-ordinated instead of a free ion by displacement of one or more H^+ ions in a manner analogous to the above reaction. In addition, the metal complex might eliminate a hydroxide group giving rise to a metal–metal bond as

$$-X - O - H + ML_n^{z+} \rightleftharpoons -X - ML_n^{(z-1)+} + OH^-$$

It should be noted that all of these reactions are equilibria for which an appropriate equilibrium constant can be defined and measured. This data can then be incorporated into the speciation models discussed in Section 4.2.5.

4.3.2.3 Ion Exchange Reactions (see also Chapter 5). Both mineral particles and particulate organic material can take up cations from solution and release an equivalent amount of another cation into solution. This process is termed cation exchange and the cation exchange capacity (CEC) for a given phase is a measure of the number of exchange sites present per 100 g of material. This is operationally defined by the uptake of ammonium ions from 1 mol L^{-1} ammonium acetate at pH 7. The specific surface area and CEC are given in Table 9 for several sorption-active materials.

There are several factors that influence the affinity of cations towards a given surface. Firstly, the surface coverage will increase as a function of the cation concentration. Secondly, the affinity for the exchange site is enhanced as the oxidation state increases. Finally, the higher the charge density of the hydrated cation, the greater will be its affinity for the exchange site. In order of increasing charge density, the group I and II cations are

$$Ba < Sr < Ca < Mg < Cs < Rb < K < Na < Li$$

4.3.2.4 Role of Surface Chemistry in Biogeochemical Cycling. Reactions at the aqueous–particle interface have several consequences for material in the marine environment, from estuaries to the deep-sea sediment–water boundary. Within estuarine waters, suspended particles experience a dramatic change in the composition and concentration of dissolved salts. A number of halmyrolysates can be formed. Clay minerals undergo cation exchange as Mg^{2+} and Na^+ replace Ca^{2+} and K^+. Alternatively, montmorillonite may take up K^+ becoming transformed to illite. Hydrogenous components, in particular Mn and Fe oxides, may be precipitated onto the surfaces of suspended particles. The particulate material generally accumulates organic coatings within estuaries, which together with an increase in the ionic strength of the surrounding solution, leads to the formation of stable colloids. Both the oxide and organic coatings can subsequently scavenge other elements in the estuary.

Table 9   *The specific surface area and cation-exchange capacities of several sorption-active materials*

Material	Specific surface area $(mg^2\ g^{-1})$	Cation-exchange capacity $(meq/100\ g)$
Calcite (<2 μm)	12.5	—
Kaolinite	10–50	3–15
Illite	30–80	10–40
Chlorite	—	20–50
Montmorillonite	50–150	80–120
Freshly precipitated $Fe(OH)^3$	300	10–25
Amorphous silicic acid	—	11–34
Humic acids from soils	1900	170–590

Source: adapted from Förstner and Wittman, 1981.[25]

Within the ocean, the exchange of material from the dissolved to the suspended particulate state influences the distribution of several elements. This scavenging process removes dissolved metals from solution and accelerates their deposition. The effectiveness of this process is obvious in the depth profiles of metals, especially those of the surface enrichment type. Furthermore, the removal can be expressed in terms of a deepwater scavenging residence time as indicated in Table 10.

The scavenging mechanism can be particularly effective in the water–sediment boundary region. The re-suspension of fine sediments generates a very large surface area for adsorption and ion exchange processes. Within the immediate vicinity of hydrothermal springs, reduced species of Mn and Fe are released and subsequently are oxidised to produce colloidal oxyhydroxides, which have a large surface area and very great sorptive characteristics. Finally, ferromanganese nodules form at the sediment–water interface and become considerably enriched in a number of trace metals *via* surface reactions.

Ferromanganese nodules result from the authigenic precipitation of Fe and Mn oxides at the seafloor. Two morphological types are recognised, depending upon the growth mechanism. Firstly, spherical encrustations produced atop oxic deep-sea sediments grow slowly, accumulating material from seawater. These seawater nodules exhibit a relatively low Mn:Fe ratio and are especially enriched with respect to Co, Fe and Pb. Secondly, discoid shaped nodules develop in nearshore environments deriving material *via* diffusion from the underlying anoxic sediments. Such diagenetic nodules grow faster than deep-sea varieties and metals tend to be in lower oxidation states. They have a high Mn:Fe ratio and enhanced content of Cu, Mn, Ni and Zn. Ferromanganese nodules have concentric light and dark bands in cross-section, related to

Table 10 *The deepwater scavenging residence times of some trace elements in the oceans*

Element	Scavenging residence time (year)	Element	Scavenging residence time (year)
Sn	10	Mn	51–65
Th	22–33	Al	50–150
Fe	40–77	Sc	230
Co	40	Cu	385–650
Po	27–40	Be	3700
Ce	50	Ni	15 850
Pa	31–67	Cd	117 800
Pb	47–54	Particles	0.365

Source: adapted from Chester, 1990.[1]

Fe and Mn oxides, respectively. Patterns of trace element enrichment in the nodules are determined by mineralogy, the controlling mechanisms being related to cation substitution in the crystal structure. Mn phases preferentially accumulate Cu, Ni, Mo and Zn. Alternatively, Co, Pb, Sn, Ti and V are enriched in Fe phases.

4.3.3 DIAGENESIS

Diagenesis refers to the collection of processes that alter the sediments following deposition. These mechanisms may be physical (compaction), chemical (cementation, mineral segregation, ion exchange reactions) or biological (respiration). The latter are of particular importance as the bacteria control pH and pε in the interstitial waters, which in turn affect a wide range of chemical equilibria. These master variables influence the composition of the interstitial water, and can exert a feedback effect on the overlying seawater. Also, they can ultimately control the mineralogical phases that are lost to the sediment sink.

Organic material accumulates with other sedimentary components at the time of deposition. High biological productivity in surface waters and rapid sedimentation ensures that most nearshore and continental margin sediments contain significant amounts of organic matter. Biochemical oxidation of this material exhausts the available O_2 creating anoxic conditions. The oxic/anoxic boundary occurs at the horizon where the respiratory consumption of O_2 balances its downward diffusion. Upon depleting the O_2, other constituents are used as oxidants leading to the stepwise depletion of NO_3^-, NO_2^- and SO_4^{2-}. Thereafter, organic matter itself may be utilised with the concurrent production of CH_4.

This series of reactions causes progressively greater reducing conditions, with consequent influences on the chemistry of several elements. Metals are reduced and so are present in lower oxidation states. In particular, Mn undergoes reductive dissolution from MnO_{2s} to Mn_{aq}^{2+}. The divalent state being much more soluble, Mn is effectively remobilised under anoxic conditions and can be released back into overlying seawater. As seen in the previous section, this is one pathway to ferromanganese nodule formation. This can also be true for other elements that had been deposited following incorporation into the Fe and Mn oxide phases. In contrast, some elements can be preserved very effectively in anoxic sediments. Interstitial waters in marine sediments, in contrast to freshwater deposits, have high initial concentrations of SO_4^{2-}. Bacterial sulfate reduction proceeds *via* the reaction

$$2CH_2O + SO_4^{2-} \rightleftharpoons H_2S + 2HCO_3^-$$

Thus, sulfide levels in interstitial waters increase. A number of elements form insoluble sulfides, which under these anoxic conditions are precipitated and retained within the sediments. A notable example is the accumulation of pyrite, FeS_2, but also Ag, Cu, Pb and Zn are enriched in anoxic sediments in comparison with oxic ones.

4.4 PHYSICAL AND CHEMICAL PROCESSES IN ESTUARIES

Rivers transport material in several phases: dissolved, suspended particulate and bed load. Physical and chemical processes within an estuary influence the transportation and transformation of this material, thereby affecting the net supply of material to the oceans. Several definitions and geomorphologic classifications of estuaries have been reviewed by Perillo (1995).[26] From a chemical perspective, an estuary is most simply described as the mixing zone between river water and seawater characterised by sharp gradients in the ionic strength and chemical composition. Geographic distinctions can be made between drowned river valleys, fjords and bar-built estuaries. They can alternatively be classified in terms of the hydrodynamic regime as:

  (i) salt wedge,
  (ii) highly stratified,
  (iii) partially stratified, and
  (iv) vertically well mixed.

The aqueous inputs into the system are the river flow and the tidal prism. The series above is ranked according to the diminishing importance of the riverine flow and the increasing marine contribution. Thus, a salt wedge estuary represents the extreme case dominated by river flow, in which very little mixing occurs. A fresh, buoyant layer flows outward over denser, saline waters. In contrast, the vertically well-mixed estuary is one dominated by the tidal prism. The inflowing river water mixes thoroughly with and dilutes the seawater, but the effective dilution diminishes with distance along the mixing zone.

The position of the mixing zone in the estuary exhibits considerable temporal variations. There can be a strong seasonal effect, largely due to non-uniform river discharge. High winter rainfall leads to a wintertime discharge maximum. However, winter precipitation as snow creates a storage reservoir, such that the river flow maximum occurs following snow melt in spring or even early summer if the catchment area is of high elevation as for the Fraser River in western Canada. On shorter timescales, the mixing zone is influenced by tidal cycles. Thus, the penetration of seawater into the estuary depends upon the spring–neap tidal cycle and the diurnal nature of the tides. Together these influences determine the geographic extent of sediments experiencing a variable salinity regime. Variations in the river discharge affect the mass loading of the discharge, both in terms of suspended sediment and bed load material. The hydrodynamic regime in the estuary influences the deposition of the riverine sedimentary material and the mixing of dissolved material.

The estuary is a mixing zone for river water and seawater, the characteristics of which differ considerably. River water is slightly acidic and of low ionic strength with a salt matrix predominantly of Ca(H-CO$_3$)$_2$ ($\sim$120 mg L^{-1}). In contrast, seawater has a higher pH ($\sim$8), higher ionic strength ($\sim$0.7) and consists primarily of NaCl ($\sim$35 g L^{-1}). As a consequence, the salt matrix within the estuary is dominated by the sea salt end member throughout the mixing zone except for a small proportion at the dilute extreme. Salinity can be used as a conservative index, although conductivity is better, not being subject to systematic conversion errors in the initial mixing region.

In a plot of concentration *vs.* some conservative index (*i.e.*, S‰, Cl‰ or conductivity), the theoretical dilution curve would comprise a straight line between the river and seawater end members. A dissolved constituent that exhibits such a distribution is said to behave conservatively in the estuary. Whereas a negative slope shows that the riverine end member is progressively diluted during mixing with seawater, a positive slope indicates that the seawater end member has the greater concentration. Conservative behaviour is exhibited, for example, by Na$^+$, K$^+$ and SO$_4$$^{2-}$. Reactive silica may at times behave conservatively. Non-conservative behaviour can result from an additional supply of material (*i.e.*, causing positive deviations from the theoretical dilution curve). Elements that may show a maximum concentration at some intermediate salinity are Mn and Ba. Alternatively, the removal of dissolved material during mixing (*i.e.*, negative deviation from the theoretical dilution curve) can be caused by biological activity or by dissolved to particulate transformations. Biological activity can cause non-conservative behaviour for nutrient elements, including reactive silica. Dissolved constituents typically transformed to the particulate phase include Al, Mn and Fe (see Figure 14) in some estuaries. The pH distribution is usually characterised by a pH minimum in the initial mixing zone, resulting from the non-linear salinity dependence of the first and second dissociation constants of H$_2$CO$_3$. Notwithstanding the obvious utility of component–conservative index plots, they can be applied and interpreted only with caution. Often it is difficult to define the exact composition of the end members. Hence, a plot that apparently denotes non-conservative behaviour could arise if temporal fluctuations in the concentration of the component of interest occur on the same relative time scales as estuarine flushing.

A component can undergo considerable physico-chemical speciation alterations in an estuary. With respect to dissolved constituents, the composition and concentration of available ligands changes. Depending upon the initial pH of the riverine water, OH^- may become markedly more important down the estuary. Similarly, chlorocomplexes for metals such as Cd, Hg and Zn become more prevalent as the salinity increases. Conversely, the competitive influence of seawater derived Ca and Mg for organic material decreases the relative importance of humic complexation for Mn and Zn.

Estuaries are particularly well known for dissolved–particulate interactions. Such changes in phase come about *via* several mechanisms. Firstly, dissolution–precipitation processes may occur. This is especially important for the authigenic precipitation of Fe and Mn oxyhydroxides. Secondly, components may experience adsorption–desorption reactions. Desorption can occur in the initial mixing zone, partly in response to the pH minimum. Adsorption, particularly in association with the Fe and Mn phases, can accumulate material within the suspended sediments. Thirdly, flocculation and aggregation processes can remove material from solution. This occurs as particulate material with negatively charged surfaces adsorb cations in the estuary. The surface charge diminishes and as the ionic strength increases, the particles experience less electrostatic repulsion. Eventually, the situation arises whereby particle collisions lead to aggregation due to weak bonding. This process can be facilitated if particle surfaces are coated with organic material (then known as flocculation rather than aggregation). The three types of processes outlined above often happen simultaneously in the estuarine environment. Dissolved Fe, as shown in Figure 14, typifies such non-conservative behaviour. The transformation of dissolved into particulate phases can then be followed by deposition to the estuarine sediments. Thus, the flux of material to the ocean can be considerably modified, particularly as such sediments may be transported landward rather than seaward.

Biological activity in the estuarine environment can also influence the speciation of constituents, notably dissolved–particulate partitioning. A complex regeneration cycle determines distributions and modifies the riverine flux. This is especially the case for nutrients, and estuaries are often termed a nutrient trap. Estuaries tend to be regions of high biological productivity as rivers have elevated nutrient concentrations. Moreover, several freshwater organisms die upon encountering brackish water with consequent cell rupture and the release of contents into solution. Regardless of the source, the nutrients stimulate phytoplankton productivity.

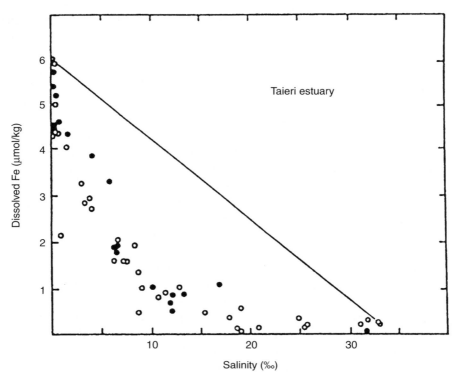

Figure 14   *The distribution of dissolved Fe vs. Salinity in the Taieri Estuary, New Zealand; • 21 October 1980, ○ 4 December 1980*
(Adapted from Hunter, 1983.[27])

Whereas some of the biological material (either transported *via* the freshwater or produced *in situ* within the estuarine zone) can be flushed out to sea, the remainder settles out of the surface water to be deposited onto the floor of the estuary. Respiration of this debris by benthic organisms regenerates nutrients and any contaminants that have been accumulated, releasing them into the bottom water. In estuaries where a two-layer flow is well defined, these nutrients are transported upriver in the salt wedge and entrained into the exiting river water, thereby adding to the available nutrient pool. Polluted rivers having a high nutrient loading are subject to eutrophication due to over-stimulation of biological activity. Anoxic conditions within the bottom waters and/or underlying sediments can result, depending upon the organic loading. Aeolian transport of nitrate to Chesapeake Bay can lead to low O_2 conditions. As mentioned previously, some fjords develop anoxic conditions when bottom waters stagnate due to limited mixing.

4.5 MARINE CONTAMINATION AND POLLUTION

Both contamination and pollution entail the perturbation of the natural state of the environment by anthropogenic activity. The two terms are distinguishable in terms of the severity of the effect, whereby pollution induces the loss of potential resources.[28] Additionally, a clear cause–effect relationship must be established for a substance to be classified as a pollutant towards a particular organism. The human-induced disturbances take many forms, but the greatest effects tend to be in coastal environments due to the source strengths and pathways. Waters and sediments in coastal regions bear the brunt of industrial and sewage discharges, and are also subject to dredging and spoil dumping. Agricultural runoff may contain pesticide residues and elevated nutrients, the latter of which may over-stimulate biological activity producing eutrophication and anoxic conditions. The deep sea has not escaped contamination. The most obvious manifestations comprise crude oil, petroleum products and plastic pollutants, but include the long-range transport of long-lived radionuclides from coastal sources. Additionally, the aeolian transport of heavy metals has enhanced natural fluxes of some elements, particularly lead. Three case studies are introduced below to illustrate diverse aspects of marine contamination and pollution.

4.5.1 Oil Slicks

Major releases of oil have been caused by the grounding of tankers (*i.e.*, *Torrey Canyon*, Southwest England, 1967; *Argo Merchant*, Nantucket Shoals, USA, 1976; *Amoco Cadiz*, Northwest France, 1978; *Exxon Valdiz*, Alaska, 1990; *Erika*, France, 1999; *Prestige*, Spain, 2002) and by the accidental discharge from offshore platforms (*i.e.*, *Chevron MP-41C*, Mississippi Delta, 1970; *Ixtox I*, Gulf of Mexico, 1979). Because oil spills receive considerable public attention and provoke substantial anxiety, oil pollution must be put into perspective. Crude oil has been habitually introduced into the marine environment from natural seeps at a rate of approximately 340×10^6 L y^{-1}. Anthropogenic activity has recently augmented this supply by an order of magnitude; however, most of this additional oil has originated from relatively diffuse sources relating to municipal runoff and standard shipping operations. Exceptional episodes of pollution occurred in the Persian Gulf in 1991 (910×10^6 L y^{-1}) and due to the *Ixtox I* well in the Gulf of Mexico in 1979 (530×10^6 L y^{-1}). In contrast to such mishaps, the *Amoco Cadiz* discharged only 250×10^6 L y^{-1} of oil in 1978 accounting for the largest spill from a tanker. The cumulative pollution from tanker accidents on an annual basis matches that emanating from natural seepage. Nevertheless, the impacts can be severe when the subsequent slick impinges on coastal ecosystems.

Regardless of the source, the resultant oil slicks are essentially surface phenomena that are affected by several transportation and transformation processes.[29] With respect to transportation, the principal agent for the movement of slicks is the wind, but length scales are important. Whereas small weather systems, such as thunderstorms, tend to disperse the slick, cyclonic systems can move the slick essentially intact. Waves and currents also affect the advection of an oil slick. To a more limited extent, diffusion can act to transport the oil.

Transformation of the oil involves phase changes and degradation. Several physical processes can invoke phase changes. Evaporation of the more volatile components is a significant loss mechanism, especially for light crude oil. The oil slick spreads as a buoyant lens under the influence of gravitational forces, but generally separates into distinctive thick and thin regions. Such pancake formation is due to the fractionation of the components within the oil mixture. Sedimentation can play a role in coastal waters when rough seas bring dispersed oil droplets into contact with suspended particulate material and the density of the resulting aggregate exceeds the specific density of seawater. Colloidal suspensions can consist of either water-in-oil or oil-in-water emulsions, which behave distinctly differently. Water-in-oil emulsification creates a thick, stable colloid that can persist at the surface for months. The volume of the slick increases and it aggregates into large lumps known as "mousse", thereby acting to retard weathering. Conversely, oil-in-water emulsions comprise small droplets of oil in seawater. This aids dispersion and increases the surface area of the slick, which can subsequently accelerate weathering processes.

Chemical transformations of oil are evoked primarily through photochemical oxidation and microbial biodegradation. Not only is the latter more important in nature, but strategies can be adopted to stimulate biological degradation, consequently termed bioremediation. All marine environments contain microorganisms capable of degrading crude oil. Furthermore, most of the molecules in crude oils are susceptible to microbial consumption. Oil contains little nitrogen or phosphorus and therefore, microbial degradation of oil tends to be nutrient limited. Bioremediation often depends upon on the controlled and gradual delivery of these nutrients, while taking care to limit the concurrent stimulation of phytoplankton activity. Approaches that have been adopted are the utilisation of slow-release fertilisers, oleophilic nutrients and a urea-foam polymer fertiliser incorporating oil-degrading bacteria. Bioremediation techniques were successfully applied in the clean up of Prince William Sound and the Gulf of Alaska following the *Exxon Valdez* accident. Alternative bioremediation procedures relying on the addition of exogenous bacteria have still to be proved. Similarly, successful bioremediation of floating oil spills has yet to be demonstrated.

Source apportionment of crude oil in seawater and monitoring the extent of weathering and biodegradation constitute important challenges in environmental analytical chemistry. As the concentration of individual compounds varies from one sample of crude oil to another, the relative amounts define a signature characteristic of the source. Compounds that degrade at the same rate stay at fixed relative amounts throughout the lifetime of an oil slick. Hence, a "source ratio", which represents the concentration ratio for a pair of compounds exhibiting such behaviour remains constant. Conversely, a "weathering ratio" reflects the concentration ratio for two compounds that degrade at different rates and consequently continually changes. Oil spill monitoring programmes conventionally determine four fractions[30]:

   (i) volatile hydrocarbons;
   (ii) alkanes;
   (iii) total petroleum hydrocarbons; and
   (iv) polycyclic aromatic hydrocarbons (PAHs).

The volatile hydrocarbons, albeit comparatively toxic to marine organisms, evaporate relatively quickly and hence serve little purpose as diagnostic aids. The alkanes and total petroleum hydrocarbons make up the bulk of the crude oil. They can be used to some extent for source identification and monitoring weathering progress. The final fraction, the PAHs, comprises only about 2% of the total content of crude oil but includes compounds that are toxic. Moreover, these components exhibit marked disparities in weathering behaviour due to differences in water solubility, volatility and susceptibility towards biodegradation. As demonstrated in Figure 15, both a source ratio (C3-dibenzothiophenes/ C3-phenanthrenes) and a weathering ratio (C3-dibenzothiophenes/ C3-chrysenes) have been defined from among such compounds that enable the extent of crude oil degradation to be estimated in the marine environment, as well as for sub-tidal sediments and soils.[30]

The discussion above focuses on petroleum hydrocarbons from oil spills, representing just one source of contamination. Marine contaminant surveys are conducted not just in response to tanker accidents, but routinely as a means to assess the quality of the marine environment. Although analytical difficulties with environmental samples persist and

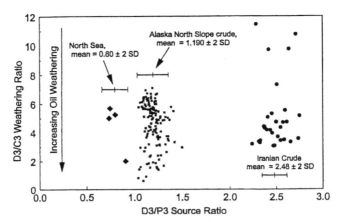

Figure 15   *Plot of weathering ratio (C3-dibenzothiophenes/C3-chrysenes) vs. source ratio*
(C3-dibenzothiophenes/C3-phenanthrenes) for fresh and degraded oil samples
from three different crude oil spills
(Adapted from Douglas *et al.*, 1996.[30])

much material can only be classified as an unresolved complex mixture (UCM), many individual aliphatic and aromatic compounds can be routinely determined. Total petroleum hydrocarbon distributions, as shown in Figure 16 for the Caspian Sea, provide evidence of pollution "hot spots".[31] In this case, the highest concentrations were observed in Azerbaijan. Given that several individual PAHs can be ascribed to different origins, plots of the relative amounts allows an assessment to be made of the relative importance of petrogenic (fossil oil), pyrolytic (combustion products) and natural (*in situ* biological activity) sources. As shown in Figure 17, most of the PAHs in the Caspian Sea sediments were derived from fossil oil.[31] This technique cannot distinguish between material derived from natural seeps or anthropogenic activity. Otherwise, the contribution of combustion products was evident in Russia and natural sources were only observed in the Iranian sector.

4.5.2   Plastic Debris

The accumulation of litter and debris along shorelines epitomises a general deterioration of environmental quality on the high seas. The material originates not only from coastal sources, but also from the ancient custom of dumping garbage from ships. Drilling rigs and offshore production platforms have similarly acted as sources of contamination. Some degree of protection in recent years has accrued from both the London Dumping Convention (LDC) and the International

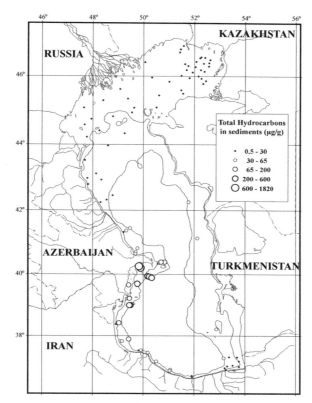

Figure 16 *Distribution of total hydrocarbons (total aliphatic and aromatic) in sediments from the Caspian Sea (μg g^{-1} dry wt)*
(From Tolosa *et al.*, 2004.[31])

Convention for the Prevention of Pollution from Ships (MARPOL) that outlaw such practices. However, the problem of seaborne litter remains global in extent and not even Antarctica has been left unaffected.[32]

The debris consists of many different materials, which tending to be non-degradable endures in the marine environment for many years. The most notorious are the plastics (*e.g.*, bottles, sheets, fishing gear, packaging materials and small pellets), but also includes glass bottles, tin cans and lumber. This litter constitutes an aesthetic eyesore on beaches, but more importantly can be potentially lethal to marine organisms. Deleterious impacts on marine birds and mammals result from entanglement and ingestion. Lost or discarded plastic fishing nets remain functional and can continue "ghost fishing" for several years. This is similarly true for traps and pots that go astray. Plastic debris settling on soft and hard bottoms can smother benthos and limit gas exchange with

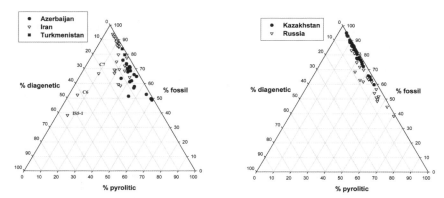

Figure 17   *Ternary diagram showing the relative amounts of PAHs from different sources (fossil oil, combustion products and natural biological activity) and distinguished on a country basis*
(From Tolosa *et al.*, 2004.[31])

pore waters. Despite the negative effects of seaborne plastic debris, this material can have positive consequences serving as new habitats for opportunistic colonisers.

4.5.3   Tributyltin

Tributyltin (TBT) provides an interesting case study of a pollutant in the marine environment.[33] Because TBT compounds are extremely poisonous and exhibit broad-spectrum biocidal properties, they have been utilised as the active ingredient in marine anti-fouling paint formulations. Its potency and longevity ensures good fuel efficiencies for ship operations and guarantees a long lifetime between repainting. TBT-based paints have been used on boats of all sizes, from small yachts to supertankers, ensuring the global dispersion of TBT throughout the marine environment, from the coastal zone to the open ocean.

Notwithstanding such benefits, the extreme toxicity and environmental persistence has resulted in a wide range of deleterious biological effects on non-target organisms. TBT is lethal to some shellfish at concentrations as low as 0.02 μg TBT-SnL^{-1}. Lower concentrations result in sub-lethal effects, such as poor growth rates and reduced recruitment leading to the decline of shellfisheries. The most obvious manifestations of TBT contamination have been shell deformation in Pacific oysters (*Crassostrea gigas*), and the development of imposex (*i.e.*, the imposition of male sex organs on females) in marine gastropods. The later effect has caused dramatic population decline of gastropods at locations throughout the world. TBT has been observed to accumulate in fish and various marine birds and mammals, with yet unknown consequences. Although it has not been shown to pose a public health risk, one recent study reported measurable butyltin concentrations in human liver.[34]

The economic consequences of the shellfisheries decline led to a rapid political response globally. The first publication[35] suggesting TBT to be the causative agent appeared in only 1982, and led to prompt legislative action. France was the first country to control the use of TBT-based paints, in the first instance on boats <25 m in length at some coastal locations. Oyster aquaculture in Arcachon Bay benefited immediately, with a notably decline in shell deformations and TBT-body burdens, and the complete recovery of production within two years (Figure 18). Several other countries introduced various controls and bans on these marine anti-foulants. Comparable improvements in oyster conditions were subsequently observed in Great Britain and Australia. Similarly, there have been many reported instances of restoration of gastropod populations at previously impacted locations. Certainly the TBT flux to the marine environment decreased, as manifested in sedimentary TBT profiles. However, most of the legislation first introduced applied only to small vessels and large ships continued to act as a source of TBT to the marine environment. Imposexed gastropods were observed at sites (*e.g.*, North Sea and Strait of Malacca) where the source of TBT could only

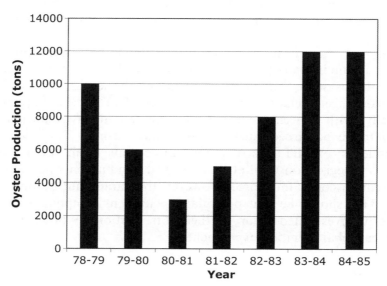

Figure 18   *Annual oyster (Crassostrea gigas) production in Arcachon Bay, 1978–1985; Restrictions on TBT use were applied starting January 1982* (Data from Alzieu, 1991.[36])

have been attributed to shipping. Global concern was sufficiently great that in 2001 the International Maritime Organisation (IMO) successfully negotiated the *International Convention on the Control of Harmful Anti-fouling Systems on Ships*. The Convention aims to ban the application of organotin-based anti-fouling paints on ships from 2003 and their presence on ship hulls from 2008, but does require ratification of 25 countries comprising 25% of the world's merchant shipping tonnage before coming into effect.

TBT exists in solution as a large univalent cation and forms a neutral complex with Cl^- or OH^-. It is extremely surface active and so is readily adsorbed onto suspended particulate material. Such adsorption and deposition to the sediments limits its lifetime in the water column. Degradation, *via* photochemical reactions or microbially mediated pathways, obeys first-order kinetics. Several marine organisms, as diverse as phytoplankton to starfish, debutylate TBT. Stepwise debutylation produces di- and mono-butyltin, which are much less toxic in the marine environment than is TBT. As degradation rates in the water column are on the order of days to weeks, they are slow relative to sedimentation. TBT accumulates in the sediments where degradation rates are much slower, with the half-life being on the order of years.[37] Furthermore, concentrations are highest in those areas, such as marinas and harbours, which are most likely to undergo dredging. The intrinsic toxicity of TBT, its persistency in the sediments and its periodic remobilisation by anthropogenic activity are likely to retard the long-term recovery of the marine ecosystem.

Although the unrestricted use of TBT should end, significant challenges remain in many parts of the world. For the most part, the coastal tropical ecosystems remain unprotected and the sensitivity of its indigenous organisms relatively poorly evaluated. TBT endures in sediments globally, with concentrations usually greatest in environments most likely to be perturbed, such as ports and marinas. The widespread introduction of TBT into seawater will continue from vessels not yet subject to legislation until at least 2008, depending on when the IMO Convention comes into force. However, the paramount lesson learned from TBT should be that potential replacement compounds must be properly investigated prior to their introduction in order to avoid another global pollution experiment.

QUESTIONS

(i) What are the main sources and sinks for dissolved and particulate metals entering the ocean?

(ii) Define residence time and outline the factors that influence the residence time of an element in the ocean. Provide examples of oceanic residence times for elements that span the time scale.

(iii) Define salinity. What are the notable variations in sea salt, in terms of concentration and composition, in the world ocean?

(iv) Using data available in Table 4, calculate the concentration of H_2CO_3 in seawater at 5°C in equilibrium with atmospheric CO_2 at 370 ppm.

(v) Calculate the speciation of H_2S in seawater at pH 8.3 at 25°C given that $pK_1 = 7.1$ and $pK_2 = 17.0$.

(vi) Explain how DMS of marine origin might affect global climate.

(vii) What are the major nutrients in the ocean? Describe their concentration profiles and account for any differences between them according to chemical behaviour.

(viii) Describe the different types of depth profiles that various metals exhibit and explain how differences in their profiles originate.

(ix) Assuming all γ's $= 1$, calculate the speciation of lead in seawater at 25°C given the following values for stability constants for the chlorocomplexes ($K_1 = 10^{7.82}$, $\beta_2 = 10^{10.88}$, $\beta_3 = 10^{13.94}$, $\beta_4 = 10^{16.30}$). Note that $[Cl^-]$ is 0.559 mmol L^{-1}.

(x) With specific reference to Question 9, outline the potential limitations of using equilibrium models to explain chemical behaviour of trace metals in seawater.

(xi) Describe the classification of marine sediments and give examples of each sediment type.

(xii) Why do suspended particles exhibit a surface charge and how could this characteristic moderate the composition of ocean's waters?

(xiii) With respect to estuarine chemistry, describe conservative and non-conservative behaviour and provide examples of cationic and anionic species for each category.

(xiv) What natural processes are responsible for weathering an oil slick?

(xv) Why is tributyltin considered such a potent pollutant in the marine environment?

REFERENCES

1. R. Chester, *Marine Geochemistry*, Unwin Hyman, London, 1990, 698.
2. P. Tchernia, *Descriptive Regional Oceanography*, Pergamon Press, Oxford, 1980, 253.

3. The Open University, *Seawater: Its Composition, Properties and Behaviour*, Open University, Oxford, 1989, 165.
4. K.S. Stowe, *Ocean Science*, Wiley, New York, 1979.
5. International Panel on Climate Change (IPCC), *Climate Change 2001: Synthesis Report*, WMO–UNEP, Third Assessment Report, vol 4, Geneva, 2001, 397.
6. D. Dyrssen and M. Wedborg, in *The Sea*, E. Goldberg (ed), Wiley, New York, 1974, 181.
7. P. Brewer, in *Chemical Oceanography*, 2nd edn, vol. 1, J.P. Riley and G. Skirrow (eds), Academic Press, London, 1975, 415.
8. K.W. Bruland, in *Chemical Oceanography*, vol. 8, J.P. Riley and R. Chester (eds), Academic Press, London, 1983, 157.
9. R.F. Weiss, *Deep-Sea Res.*, 1970, **17**, 721.
10. D. Kester, *Chemical Oceanography*, 2nd edn, vol 1, J.P. Riley and G. Skirrow (eds), Academic Press, London, 1975, 497.
11. W.S. Broecker and T.H. Peng, *Tracers in the Sea*, Lamont-Doherty Geological Observatory, Palisades, 1982, 690.
12. P.S. Liss, in *Air–Sea Exchange of Gases and Particles*, P.S. Liss and W.G.N. Slinn (eds), Reidel, Dordrecht, 1983, 241.
13. K.W. Bruland, *Earth Planet. Sci. Lett.*, 1980, **47**, 176.
14. W. Stumm and J.J. Morgan, *Aquatic Chemistry*, 3rd edn, Wiley, New York, 1996, 1022.
15. J. Lovelock, *Gaia. A New Look at Life on Earth*, Oxford University Press, Oxford, 1979.
16. R.J. Charlson, J.E. Lovelock, M.O. Andreae and S.G. Warren, *Nature*, 1987, **326**, 655.
17. M.O. Andreae, in *The Role of Air–Sea Exchange in Geochemical Cycling*, P. Buat-Ménard (ed), Reidel, Dordrecht, 1986, 331.
18. A.C. Redfield, *Am. J. Sci.*, 1958, **46**, 205.
19. E. Goldberg, M. Koide, J.S. Yang and K.K. Bertine, *Metal Speciation: Theory, Analysis and Applications*, J.R. Kramer and H.E. Allen (eds), Lewis Publishers, Chelsea, 1988.
20. L. Öhman and S. Sjöberg, in *Metal Speciation: Theory, Analysis and Applications*, J.R. Kramer and H.E. Allen (eds), Lewis Publishers, Chelsea, 1988, 1.
21. S.J. de Mora and R.M. Harrison, in *Hazard Assessment of Chemicals, Current Developments*, vol 3, J. Saxena (ed), Academic Press, London, 1984, 1.
22. P. Craig, *Organometallic Compounds in the Environment*, P.S. Craig (ed), Longman, Harlow, 1986, 1.

23. R.M. Harrison and S.J. de Mora, *Introductory Chemistry for the Environmental Sciences*, 2nd edn, Cambridge University Press, Cambridge, 1996, 373.

24. E.K. Berner and R.A. Berner, *The Global Water Cycle*, Prentice-Hall, NJ, 1987, 397.

25. U. Förstner and G.T.W. Wittman, *Metal Pollution in the Aquatic Environment*, 2nd edn, Springer, Berlin, 1981, 486.

26. G.M.E. Perillo, in *Geomorphology and Sedimentology of Estuaries*, G.M.E. Perillo (ed), Elsevier Science, Amsterdam, 1995, 17.

27. K. Hunter, *Geochim. Cosmochim. Acta*, 1983, **47**, 467.

28. E. Goldberg, *Mar. Pollut. Bull.*, 1992, **25**, 45.

29. S. Murray, in *Pollutant Transfer and Transport in the Sea*, vol. 2, G. Kullenberg (ed), CRC Press, Boca Raton, 1982, 169.

30. G.S. Douglas, A.E. Bence, R.C. Prince, S.J. McMillen and E.L. Butler, *Environ. Sci. Technol.*, 1996, **30**, 2332.

31. I. Tolosa, S.J. de Mora, M.R. Sheikholeslami, J.-P. Villeneuve, J. Bartocci and C. Cattini, *Mar. Pollut. Bull.*, 2004, **48**, 44–60.

32. M.R. Gregory and P.G. Ryan, in *Marine Debris: Sources, Impacts and Solutions*, J.M. Coe and D.B. Rogers (eds), Springer, New York, 1996, 49.

33. S.J. de Mora (ed.), *Tributyltin: Case Study of an Environmental Contaminant*, Cambridge University Press, Cambridge, 1996, 301.

34. K. Kannan and J. Falandysz, *Mar. Pollut. Bull.*, 1997, **34**, 203.

35. C. Alzieu, M. Heral, Y. Thibaud, M.J. Dardignac and M. Feuillet, *Rev. Trav. Inst. Marit.*, 1982, **45**, 100.

36. C. Alzieu, *Mar. Environ. Res.*, 1991, **32**, 7.

37. C. Stewart and S.J. de Mora, *Environ. Technol.*, 1990, **11**, 565.

CHAPTER 5

The Chemistry of the Solid Earth

IAN D. PULFORD

Chemistry Department, University of Glasgow, Joseph Black Building, Glasgow, G12 8QQ, UK

5.1 INTRODUCTION

Soil is the product formed when the rocks of the earth's crust are exposed at the surface and are subjected to various physical, chemical, and, eventually, biological weathering processes. The minerals in these rocks are predominantly silicates, which dominate the characteristics of most soils. Table 1 shows those elements that are found in the crust above an average concentration of 1% and their corresponding soil content. The importance of aluminosilicates in soil is clear from the enrichment factors of approximately 1 for O, Si, and Al. Some loss occurs of K, Fe, Ca, Na, and Mg as a result of soil processes. But two elements, C and N, show considerable enrichment in soil because of the crucial role played by organic matter.

Table 1 *Average elemental concentrations in the earth's crust and in soil, and the enrichment factor in soil*

Element	Average concentration in crust (%)	Average concentration in soil (%)	Soil:crust ratio
Oxygen	47	49	1.0
Silicon	28	31	1.1
Aluminium	8.2	7.2	0.88
Iron	4.1	2.6	0.63
Calcium	4.1	2.4	0.59
Magnesium	2.3	1.2	0.52
Sodium	2.3	1.2	0.52
Potassium	2.1	1.5	0.71
Carbon	0.05	2.5	50
Nitrogen	0.0025	0.20	80

Source: adapted from Sposito, *The Chemistry of Soils*, Oxford University Press, 1989.

Soil is formed not just by the weathering of minerals derived from rocks, but also by the input of organic matter from the decomposition of plants and, to a lesser extent, the decomposition and waste products of animals and microorganisms. The relative amounts of the mineral and organic components can vary widely, from almost all mineral materials, such as in desert sand, to highly organic soil such as peat. It is the combination of the mineral and organic components that gives soil its characteristics and makes it a valuable resource, which can be used to grow crops for a number of uses, such as food, fibre, and energy. These two aspects – soil development and soil use – are reflected in two definitions of soil given by the Soil Science Society of America (http://www.soils.org/sssagloss/):

The unconsolidated mineral or organic material on the immediate surface of the earth that serves as a natural medium for the growth of land plants.

The unconsolidated mineral or organic material on the surface of the earth that has been subjected to and shows the effects of genetic and environmental factors of: climate (including water and temperature effects), and macro- and microorganisms, conditioned by relief, acting on parent material over a period of time. A product – soil differs from the material from which it is derived in many physical, chemical, biological, and morphological properties and characteristics.

The five points underlined in the second definition are known as the *soil-forming factors*, which interact to produce a soil of particular characteristics in any given place. The parent material is acted upon by a series of weathering processes, outlined in Figure 1. Initially these are mainly physical processes, which bring about a decrease in particle size, and hence an increase in surface area. Weathering is strongly influenced by climate, and, once living organisms colonize the developing soil, by biological processes. Climate and relief interact to influence the movement of water and weathering products through the soil. Most of these physical, chemical, and biological processes proceed slowly, and so soil is evolving with time.

Soil is a combination of all of the major components of the surface environment: the atmosphere, hydrosphere, lithosphere, and biosphere (Figure 2). It is the mix and interaction of these four components that result in the particular properties of a specific soil. The solid components (mineral and organic matter) make up approximately 50% of soil by volume. Air and water occupy the pore space between the solid phase, and their relative amounts can vary considerably, resulting in important effects on the chemical and biological processes in a soil. Various measurements can be made to express the air–water balance in a soil (Table 2).

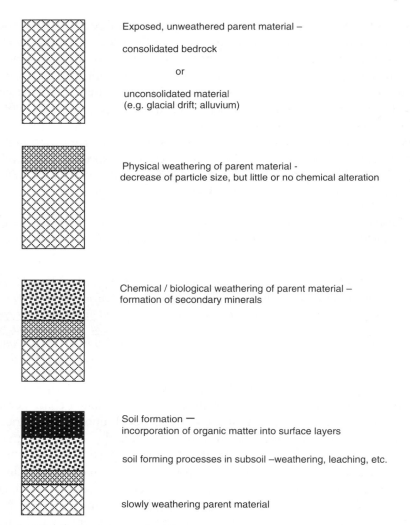

Exposed, unweathered parent material –

consolidated bedrock

or

unconsolidated material
(e.g. glacial drift; alluvium)

Physical weathering of parent material -
decrease of particle size, but little or no chemical alteration

Chemical / biological weathering of parent material –
formation of secondary minerals

Soil formation ⏤
incorporation of organic matter into surface layers

soil forming processes in subsoil –weathering, leaching, etc.

slowly weathering parent material

Figure 1 *Weathering sequence in soil*

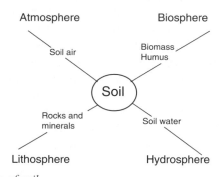

Atmosphere   Biosphere

Soil air   Biomass
                                     Humus

Soil

Rocks and
minerals   Soil water

Lithosphere   Hydrosphere

Figure 2 *Components of soil*

Worked example 5.1 – soil water and air contents, and bulk density
A soil core of 5 cm diameter and 8 cm height contains 257 g of fresh soil. After drying the soil at 110°C, the soil weighed 196 g. Calculate (i) the gravimetric soil water content, (ii) the volumetric soil water content, (iii) the soil bulk density, (iv) the % pore space, (v) the % water-filled pore space, and (vi) the % air-filled pore space.
Density of water $= 0.997$ g cm^{-3}
Volume of soil core $= \pi.r^2.h = 3.14 \times (2.5)^2 \times 8 = 157$ cm^3
(i) Gravimetric soil water content $= [(257-196 \text{ g})/196 \text{ g}] \times 100 = 31\%$
(ii) Volumetric soil water content $= [(257-196 \text{ g})/(0.997 \text{ g cm}^{-3} \times 157 \text{ cm}^3)] \times 100 = 39\%$
(iii) Soil bulk density $= 196$ g/157 cm$^3 = 1.25$ g cm^{-3}
Assuming a soil particle density of 2.6 g cm^{-3}, the volume of particles in the soil core $= 196$ g/2.6 g cm$^{-3} = 75.4$ cm^3
(iv) % pore space $= [(157 \text{ cm}^3 - 75.4 \text{ cm}^3)/157 \text{ cm}^3] \times 100 = 52\%$
(v) % water-filled pore space $=$ volumetric water content $= 39\%$
(vi) % air-filled pore space $= 52\% - 39\% = 13\%$
What is the weight of this soil in 1 ha to a depth of 20 cm?
1 ha $= 10,000$ m^2, therefore volume of soil to 20 cm $= 10,000$ m$^2 \times 0.2$ m $= 2000$ m^3
Bulk density of soil $= 1.25$ g cm$^{-3} \equiv 1.25$ t m^{-3}
2000 m$^3 \times 1.25$ t m$^{-3} = 2500$ t of soil.

Table 2   *Determination of soil water and air contents and bulk density*

Gravimetric soil water content (%)	(Mass of water/mass of oven dry soil) $\times$ 100
Volumetric soil water content (%)	(Volume of water/volume of soil core) $\times$ 100
Soil bulk density (g cm^{-3})	Mass of oven dry soil/volume of soil core
Soil particle-specific gravity (g cm^{-3})	Mass of oven dry soil solids/volume of soil solids
Pore space (%)	((Soil core volume–volume of soil particles in core)/soil core volume) $\times$ 100
Solid material (%)	(Volume of soil particles in core/total core volume) $\times$ 100
Water-filled pore space (%)	Volumetric soil water content
Air-filled pore space (%)	% pore space–% water-filled pore space

5.2 MINERAL COMPONENTS OF SOIL

5.2.1 Inputs

The nature of the parent material is the most important factor influencing the mineral components of a soil. In particular, the textural (particle size) properties and inherent fertility are directly affected by the types of rocks and minerals found in the parent material.

Parent material can be consolidated rock (igneous, sedimentary, or metamorphic) or unconsolidated superficial deposits that have been transported by some agency, such as wind, ice, or water. Igneous rocks are formed from molten magma from the earth's crust and are the ultimate source of all rocks. Extruded igneous rock is formed when molten magma appears at the surface and cools quickly, resulting in small grain size (*e.g.* basalt). Intruded igneous rock cools slowly on its way to the surface and results in large grain size (*e.g.* granite). Sedimentary rocks are formed from the weathering products of igneous, metamorphic, or older sedimentary rocks. Metamorphic rocks are formed by changes in igneous or sedimentary rocks owing to high temperature and/or high pressure, which usually has the effect of making the rock more resistant to weathering. Unconsolidated materials are surface deposits of partially weathered rock that have been transported and deposited in various ways. The most important groups of such materials, from which much of the world's major soils are formed, are: glacial deposits, resulting from the action of ice; alluvium, deposited from water; and aeolian or wind-blown deposits.[†]

5.2.2 Primary Minerals

The primary soil minerals, sometimes called 'inherited' minerals, are those that are derived from the rocks of the parent material. They are distinguished from the secondary, or 'pedogenic' minerals, which are formed during the process of soil formation (see next section). The main primary minerals are silicates (Table 3), all based on the SiO_4^{4-} tetrahedron, in which three oxygen atoms are in a triangular plane below the silicon atom and one oxygen is above the silicon (Figure 3a). There may be isomorphous substitution of silicon by aluminium. The charge is balanced by cations such as K^+, Na^+, Ca^{2+}, Mg^{2+}, and Fe^{2+}.

The silica tetrahedra can be linked in a number of ways to form silicates of differing structures, and hence susceptibility to weathering

---

[†] A description of the origins and morphology of specific soil types is outwith the scope of this chapter. An excellent description of the major soils of the world can be found at http://soils.usda.gov/education/.

Table 3 *Structures of common primary soil minerals*

Quartz	SiO_2
Feldspars	
Microcline	$KAlSi_3O_8$
Albite	$NaAlSi_3O_8$
Anorthite	$CaAlSi_2O_8$
Micas	
Muscovite	$KAl_2(Si_3Al)O_{10}(OH)_2$
Biotite	$K(Mg, Fe)_2(Si_3Al)O_{10}(OH)_2$
Amphiboles	
Hornblende	$(Na, Ca)_2(Mg, Fe, Al)_5(Si, Al)_8O_{22}(OH)_2$
Pyroxenes	
Enstatite	$MgSiO_3$
Augite	$Ca(Mg, Fe, Al)(Si, Al)_2O_6$
Olivine	$(Mg, Fe)_2SiO_4$

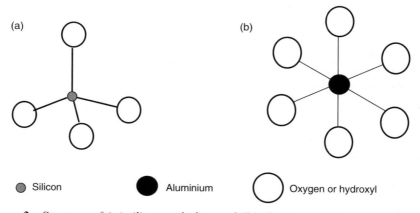

Figure 3 *Structure of (a) silica tetrahedron and (b) alumina octahedron*

(Figure 4). Quartz and the feldspars are *framework silicates*, in which adjacent tetrahedra share all four oxygens. The micas form *sheet silicates* by the sharing of three oxygens by each silica tetrahedron. The amphiboles and pyroxenes are *chain silicates*: amphiboles are double chains, with the tetrahedra sharing two or three oxygens alternately; pyroxenes are single chains with sharing of two oxygens. Olivine is an isolated tetrahedon (or orthosilicate).

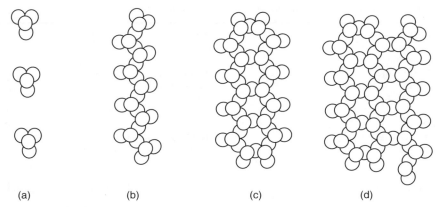

Figure 4   *Structures of (a) ortho (isolated); (b) single chain; (c) double chain; and (d) sheet silicates*

In Table 3, susceptibility to weathering increases down the list as fewer silicon–oxygen bonds need to be broken to release silicate. Consequently, quartz and feldspars especially, but also mica in temperate soils, are common inherited minerals in the coarse particle size fractions of soil (the silt and sand fractions, 0.002–2 mm). The amphiboles, pyroxenes, and olivine are much more easily weathered. Thus, soils derived from parent material with rock containing a predominance of framework silicates (*e.g.* granite, sandstone) tend to be more sandy, while those derived from rocks containing the more easily weathered minerals tend to be more clayey.

The chemical composition of the silicate minerals affects the inherent nutrient properties of soil. Coming down the list of silicates in Table 3, there is a change from the dominance of Si, Al, and Na in quartz and the feldspars to greater contents of Ca, Mg, and Fe in the amphiboles, pyroxenes, and olivine. These latter make up the so-called 'ferromagnesian' group of minerals, and soils in which they are an important component of the parent material have better inherent nutrient properties. In particular, such soils are better buffered against the natural acidification process that occurs due to loss of exchangeable cations, especially Ca^{2+} and Mg^{2+}, by leaching (see Sections 5.4 and 5.5).

5.2.3   Secondary Minerals

Secondary minerals (which dominate the clay-size fraction of soil, <0.002 mm) are formed in soil by the action of various weathering processes on

Table 4 *Structures of common secondary soil minerals*

Clay minerals (phyllosilicates)

Kaolinite	$Al_4Si_4O_{10}(OH)_8$
Illite	$K_{1.5}(Al_{3.5}Mg_{0.5})(Si_7Al)O_{20}(OH)_4$
Vermiculite	$M_{0.7}Al_2(Al_{0.7}Si_{3.3})O_{10}(OH)_2$
	$M=Ca^{2+}, Mg^{2+}, K^+, etc.$
Smectite	$M_{0.3}Al_2(Al_{0.3}Si_{3.7})O_{10}(OH)_2$
	$M=Ca^{2+}, Mg^{2+}, K^+, etc.$
Chlorite	$(Mg, Fe)_3(Si, Al)_4O_{10}(OH)_2(Mg, Fe)_3(OH)_6$

Amorphous aluminosilicates

Allophane	$xSiO_2 \cdot Al_2O_3 \cdot yH_2O$ (x=0.8–2, $y \geq 2.5$)
Imogolite	$SiO_2 \cdot Al_2O_3 \cdot 2.5H_2O$

Hydrous oxides

Goethite	$FeOOH$
Haematite	Fe_2O_3
Ferrihydrite	$Fe_5O_7(OH) \cdot 4H_2O$
Gibbsite	$Al(OH)_3$
Birnessite	$(Na, Ca)(Mn^{3+}, Mn^{4+})_7O_{14} \cdot 2.8H_2O$

Carbonates

Calcite	$CaCO_3$
Dolomite	$CaMg(CO_3)_2$

Others

Gypsum	$CaSO_4 \cdot 2H_2O$

the primary minerals derived from the parent material (see section 5.2.4). These may be formed by alteration of a primary mineral – *e.g.* the conversion of mica into a clay mineral – or by reactions of soluble products released into the soil environment – *e.g.* precipitation of iron oxide. The most important groups of secondary minerals are: (a) the aluminosilicate clay minerals; (b) short-range order (amorphous) aluminosilicates; and (c) the hydrous oxides of Al, Fe, and Mn (Table 4).

In the study of soil science, most attention has historically been paid to the aluminosilicate clays, which dominate the properties of temperate soils, the first to be scientifically studied. More recently, the importance of the amorphous aluminosilicates has been shown in young soils, in soils derived from volcanic ash and in leached, acidic soil (*e.g.* podzols or spodosols). The hydrous oxides are especially important components of old, highly weathered soils, such as those found in the tropics (*e.g.* oxisols). This is an important distinction as the charge on the aluminosilicate clays is predominantly a permanent negative charge, while the amorphous aluminosilicates and hydrous oxides have a variable, pH-dependent charge. This results in very different properties of tropical soils compared to temperate soils.

5.2.3.1 Aluminosilicate Clay Minerals (Phyllosilicates).

The aluminosilicate clay minerals are sheet silicates. They are sometimes referred to as phyllosilicates, but strictly this term should also include the micas, which are primary minerals.

The basic building blocks of all the aluminosilicate clay minerals are the silica tetrahedron (Figure 3a) and alumina octahedron (Figure 3b). These build up into tetrahedral sheets by the sharing of oxygens between adjacent silicons, and into octahedral sheets by the sharing of oxygens or hydroxyls between adjacent aluminiums. The sharing of oxygens between tetrahedral and octahedral sheets leads to the formation of clay *unit layers*, of which there are two basic types (Table 5): 1:1, one tetrahedral Si sheet and one octahedral Al sheet, and 2:1, two tetrahedral Si sheets sandwiching one octahedral Al sheet.

Unit layers are held together in various ways to produce a clay crystal. These are regular, rigid systems and the distance between equivalent points in adjacent unit layers, the c spacing or basal spacing, can be measured by X-ray diffraction and is used to identify the clay minerals (Figure 5).

Within the sheets, Si or Al may be replaced by a different element by the process of *isomorphous substitution*. Common replacements in the clay minerals are Al^{3+} for Si^{4+} in the tetrahedral sheet and Mg^{2+} or Fe^{2+} for Al^{3+} in the octahedral sheet. Substitution of an element by one of a lower valency results in a deficit of positive charge, and hence the development of a *permanent negative charge* on the clay mineral. In soils, this charge is neutralized by cations in solution being attracted to the negatively charged surface, and is called the *cation-exchange capacity*

Table 5 *Components of soil clays and how they combine to form a clay mineral*

Basic building blocks	Si tetrahedra, Al octahedra
Sheets	Tetrahedral sheet, made up of Si tetrahedra sharing oxygen atoms; octahedral sheet, made up of Al octahedra sharing oxygens or hydroxyls
Unit layers	Sheets combined in the ratio of one tetrahedral:one octahedral (1:1), or two tetrahedral:one octahedral (2:1)
Clay crystal	Stacks of unit layers held together by various bonding mechanisms, depending on the type of clay

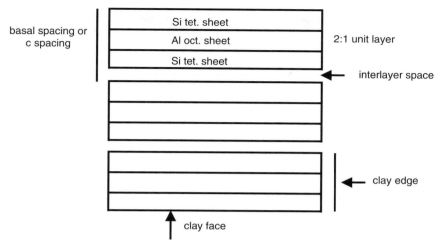

Figure 5 *Structure of a 2:1 aluminosilicate clay mineral*

(CEC). This permanent negative charge is the major characteristic of temperate soils, in which the aluminosilicate clays dominate the reactive fraction. The consequences of this are discussed in relation to ion-exchange processes in Section 5.5.

Kaolinite is the commonest 1:1 clay mineral. There is a small amount of isomorphous substitution of Al for Si in the tetrahedral sheet, resulting in a charge of ≤ 0.005 mol negative charge per unit cell. The unit layers are bound by hydrogen bonds between the oxygen atoms on the tetrahedral face and the hydroxyls on the adjacent octahedral face, resulting in a low CEC and surface area (Table 6) and a fixed c spacing of 0.7 nm. Unit layers stack up to form hexagonal crystals typically 0.05–2 μm thick. This means that the variable, pH-dependent charge on the clay edge is relatively more important in kaolinite than the other aluminosilicate clays.

Illite (a hydrous mica) is a 2:1 clay mineral with isomorphous substitution occurring mainly in the tetrahedral sheet (Al^{3+} for Si^{4+}), although a small amount also occurs in the octahedral sheet (Mg^{2+} or Fe^{2+} for Al^{3+}), resulting in a charge of 1.5 mol negative charge per unit cell. Most of this charge is neutralized by K^+ ions in the interlayer space. The geometrical arrangement of oxygen atoms on the tetrahedral face allows the K^+ ions to sit very close to the clay surface, forming a very strong bond between unit layers. This results in a low CEC and surface area (Table 6) and a fixed c spacing of 1.0 nm.

Vermiculite, a 2:1 clay mineral, is also a hydrous mica, with isomorphous substitution in the tetrahedral sheet, resulting in a charge of between 1.2 and 1.9 mol negative charge per unit cell. In this case the

Table 6 *Properties of the common aluminosilicate clay minerals*

Clay mineral	Type	CEC ($cmol_c$ kg^{-1})	Surface area (m^2 g^{-1})	Expanding/ non-expanding	c spacing nm	Interlayer binding
Kaolinite	1:1	3–20	5–100	Non-expanding	0.7	Hydrogen bonds
Illite	2:1	10–40	100–200	Non-expanding	1.0	K^+ ions
Vermiculite	2:1	80–150	300–500	Slight expansion	1.4	K^+, Ca^{2+}, Mg^{2+} ions
Smectite	2:1	80–120	700–800	Fully expanding	1.4–1.9	Interlayer cations
Chlorite	2:1:1	10–40	300–500	Non-expanding	1.4	Brucite sheet and H bonds

interlayer cations are mainly Mg^{2+} with some Ca^{2+}, which are more strongly hydrated cations than K^+ (see Section 5.5). The water of hydration associated with these cations widens the c spacing to 1.4 nm, and allows some degree of expansion.

Smectites are a group of 2:1 clay minerals that have a low degree of isomorphous substitution and hence a low-layer charge. The commonest smectite found in soil is montmorillonite, in which substitution occurs in the octahedral layer giving 0.7 mol negative charge per unit cell. Because the charge originates further from the clay surface, interlayer bonding is weak, allowing expansion of the clay lattice and easy entry of cations and water molecules into the interlayer space. Therefore, the CEC and surface area are both high (Table 6). The c spacing is variable, up to 1.9 nm when Ca^{2+} or Mg^{2+} are the dominant interlayer cations. If sodium is the dominant cation, the smectites can be fully dispersed in suspension.

Chlorite is a 2:1:1 clay mineral, with the usual 2:1 unit layer of two tetrahedral sheets and one octahedral held together by a sheet of brucite. Brucite is a magnesium hydroxide in which about two thirds of the Mg^{2+} ions are substituted by Al^{3+}, resulting in a positive charge. This neutralizes much of the negative charge that arises due to isomorphous substitution in the tetrahedral sheets (2 mol negative charge per unit cell) and forms a strong electrostatic bond between unit layers. There is also hydrogen bonding between hydroxyls on the brucite sheet and oxygens on adjacent tetrahedral sheets. As a result of this strong interlayer bonding, the CEC and surface area are low, and the c spacing is fixed at 1.4 nm.

In reality, soil clays are often not so well defined. Interstratification is common, with, for example, intergrades of vermiculite and illite or smectite and illite being found. This means that some layers in the clay mineral are illitic in nature, while others more closely resemble vermiculite or smectite. Weathering at the edges of the clay mineral can also lead to changes. For example, at the edge of an illite crystal, there may be some loss of interlayer K^+ ions and their replacement by other cations. This results in a widening of the interlayer space to form the so-called wedge site (Figure 6).

5.2.3.2 Short-Range Order (Amorphous) Aluminosilicates. This is a group of minerals that have small particle size and no regular crystal structure, and as a result produce poorly defined X-ray diffraction patterns. They do, however, show some regular short-range structure. Thus, there has been a tendency over recent years to describe them as short-range order, rather than amorphous, aluminosilicates. They have a very high surface area, making them highly reactive. They have both a

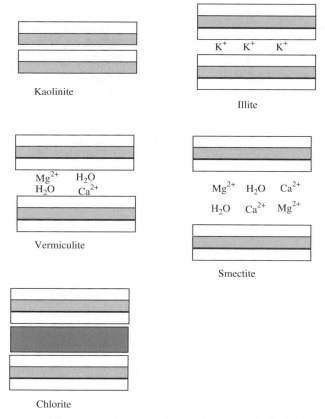

Figure 6 *Structures of common clay minerals* ☐ *Silica tetrahedral sheet;* ▨ *alumina octahedral sheet;* ■ *brucite sheet.*

fixed negative charge, arising from isomorphous substitution in Si tetrahedral sheets, and a variable, pH-dependent charge similar to that on clay edges and hydrous oxide surfaces. The two main minerals in this group, allophane and imogolite, are found in young soils and soils of volcanic origin.

5.2.3.3   Hydrous Oxides.   This term is generally taken to include the oxides, hydroxides, and oxyhydroxides of aluminium, iron and manganese, which form in soil when these elements are released from primary minerals by weathering. They exist mainly as small particles in the clay-sized fraction of a soil (<2 µm), and also as coatings on other soil minerals or as components of larger aggregates.

The most common aluminium oxide found in soil is gibbsite ($\gamma Al(OH)_3$), which may be the final product resulting from the precipitation of amorphous aluminium hydroxide. A number of iron oxides are found in soil, the commonest being ferrihydrite, goethite, and haematite. Iron released from weathering of the ferromagnesian minerals precipitates out of solution as either ferrihydrite or goethite depending on soil conditions (low organic matter and a high rate of Fe release favour ferrihydrite formation). Ferrihydrite is a poorly ordered oxide of small particle size, whereas goethite is a well-structured mineral and is the commonest of the iron oxides. Haematite is derived by structural changes to ferrihydrite, and is commonly found in tropical soils. Manganese oxides occur in a large number of forms, often with variable valency, for example, birnessite, and in mixed oxides, especially with iron. They are the least well understood of this group of minerals.

The hydrous oxides have a variable, pH-dependent charge and are important in soils as aggregating agents and as adsorptive surfaces (see Section 5.5) (Figure 7).

5.2.4   Weathering Processes (See Also Chapter 3)

The processes that cause changes to the rocks and minerals of the soil parent material are collectively known as weathering. In the early stages, physical weathering dominates and results in fragmentation of rock, a decrease in particle size, and increase in surface area, but no chemical change. It is brought about by a number of actions:

- Abrasion of the rock caused by the action of ice, water, and wind leads to physical disintegration.
- Temperature change due to diurnal temperature variation and by freeze–thaw cycles can create significant pressures that will cause rock to split along lines of weakness.

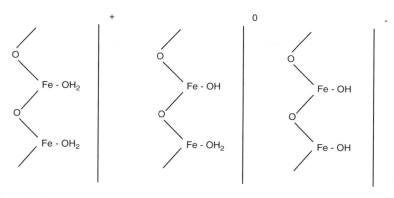

Figure 7   *Variable, pH-dependent charge on an iron oxide surface*

- Roots tend to grow into existing fissures and channels and can exert significant pressure.
- Crystallization of some salts can also cause swelling, which can exert pressure.

The key factor about physical weathering is the increase in surface area of rock that is then available for attack by various chemical-weathering processes. Chemical weathering is the reaction with natural waters, and dissolved components, resulting in partial or complete dissolution of the rock and formation of a new mineral phase. In most cases, it is the interaction of water with the minerals in the rock that is the crucial issue.

5.2.4.1 Dissolution.   Some minerals, such as gypsum ($CaSO_4.2H_2O$), are readily soluble in water and so can persist only in soils of arid and semi-arid regions, where water is limiting. In temperate and tropical regions sufficient water will pass through the soil to cause dissolution and loss of gypsum:

$$CaSO_4.2H_2O \leftrightharpoons Ca^{2+} + SO_4^{2-} + 2H_2O \qquad (5.1)$$

Quartz is a much less soluble mineral than gypsum, but even this is slightly soluble:

$$SiO_2 + 2H_2O \rightleftharpoons H_4SiO_4 \qquad (5.2)$$

This reaction is driven by the loss of the soluble silicic acid from the soil. In temperate regions insufficient water passes through the soil, and weathering has proceeded over a relatively short time since the last ice age, for there to have been significant loss of silica, and so quartz is a dominant mineral in the coarse fraction of soils. In the humid tropics, however, the much larger volume of water passing through the soil, over a much longer timescale, has resulted in significant, if not complete, loss of silica.

5.2.4.2   Acid Hydrolysis.   The water that enters soil as rain or snow is in equilibrium with CO_2 in the atmosphere, which dissolves to form carbonic acid. Unpolluted rainwater has a pH of approximately 5.7, whereas water in soil pores may be exposed to air containing a higher partial pressure of CO_2 than the free atmosphere, and hence soil water may be more acidic (see Section 5.4). It is the attack on soil minerals by this weak carbonic acid that is the major chemical weathering process in most soils. For example, acid hydrolysis of calcium carbonate yields calcium and bicarbonate ions:

$$CaCO_3 + H_2CO_3 \rightleftharpoons Ca^{2+} + 2HCO_3^- \qquad (5.3)$$

Acid hydrolysis of the primary mineral microcline feldspar results in release of some soluble components (silicic acid, potassium ions, and bicarbonate ions) and alteration of the solid phase to kaolinite:

$$4KAlSi_3O_8 + 22H_2O + 4CO_2 \rightleftharpoons Al_4Si_4O_{10}(OH)_8 + 8H_4SiO_4 \quad (5.4)$$
$$+ 4K^+ + 4HCO_3^-$$

5.2.4.3   Oxidation.   For those elements that can exist in more than one valence state oxidation, and indeed reduction, may be a major reaction in the chemical-weathering process. Iron and manganese are the most important elements that behave in this way. For example, iron in the ferromagnesian minerals is in the Fe(II) state, which is oxidized to Fe(III) when released from the mineral. This can cause changes in the charge balance, requiring other ions to be lost. Formation of iron oxide can cause physical disruption to the mineral.

5.2.4.4   Chelation.   Once biological activity is established, an important reaction is the formation of a stable complex between a metal ion and an organic molecule due to release of organic compounds by colonizing organisms and plant roots.

5.3   ORGANIC COMPONENTS OF SOIL

It cannot be emphasized enough that soil is formed by the interaction of the mineral weathering product of rocks and the organic material introduced as a result of biological activity. Much of this organic material is plant-derived litter and the amount added to soil per year depends very much on the dominant vegetation type (Table 7). These figures do not include inputs from roots, which may be significant but difficult to quantify, or more resistant inputs, such as wood in forests. In addition, animal and microbial excretion and decomposition must be taken into account. The values in Table 7 do, however, give a reasonable estimate of what may be described as degradable carbon entering soils per year.

The organic materials added to soil have varying degrees of susceptibility to degradation. Simple compounds, such as sugars and amino acids, are rapidly decomposed by soil microorganisms, whereas complex material, such as lignin and hemicellulose, tends to persist in an undegraded or partially degraded form. About 70% of the carbon added to soil is lost as carbon dioxide, produced mainly by microbial respiration. The other 30% is incorporated into the microbial biomass and soil humus, a resistant material, which degrades only very slowly. Figure 8 shows typical values for the annual turnover of organic matter in soil.

Humus formation proceeds broadly along two pathways (Figure 9). Readily decomposed material releases low molecular weight compounds, especially phenolic and amino compounds, which react and polymerize to form humic material. Alternatively, the more resistant organic material can undergo partial degradation, and this partially degraded material may also react with the low molecular weight compounds released into the soil, again leading to the formation of humus. It is likely that both of these processes operate in soils.

The terms humus and humic substances are used interchangeably to refer to the resistant organic material that persists in soil over relatively long timescales (100s–1000s of years). Although the humic substances may be chemically diverse, being dependent on the dominant type of plant tissue input, they do have some overall similarities. They are amorphous, colloidal polymers built up mainly from aromatic units. There is a wide range of molecular weights, from 100s to 100,000s Da.

Table 7 *Inputs of carbon to soil by various types of land use*

Dominant land use	Tonnes carbon added to soil ha^{-1} $year^{-1}$
Alpine and arctic forest	0.1–0.4
Arable agriculture	1–2
Coniferous forest	1.5–3
Deciduous forest	1.5–4
Temperate grassland	2–4
Tropical rain forest	4–10

Source: adapted from White, *Principles and Practice of Soil Science*, Blackwell Science, 1997.

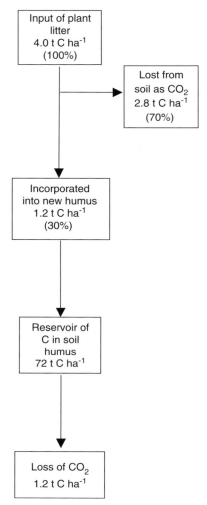

Figure 8 *Typical annual turnover of organic matter in a temperate grassland soil at steady state receiving 4 t C ha^{-1}*

All of these compounds do, however, have similar overall properties due to the preponderance of carboxylic and phenolic functional groups on the humic polymer. Both of these groups can ionize by losing a hydrogen ion to give a negatively charged group. This is a variable charge, the magnitude of which varies with pH, but is always negative and forms part of the CEC of a soil (see Section 5.5).

$$R\text{-}COOH \rightleftharpoons R\text{-}COO^- + H^+ \qquad (5.5)$$

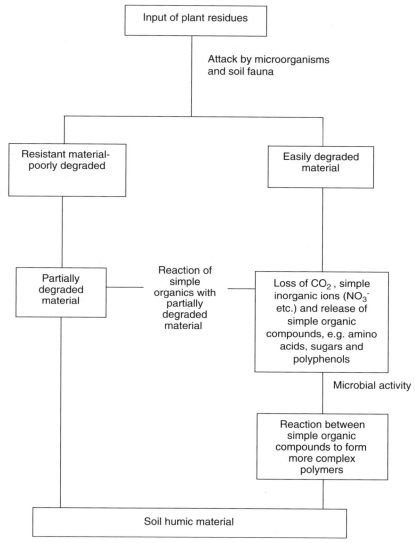

Figure 9 *Possible pathways for formation of humic material in soil from inputs of plant residues*

$$\langle \!\!\!\bigcirc\!\!\!\rangle\ OH\ \rightleftharpoons\ \langle \!\!\!\bigcirc\!\!\!\rangle\ O^- + H^+ \qquad (5.6)$$

The pK of Equation (5.5) is in the range 3–5, and of Equation (5.6) in the range 7–8. Thus, carboxyl groups are important in acid soils, while phenolic groups become important above pH 7.

Soil organic matter has been traditionally fractionated into a number of subgroups (Figure 10).

Humin and humic acid are high molecular weight fractions ($> 10,000$ Da) and it is thought that humin is not extracted by alkali because it is too strongly bound to the soil mineral component. Fulvic acid is the low molecular weight fraction ($< 10,000$ Da). There are some chemical differences between humic acid and fulvic acid, the latter being more acidic and having a higher oxygen content and lower carbon content than humic acid. This reflects the greater number of carboxylic and phenolic groups in the fulvic acid fraction. It is speculated that humic acid forms the less reactive backbone or core of the humic material, while the more reactive fulvic acid is present as side chains branching from the core.

Although there is considerable variation between soils, the ratio of the major elements, C:N:P:S, in humic material is approximately 100:10:2:1, forming a major reservoir of N, P, and S that is made available to plants as soil organic matter, is broken down by soil micoorganisms.

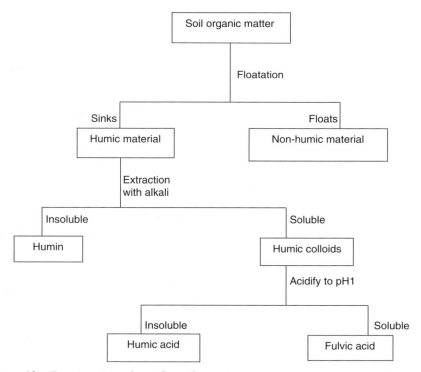

Figure 10 *Fractionation scheme for soil organic matter*

Identification of N-, P-, and S-containing compounds is difficult, and those that are identified tend to be metabolic products, such as nucleotides and vitamins, released into the soil following the death of cells. Broadly, nitrogen is an integral part of the humic molecule, and is released as NH_4^+ ions when the humic material is degraded. Phosphorus and sulfur are more commonly found as P and S esters, which can be released as orthophosphate and sulfate ions by the action of phosphatase and sulfatase enzymes, respectively.

Worked example 5.2 – organic carbon content of soil

Organic carbon in soil can be measured by oxidation with acid dichromate solution:

$$2K_2Cr_2O_7 + 3C + 8H_2SO_4 \rightleftharpoons 2Cr_2(SO_4)_3 + 3CO_2 \qquad (5.7)$$
$$+ 2K_2SO_4 + 8H_2O$$

The excess dichromate is measured by titration with ferrous sulfate

$$Cr_2O_7^{2-} + 6Fe^{2+} + 14H^+ \rightleftharpoons 2Cr^{3+} + 6Fe^{3+} + 7H_2O \quad (5.8)$$

0.20 g soil was oxidized with 20 cm^3 of $M/6 K_2CrO_7$ in sulfuric acid. The excess dichromate was determined by titration with 0.5 M $FeSO_4$ solution. A blank, no soil, control was carried out to assess loss of dichromate by thermal decomposition. The titration values obtained were 17.5 cm^3 for the soil sample and 39.0 cm^3 for the blank. What is the % organic carbon in the soil?

Moles of $Cr_2O_7^{2-}$ added initially $= 20 \times 1/6 = 3.334$ mmol

Moles of $Cr_2O_7^{2-}$ remaining after reaction:

$17.5 \times 0.5 = 8.75$ mmol Fe^{2+}

From Equation (5.8), 1 mol of $Cr_2O_7^{2-}$ reacts with 6 mol of Fe^{2+}

Therefore, $8.75/6 = 1.458$ mmol of $Cr_2O_7^{2-}$ unused.

Blank titration

$39.0 \times 0.5 = 19.5$ mmol $Fe^{2+} = 3.25$ mmol of $Cr_2O_7^{2-}$

Therefore, $3.334–3.25 = 0.084$ mmol of $Cr_2O_7^{2-}$ lost by thermal decomposition.

Moles of $Cr_2O_7^{2-}$ used in oxidation $= 3.334–1.458 = 1.876$ mmol

Subtract $Cr_2O_7^{2-}$ lost by thermal decomposition $= 1.876–0.084 = 1.792$ mmol of $Cr_2O_7^{2}$

From Equation (5.7) 2 mol of $Cr_2O_7^{2-}$ react with 3 mol of C

$1.792 \times 3/2 = 2.688$ mmol C

$2.688 \times 12 = 32.26$ mg carbon

$(32.26$ mg$/200$ mg$) \times 100 = 16.13\%$ C in the soil.

5.4 SOIL pH AND REDOX POTENTIAL

Soil pH, which is a measure of the acidity or alkalinity, and redox potential, a measure of the aeration status, are the two crucial factors that control many of the chemical and biological processes in soil.

5.4.1 pH and Buffering

pH is defined as $-\log[H^+]$, *i.e.* the negative logarithm of the hydrogen ion concentration (or strictly speaking its activity) in solution, and is based on the concept of the partial ionization of water. In pure water, the H^+ ion concentration is 10^{-7} mol L^{-1} and so the pH is 7. If the concentration of H^+ ions increases, the pH falls below 7 (acid conditions) and if the concentration decreases, the pH rises above 7 (alkaline conditions). In soil, however, factors other than the H^+ ion concentration in solution affect the pH.

Measurement of pH in a soil is usually done using a glass pH electrode in a soil suspension. It is important to state the ratio of the soil weight to the solution volume used, because pH varies with dilution. Common ratios are 1:1 (10 g:10 cm^3), 1:2.5 (10 g:25 cm^3), and 1:5 (10 g:50 cm^3). It is also necessary to state the nature of solution used, commonly deionized water or a dilute salt solution (*e.g.* 0.01 M CaCl$_2$). What is actually being measured is the pH of a solution in equilibrium with negatively charged soil particles – thus, some H^+ ions are in solution and some on soil surfaces (see ion exchange, Section 5.5). The pH measured depends on the ratio of these two pools of ions, and is affected by dilution and the concentration of other ions in solution. If care is taken in measuring pH, it is a very useful parameter from which a lot can be deduced about a soil. However, it must be borne in mind that in the field soil pH can vary significantly over short distances and over relatively short timescales.

H^+ and Al^{3+} ions in solution and on surface-exchange sites constitute the *solution and exchangeable acidity* in a soil, also called *the active acidity*. The *non-exchangeable*, or *reserve*, *acidity* is associated with carboxyl and phenolic groups on humic material, OH groups on hydrous oxide surfaces and clay edges, and polymeric aluminium hydroxides. The balance between solution and exchangeable acidity and non-exchangeable acidity can shift in response to a change in pH:

Non-exchangeable acidity $\rightleftharpoons$ Exchangeable and solution acidity (5.9)

If soil pH is increased (H^+ in solution decreased), the equilibrium shifts to the right – some non-exchangeable acidity released. If soil pH is decreased (H^+ in solution increased), the equilibrium shifts to left – some acidity goes on to non-exchangeable acidity. This represents the *buffer capacity* of a soil, and acts to avoid large changes in pH. Acidity in soil is normally corrected by addition of a liming material such as $CaCO_3$. The amount required can be calculated by measuring the buffering of a soil by titration with a dilute solution of an alkali (often $Ca(OH)_2$ is used). The amount of alkali needed to reach the required pH can be found from the buffer curve.

Worked example 5.3 – soil buffering and liming

To raise the pH of 5 g of the soil in Worked example 5.1 to 6.5, required addition of 4 cm^3 of 0.005 mol L^{-1} $Ca(OH)_2$ solution. How much calcium carbonate would be required to be added to 1 ha? (Mol wt $CaCO_3$ = 100).

$$4\,cm^3 \text{ of } 0.005\,mol\,L^{-1} = 0.02\,mmol\,Ca(OH)_2 \text{ solution for } 5\,g\,soil$$
$$= 0.004\,mmol\,g^{-1}$$
$$= 0.4\,mg\,CaCO_3\,g^{-1}$$
$$(1\,mmol\,CaCO_3 = 100\,mg)$$
$$= 0.4\,kg\,CaCO_3\,t^{-1}$$

From Worked example 5.1, there are 2500 t ha^{-1} in the top 20 cm, which approximates to rooting depth.

0.4 kg $CaCO_3$ t^{-1} × 2500 t = 1000 kg $CaCO_3$ required ha^{-1}.

5.4.2 Soil Acidity

The atmosphere is a major source of soil acidity. Even in unpolluted environments rainwater is slightly acidic, having a pH of about 5.7 due to the dissolution of atmospheric CO_2 to form the weak carbonic acid (see Worked example 5.4). The CO_2 concentration in the partially enclosed soil pore system can be significantly higher (typically up to about 10 times) than in the free atmosphere due to respiration of soil microorganisms and plant roots. This results in a lower pH. In areas affected by industrial pollution, sulfur dioxide and nitrogen oxides dissolve in rainwater to produce sulfuric and nitric acids (acid rain), which are both strong acids and cause even more acidity.

Worked example 5.4 – pH of soil solution
Carbon dioxide dissolves in water to form carbonic acid, which is a weak acid.

$$CO_2 + H_2O \rightleftharpoons H_2CO_3 \quad K = 10^{-1.5} \tag{5.10}$$

$$[H_2CO_3]/pCO_2 = 10^{-1.5} \text{ mol L}^{-1} \text{ atm}^{-1}$$

Carbonic acid dissociates to produce bicarbonate and hydrogen ions

$$H_2CO_3 \rightleftharpoons HCO_3^- + H^+ \quad K = 10^{-6.4} \tag{5.11}$$

$$\frac{[HCO_3^-][H^+]}{[H_2CO_3]} = 10^{-6.4} \text{ mol L}^{-1}$$

$$[HCO_3^-] = [H^+]$$

$$[H^+]^2 = 10^{-6.4} [H_2CO_3] = 10^{-6.4} \times 10^{-1.5} \times pCO_2$$

pCO_2 in the atmosphere $= 10^{-3.5}$ atm

$$[H^+]^2 = 10^{-6.4} \text{ mol L}^{-1} \times 10^{-1.5} \text{ mol L}^{-1} \text{ atm}^{-1}$$
$$\times \ 10^{-3.5} \text{ atm} = 10^{-11.4}$$
$$[H^+] = 10^{-5.7} \text{ mol L}^{-1}$$
$$pH = 5.7$$

If the partial pressure of CO_2 in the soil atmosphere is considered to be 10 times greater than the free atmosphere,

$$[H^+]^2 = 10^{-6.4} \text{ mol L}^{-1} \times 10^{-1.5} \text{ mol L}^{-1} \text{ atm}^{-1}$$
$$\times \ 10^{-2.5} \text{ atm} = 10^{-10.4}$$
$$[H^+] = 10^{-5.2} \text{ mol L}^{-1}$$
$$pH = 5.2$$

Sulfur dioxide dissolves to form the strong acid, sulfurous acid, so is strongly acidifying at very low concentrations. For example, for $pSO_2 = 10^{-7}$ atm (a value much higher than concentrations now occurring owing to air pollution control measures in developed countries. Note, however, that low-pH rainfall still occurs as a result of incorporation of sulfuric and nitric acids).

$$SO_2 + H_2O \rightleftharpoons H_2SO_3 \quad K = 1 \tag{5.12}$$

$$[H_2SO_3]/pSO_2 = 1 \text{ mol L}^{-1} \text{ atm}^{-1}$$

$$H_2SO_3 \rightleftharpoons HSO_3^- + H^+ \quad K = 1.7 \times 10^{-2} \qquad (5.13)$$

$$\frac{[HSO_3^-][H^+]}{[H_2SO_3]} = 1.7 \times 10^{-2}\, mol\, L^{-1}$$

$$[H^+]^2 = 1.7 \times 10^{-2}\, mol\, L^{-1} \times 1\, mol\, L^{-1}\, atm^{-1}$$
$$\times 10^{-7}\, atm = 1.7 \times 10^{-9}$$

$$[H^+] = 4.12 \times 10^{-5}$$

$$\log[H^+] = -4.4$$

$$pH = 4.4$$

Various soil processes also contribute to soil acidity. Decaying organic matter releases a number of organic acids, as indeed do plant roots. Nitrification, the microbial oxidation of ammonium ions to nitrate by *Nitrosomonas* and *Nitrobacter*, occurs in slightly acidic to neutral soils and releases H^+ ions:

$$2NH_4^+ + 3O_2 \rightarrow 2NO_2^- + 4H^+ + 2H_2O \qquad (5.14)$$

$$2NO_2^- + O_2 \rightarrow 2NO_3^- \qquad (5.15)$$

Sulfide oxidation, another microbially mediated process, also results in the production of acidity:

$$2FeS_2 + 2H_2O + 7O_2 \rightleftharpoons 2FeSO_4 + 2H_2SO_4 \qquad (5.16)$$

$$4FeSO_4 + 10H_2O + O_2 \rightleftharpoons 4Fe(OH)_3 + 4H_2SO_4 \qquad (5.17)$$

Note that in this case much of the acidity is due to the precipitation of iron(III) oxide.

Leaching, the passage of water through the soil profile, is a process that occurs in soils in areas of moderate to high rainfall, and which results in the loss of exchangeable cations (such as Ca^{2+}, Mg^{2+}, K^+, and Na^+ – see Section 5.5). Unless the rate of weathering of soil minerals is sufficiently rapid to replace these losses, the exchange sites will become occupied by H^+ and Al^{3+}, which are acidic cations.

5.4.3 Soil Alkalinity (see also Section 3.2.4)

Alkaline soils (pH > 7) contain solid phase carbonate, and bicarbonate is the dominant anion in solution. Calcium and magnesium carbonates produce a soil of pH between 7 and 8.5, depending upon the concentrations of CO_2 and Ca^{2+} or Mg^{2+} ions. In arid regions, the lack of rainfall, and hence leaching, allows the more soluble Na_2CO_3 to accumulate and the pH rises to about 10.5.

5.4.4 Influence of pH on Soils

A major influence of pH on soil is its effect on biological activity. Many of the soil microorganisms function within a narrow optimal range of pH and their activity is inhibited in more acidic or alkaline conditions. For example, the nitrifying bacteria mentioned above, *Nitrosomonas* and *Nitrobacter*, have a pH optimum in the range 6–8, and are severely inhibited below pH 5.5. Conversely, the iron- and sulfur-oxidizing bacteria of the genus *Thiobacillus* are active only under acid conditions, and are inhibited at pH > 5. In general, bacteria are less tolerant of acid conditions than fungi. Soil animals too are affected by pH; earthworms, for example, cannot survive below a pH of about 4.5.

Certain soil components have a variable, pH-dependent charge. The charge on the humified soil organic matter is negative overall due to dissociation of carboxyl and phenolic groups, but its magnitude varies, being greater at high pH (see Section 5.3). The hydrous oxides and edges of clay minerals are positively charged at low-pH values and negative at

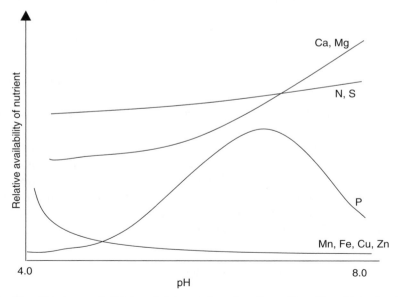

Figure 11   *Relative availabilities of plant nutrients in soil as affected by pH (only relative amounts of individual nutrients should be compared)*

high pH (see Section 5.2). In general, therefore, low pH results in a lessening of the negative charge on the surfaces of soil particles.

pH has a big influence on the solubility of many important elements in soil, and can have a big influence on the availability of plant nutrients (Figure 11). Aluminium, which is not required by plants, is relatively insoluble except under acid conditions when toxicity problems may occur. Using gibbsite, $Al(OH)_3$, which is a common Al mineral in many soils, as an example, the effect of pH on the concentration of aluminium in solution can be modelled.

Worked example 5.5 – effect of pH on concentration of aluminium in soil solution

What is the concentration of Al^{3+} ions in soil solution at the pH values calculated in Worked example 5.4 for soil solution at (a) $pCO_2 = 10^{-2.5}$ atm and (b) $pSO_2 = 10^{-7}$ atm?

Using the dissolution of gibbsite as a model,

$$Al(OH)_3 \rightleftharpoons Al^{3+} + 3OH^- \quad \text{Solubility product,} \quad (5.18)$$
$$K_{SP} = 5 \times 10^{-33}$$

$$[Al^{3+}][OH^-]^3 = 5 \times 10^{-33}$$
$$\text{(a) At } pCO_2 = 10^{-2.5} \text{ atm}$$
$$pH = 5.2, pOH = 8.8$$

$$[OH^-] = 1.58 \times 10^{-9}$$
$$[OH^-]^3 = 3.94 \times 10^{-27}$$
$$[Al^{3+}] = (5 \times 10^{-33})/(3.94 \times 10^{-27})$$
$$= 1.27 \times 10^{-6} \, mol \, L^{-1}$$

$$\text{(b) At } pSO_2 = 10^{-7} \text{ atm}$$
$$pH = 4.4, pOH = 9.6$$

$$[OH^-] = 2.51 \times 10^{-10}$$
$$[OH^-]^3 = 1.58 \times 10^{-29}$$
$$[Al^{3+}] = (5 \times 10^{-33})/(1.58 \times 10^{-29})$$
$$= 3.2 \times 10^{-4} \, mol \, L^{-1}.$$

5.4.5 Redox Potential

Redox potential is the measure of the oxidation–reduction state of a soil, and is determined by redox reactions involving the transfer of electrons from one chemical species to another. A generalized redox reaction can be written as

$$Ox + mH^+ + ne^- \rightleftharpoons Red \tag{5.19}$$

where 'Ox' and 'Red' are the oxidized and reduced species, respectively. Note also that H^+ ions are involved in the reaction, and so pH affects the redox potential.

The potential produced is a consequence of the ratio of oxidized to reduced species, and is expressed by the Nernst equation

$$E = E^0 - \frac{RT}{nF} \ln \frac{[Red]}{[Ox]} \tag{5.20}$$

where

E = potential in volts
E^0 = standard electrode potential
R = universal gas constant ($8.314 \ J \ mol^{-1} \ K^{-1}$)
T = absolute temperature in K
n = number of electrons involved
F = Faraday constant $96,487 \ C \ mol^{-1}$ (the charge when 1 mole of [Ox] is reduced).

If a temperature of 25°C (298 K) is assumed, the values of the constants R and F are used and natural logs converted to $\log_{10}$, the equation becomes

$$E = E^0 - \frac{0.0591}{n} \log_{10} \frac{[Red]}{[Ox]} \tag{5.21}$$

Redox potential is measured using an inert platinum electrode, which acquires the electric potential (E_H) of the soil when placed into the soil or a soil suspension. This potential is measured relative to a reference electrode with a known potential, such as the calomel electrode (E_{cal}), which has a potential of 0.248 V at 25°C. This potential has to be added to the measured value

$$E_H = E_{cal} + 0.248 \ V \tag{5.22}$$

E_H is measured at the pH of the sample, but is often expressed corrected to pH 7 by subtracting 0.0591 V per unit pH up to 7 for samples with a pH below 7 and adding 0.0591 V per unit pH down to 7 for samples above pH 7.

E_H can also be expressed as $p\varepsilon$, which is the negative log of the electron activity (analogous with pH) (see also Chapter 3). This approach considers electrons as a reactant or product in the reaction. The two are related by

$$E_H \text{ (in volts)} = 0.0591\, p\varepsilon \qquad (5.23)$$

Worked example 5.6 – redox limits in aqueous systems (see also Section 3.2.4.2)

The limits of redox potential in aqueous systems are determined by the oxidation or reduction of water.

Oxidation of water

$$2H_2O \rightleftharpoons O_2 + 4H^+ + 4e^- \quad \log K = -83.1 \qquad (5.24)$$

$$\log K = \log pO_2 + 4\log H^+ + 4\log e^-$$
$$-83.1 = \log pO_2 + 4\log H^+ + 4\log e^-$$
$$83.1 = -\log pO_2 + 4pH + 4p\varepsilon$$
$$\text{For } pO_2 = 1\,\text{atm}, \log pO_2 = 0$$
$$83.1 = 4pH + 4p\varepsilon$$
$$p\varepsilon = 20.78 - pH$$

Reduction of water

$$2H^+ + 2e^- \rightleftharpoons H_2 \quad \log K = 0 \qquad (5.25)$$

$$pK = -\log pH_2 - 2pH - 2p\varepsilon$$
$$\text{For } pH_2 = 1\,\text{atm}, \log pH_2 = 0$$
$$p\varepsilon = -pH$$

At pH 7, this gives theoretical limits for the soil environment at E_H = approximately +0.8 V (oxidizing) and approximately −0.4 V (reducing). In practice soil atmospheres will not contain 1 atm of oxygen or hydrogen.

5.4.6   Reduction Processes in Soil

Reduction is caused by a decrease in oxygen concentration in the soil atmosphere; for example, by waterlogging or compaction. Oxygen diffuses through water 10^4 times more slowly than through air, and so oxygen concentration falls. The microbial population changes in response to the decreased oxygen, from *aerobic organisms*, requiring oxygen for respiration, to *facultative anaerobes*, which can use sources of electrons other than oxygen, to *obligate anaerobes*, which cannot survive if oxygen is present. As the microbial population changes, their use of alternative electron sources to oxygen causes a series of important redox changes in soil (Table 8).

As each of these reduction processes occurs, the E_H is buffered, or poised, at a particular value. So, for example, while there is nitrate present, the system will be poised at around +0.22 V; once the nitrate has been used up, reduction of Mn^{IV} will poise the system at around +0.2 V, and so on. E_H in oxidized systems is difficult to measure because there is not a specific reaction controlling it. Generally, when oxygen is present, the E_H is greater than +0.3 V.

The phase changes that occur are important: the gaseous products can be lost from the soil to the atmosphere; Mn and Fe become soluble and can be moved within or out of the system.

Reduction of NO_3^-, Mn^{IV}, and Fe^{III} can occur under moderately reduced conditions, such as intermittently waterlogged soil or sediment. Reduction of nitrate causes the loss of a major nutrient. Often the reduction does not proceed through to the formation of N_2 gas, and nitrous oxide (N_2O) is formed, which is a potent greenhouse gas and can

Table 8 *Redox reactions in soil*

Redox reaction	Range of soil E_H values (V)
$O_2 + 4H^+ + 4e^- \rightleftharpoons 2H_2O$	Disappearance of oxygen 0.6–0.4
$2NO_3^- + 12H^+ + 10e^- \rightleftharpoons N_2 + 6H_2O$ solution gas	Disappearance of nitrate 0.5–0.22
$MnO_2 + 4H^+ + 2e^- \rightleftharpoons Mn^{2+} + 2H_2O$ solid solution	Appearance of Mn^{II} 0.4–0.2
$Fe(OH)_3 + 3H^+ + e^- \rightleftharpoons Fe^{2+} + 3H_2O$ solid solution	Appearance of Fe^{II} 0.3–0.1
$SO_4^{2-} + 10H^+ + 8e^- \rightleftharpoons H_2S + 4H_2O$ solution gas	Disappearance of sulfate 0 to −0.15
$CO_2 + 8H^+ + 8e^- \rightleftharpoons CH_4 + 2H_2O$ gas gas	Appearance of methane −0.15 to −0.22

Source: adapted from Bohn, McNeal and O'Connor, *Soil Chemistry*, Wiley Interscience, 1985.

cause ozone depletion in the atmosphere. Redox changes of Mn and Fe are the main factor in the formation of gleys (waterlogged soils). Re-oxidation to Mn^{IV} and Fe^{III} results in precipitation of highly reactive oxides, which can sorb significant amounts of anions such as phosphate and heavy metals. Sulfide and methane production occur only under highly reduced conditions, such as permanently waterlogged and peat soils. Sulfide may be lost as hydrogen sulfide gas, but often metals sulfides (*e.g.* iron sulfide (FeS)) are precipitated. Methane is a potent greenhouse gas.

5.5 CHEMICAL REACTIONS IN SOIL

Two of the important chemical functions of a soil are:

(i) to supply nutrients to plants and
(ii) to act as a filter or sink for pollutants.

In both cases, the crucial chemistry governing the process occurs at the interface between soil solution and the solid phase. Various mechanisms control the equilibrium between chemical species in the two phases, with movement between them being controlled by changes in solution concentration.

5.5.1 Reactions in Soil Solution

The soil solution contains dissolved gases (carbon dioxide being particularly important – see Section 5.4), soluble organic molecules from soil organic matter or exuded by plants and microorganisms, and inorganic ions released from minerals by weathering or, in agricultural soils, added as fertilizer. Typical ranges of ionic concentrations are given in Table 9.

Inorganic ions can exist as hydrated ions in solution, with the form of the ion being determined by its charge (z) and size (r), expressed as the ionic potential z^2/r. Ions with a small ionic potential ($<$approximately six) are present in solution as the hydrated ion (*e.g.* in soil solution, K^+

Table 9 *Typical ionic concentrations in soil solution at pH 6–7*

Ions	Concentration range
Ca^{2+}, NO_3^-	10^{-2}–10^{-3} mol L^{-1}
Mg^{2+}, K^+, Na^+, Cl^-, HCO_3^-	10^{-3}–10^{-4} mol L^{-1}
$H_2PO_4^-$	10^{-5}–10^{-6} mol L^{-1}
Trace elements (Zn, Cu, *etc.*)	$<10^{-6}$ mol L^{-1}

and Na^+ are weakly hydrated cations, while Ca^{2+} and Mg^{2+} are strongly hydrated). Ions with a medium ionic potential (10–20) tend to hydrolyze and precipitate as oxides and hydroxides (*e.g.* Fe^{III}, Al^{III}, Cr^{III}). Those with a high-ionic potential (>20) form soluble oxyanions (*e.g.* phosphate, nitrate, sulfate).

The chemical speciation of the ions is highly dependent on pH and redox potential. Phosphorus, for example, exists as orthophosphate anions:

$$H_3PO_4 \rightleftharpoons H_2PO_4^- + H^+ \qquad pK \quad 2.15 \qquad (5.26)$$

$$H_2PO_4^- \rightleftharpoons HPO_4^{2-} + H^+ \qquad pK \quad 7.20 \qquad (5.27)$$

$$HPO_4^{2-} \rightleftharpoons PO_4^{3-} + H^+ \qquad pK \quad 12.35 \qquad (5.28)$$

Over the pH range of most soils, the dominant orthophosphate anion in soil solution will be $H_2PO_4^-$ (Figure 12). Only in alkaline soils will HPO_4^{2-} be significant, and PO_4^{3-} ions are not important in soil solution.

Interactions between ions in solution can be considered in two ways. Long-range interactions, those >0.5 nm, are accounted for by the

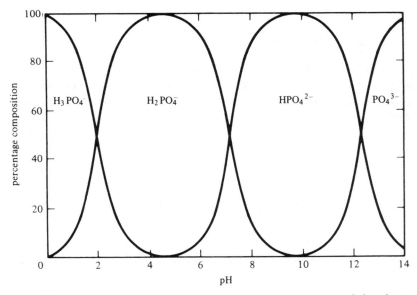

Figure 12 *A distribution diagram showing the percentage composition of phosphoric acid specified as a function of pH*
(Reproduced from Harrison, R.M. et al., Introductory Chemistry for the Environmental Sciences, Cambridge University Press, Cambridge, 1991.)

concept of ion activity (see also Chapter 3)

$$\{i\} = \gamma_i[i] \tag{5.29}$$

where
  $\{i\}$ = activity of the ion
  $[i]$ = concentration of ion (mol L^{-1})
  γ_i = activity coefficient.
  At infinite dilution, $\{i\} = [i]$, *i.e.* $\gamma_i = 1$. Ionic interaction increases with concentration and the square of the ionic charge, which is described by the ionic strength of the solution (I):

$$I = \frac{1}{2}\sum [i]\, z_i^2 \tag{5.30}$$

where

I = ionic strength
$[i]$ = molarity of each ion in soilution
z_i = charge on each ion in solution.
If all the ions in soil solution can be accounted for, I can be calculated, and then γ.

 The relationship between the activity coefficient of an ion and the ionic strength of the solution is given by the Debye–Hückel limiting law

$$\text{Log}_{10}\,\gamma_i = -0.5\, z_i^2\, \sqrt{I} \tag{5.31}$$

For soil solutions, the Davies equation, which is a modification of the Debye–Hückel equation, is commonly used, and is applicable to solutions up to approximately $I = 0.7$ mol L^{-1} (see also Section 3.2.1):

$$\text{Log}_{10}\,\gamma_i = -0.5\, z_i^2\left[\left(\sqrt{I}/\left(1 + \sqrt{I}\right)\right) - 0.3I\right] \tag{5.32}$$

Worked example 5.7 – soil solution chemistry: ionic strength and activity coefficients
The following analytical data were obtained for an isolated soil solution:
Ca^{2+}, 5.05 mmol L^{-1}; Mg^{2+}, 3.55 mmol L^{-1}; K^+, 0.68 mmol L^{-1}; Na^+, 1.80 mmol L^{-1}; NO_3^-, 3.70 mmol L^{-1}; HCO_3^-, 1.80 mmol L^{-1}; SO_4^{2-}, 6.25 mmol L^{-1}; and Cl^-, 1.40 mmol L^{-1}.

How can we tell from these figures that the main ionic species have been accounted for?
There are 19.68 mmol L^{-1} of monovalent positive charge and 19.40 mmol L^{-1} of monovalent negative charge.
What is the ionic strength of this solution?

$$I = 1/2[(5.05 \times 2^2) + (3.55 \times 2^2) + (0.68 \times 1^2) + (1.80 \times 1^2)$$
$$+ (3.70 \times 1^2) + (1.80 \times 1^2) + (6.25 \times 2^2) + (1.40 \times 1^2)]$$
$$= 1/2 \times 68.78$$
$$= 34.39 \text{ in terms of concentrations in mmol } L^{-1}$$

Ionic strength is expressed in terms of $mol\,L^{-1}$

$I = 0.0344$.
What is the activity coefficient of each ion?
By substituting $I = 0.0344$ into the Davies equation,
γ monovalent ions $= 0.843$,
γ divalent ions $= 0.504$.

Short-range (<0.5 nm) interactions between ions in solution result in the formation of ion pairs or complexes with other ions or with organic molecules. An ion pair is a transient entity formed by coulombic attraction between ions of opposite charge, with each ion retaining its own water of hydration. In a complex, the ions form covalent or coordinate bonds to form an entity with its own hydration shell.

An example of an ion pair could be between calcium and sulfate ions in solution:

$$Ca^{2+} + SO_4^{2-} \rightleftharpoons CaSO_4^0 \tag{5.33}$$

An example of a complex could be between copper ions and an organic molecule:

$$\tag{5.34}$$

Worked example 5.8 – activity of ions pairs in soil solution

In a soil solution the activity of $Ca^{2+} = 10^{-3}$ mol L^{-1} and SO_4^{2-} ions $= 5 \times 10^{-4}$ mol L^{-1}, what is the activity of the ion pair $CaSO_4^0$?

$$CaSO_4^0 \rightleftharpoons Ca^{2+} + SO_4^{2-} \quad K = 5.25 \times 10^{-3} \quad (5.35)$$

$$\frac{[Ca^{2+}][SO_4^{2-}]}{[CaSO_4^0]} = 5.25 \times 10^{-3}$$

$$[CaSO_4] = \frac{(10^{-3})(5 \times 10^{-4})}{5.25 \times 10^{-3}} = 9.5 \times 10^{-5} mol L^{-1}$$

Alternatively, this calculation can be done using logarithms.

$$CaSO_4^0 \rightleftharpoons Ca^{2+} + SO_4^{2-} \quad \log K = -2.28 \quad (5.36)$$

$$\log[CaSO_4] = \log[Ca^{2+}] + \log[SO_4^{2-}] - (-2.28)$$
$$= (-3.00) + (-3.30) - (-2.28) = -4.02$$
$$[CaSO_4] = 9.5 \times 10^{-5} mol L^{-1}$$

5.5.2 Ion Exchange (Physisorption)

Ion exchange is the process where ions held at a charged surface by coulombic bonding are exchangeable with other ions of the same charge in solution in contact with the surface. The reaction of major importance in soils is *cation exchange – i.e.* positively charged ions (cations) held at negatively charged surfaces. Anion exchange can occur, but is of minor importance.

The solid phase surfaces on which cation exchange occurs are:

Silicate clays	permanent negative charge on clay faces pH-dependent charge on clay edges
Humified organic matter	pH-dependent charge (always overall negative)
Hydrous oxides	pH-dependent charge (overall positive or negative)

The strength of the attractive force between the cation in solution and the negatively charged surface of the soil component is governed by Coulomb's law:

$$F = \frac{q'.q'' \text{ (constant)}}{D.r^2} \qquad (5.37)$$

Where

F is the attractive (or repulsive) force
q' and q'' are the charges on the ion and the surface
D the dielectric constant and
r the distance of separation between the two charges.

Thus, more highly charged ions are held at exchange surfaces in preference to lesser charged ions. Trivalent ions such as Al^{3+} and Fe^{3+} are most strongly held, then divalent ions such as Ca^{2+} and Mg^{2+}, with monovalent ions being the least strongly held. For ions of the same charge, the strength of binding depends on the size of the hydrated ion, which is the species that exists in soil solution, and hence how close to the surface it can approach. So for monovalent ions, the relative strength of binding, or affinity for the surface, is $Cs^+ > Rb^+ > K^+ > Na^+ > Li^+$. Ionic concentration must also be taken into account, as these preferences can be overcome by high concentrations of an ion in soil solution. For example, sodium ions would dominate the exchange sites in soil flooded by seawater.

Ions are considered to be held at the charged surface in a *diffuse double layer* (the Guoy–Chapman model) in which the concentration of cations falls, and that of anions increases, with distance from the surface. In a refinement of this, the Stern model introduces a layer of cations held directly on the surface of the soil component (Figure 13). Small cations, or dehydrated ions having lost their water of hydration, can sit at the surface and form a strong coulombic bond. This allows for the specificity described above.

The exchange process can be written as a chemical reaction, so, for example, for K^+ ions in soil solution exchanging with Ca^{2+} on the surface (X)

$$CaX + K^+ \rightleftharpoons KX + \frac{1}{2}Ca^{2+} \qquad (5.38)$$

At equilibrium, the ratio of the ions on the surface is related to their ratio in solution:

$$\frac{KX}{CaX} = k.\frac{(K^+)}{\sqrt{Ca^{2+}}} \qquad (5.39)$$

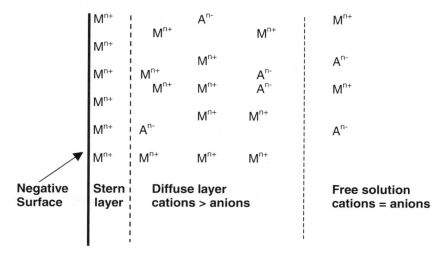

Figure 13 *The distribution of ions at a negatively charged surface of a soil component*

The exchange is between equivalents of charge. Strictly, ionic activity, not concentration, should be considered, but as ionic concentration in soil solution is so low, it is commonly used. This is known as the *Gapon equation*. This approach stresses the importance of the ion activity ratio, which tends to remain constant. As a result, monovalent ions are lost from a soil by leaching in preference to divalent and trivalent ions. A consequence is the acidifcation of soil by leaching (see Section 5.4).

5.5.2.1   Cation-Exchange Capacity (See Also Chapter 4).   The CEC of a soil is a measure of its ability to hold cations at negative sites by coulombic bonds. Because the exchange is by equivalents of charge, the units used for CEC are centimoles of monovalent charge per kilogram of soil ($cmol_c$ kg^{-1}). (Previously the units of milliequivalents per 100 g soil were used (meq 100 g^{-1}); the numerical value of CEC is the same whether expressed as $cmol_c$ kg^{-1} or meq 100 g^{-1}.) Typical values of CEC for the aluminosilicate clays given in Section 5.2 are shown in Table 6. These values for clay minerals are accounted for mainly by the permanent negative charge formed owing to isomorphous substitution, with only a minor component coming from the variable, pH-dependent charge on the clay edge. Humified organic matter has a CEC typically between 150 and 300 as $cmol_c$ kg^{-1}, all of which is pH dependent; the hydrous oxides also have a pH-dependent charge, but at most soil pHs the negative charge contributing to the CEC is small.

In a soil of between pH 6 and 7, about 80% of the cation-exchange sites hold Ca^{2+}, about 15% Mg^{2+}, K^+, and Na^+ account for up to 5%, with small amounts of NH_4^+ and H^+. As the pH falls, Al^{3+}, Mn^{2+}, and H^+ become relatively more important.

The *% base saturation* is a measure of the amount of the CEC of a soil holding the major cations (also called exchangeable bases) Ca^{2+}, Mg^{2+}, K^+, and Na^+.

$$\% \text{ base saturation} = \frac{\sum Ca^{2+}, Mg^{2+}, K^+, Na^+}{CEC} \times 100 \qquad (5.40)$$

where the units of exchangeable cations and CEC are $cmol_c \ kg^{-1}$.

% base saturation is high (approximately 80%) in fertile soils with basic minerals in the parent material and low (<30%) in highly leached acidic soils.

Worked example 5.9 – cation-exchange capacity and % base saturation
One method to measure the CEC of a soil is to leach a column containing a known weight of soil with a solution of a salt so that the exchange sites are saturated with the cation (index ion). After washing out the excess salt solution, a different salt solution is leached through the column and the exchangeable ions displaced, collected in a known volume and their concentration measured.

10 g soil was leached with 200 cm^3 of 1 mol L^{-1} CH_3COOK solution in order to saturate the exchange sites with K^+ ions. After washing out the excess CH_3COOK solution, 200 cm^3 of 1 mol L^{-1} CH_3COONH_4 solution was passed through to displace the K^+ ions by NH_4^+. The solution that leached through the soil was made up to 250 cm^3 in a volumetric flask. Ten centimetre cube of this solution was diluted to 100 cm^3 and the K^+ concentration in the diluted solution measured by flame photometry. If the K^+ concentration in this solution is 41 $\mu g \ cm^{-3}$, what is the CEC of the soil in $cmol_c \ kg^{-1}$?

Concentration of K^+ in the diluted solution

$= 41 \ \mu g \ cm^{-3}$

$= 410 \ \mu g \ cm^{-3}$ in the original 250 cm^3 of solution.

Weight of K in this solution $= 410\,\mu g\,cm^{-3} \times 250\,cm^3 = 102{,}500\,\mu g$ or 102.5 mg

$$\text{Atomic weight of K} = 39.1$$
$$102.5/39.1 = 2.62\,\text{mmol K in 10 g soil}$$
$$\text{or } 262\,\text{mmol kg}^{-1}$$
$$= 26.2\,\text{cmol kg}^{-1}$$

As K^+ is monovalent, this is 26.2 cmol of monovalent charge (cmol_c) kg^{-1} and equal to the CEC of the soil.

Concentrations of exchangeable cations in the above soil were: 3480 mg Ca kg^{-1}; 474 mg Mg kg^{-1}; 430 mg K kg^{-1}; and 161 mg Na kg^{-1}. What is their concentration in $\text{cmol}_c\,kg^{-1}$, and what is the % base saturation of the soil?

Atomic weights: Ca, 40; Mg, 24.3; K, 39.1; and Na, 23.

$$\text{Ca}: 3480/40 = 87\,\text{mmol Ca kg}^{-1}$$
$$= 8.7\,\text{cmol Ca kg}^{-1} = 17.4\,\text{cmol}_c\,\text{kg}^{-1}$$
$$\text{Mg}: 474/24.3 = 19.5\,\text{mmol Mg kg}^{-1}$$
$$= 1.95\,\text{cmol Mg kg}^{-1} = 3.9\,\text{cmol}_c\,\text{kg}^{-1}$$
$$\text{K}: 430/39.1 = 11\,\text{mmol K kg}^{-1}$$
$$= 1.1\,\text{cmol K kg}^{-1} = 1.1\,\text{cmol}_c\,\text{kg}^{-1}$$
$$\text{Na}: 161/23 = 7\,\text{mmol Na kg}^{-1}$$
$$= 0.7\,\text{cmol Na kg}^{-1} = 0.7\,\text{cmol}_c\,\text{kg}^{-1}$$
$$\% \text{ base saturation}$$
$$= [(17.4 + 3.9 + 1.1 + 0.7)/26.2] \times 100$$
$$= 88\%.$$

5.5.3 Ligand Exchange (Chemisorption)

Ligand exchange is the process by which ions are held at a solid phase surface by covalent bonding. The term *chemisorption* is often used to describe this process in order to distinguish it from *physisorption*, where coulombic bonds are involved and the process is ion exchange (see previous section).

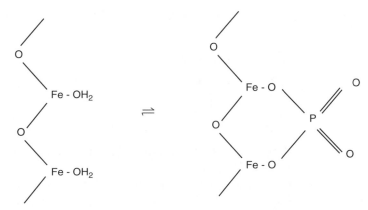

Figure 14 *Chemisorption of a phosphate anion on to an iron oxide surface by ligand exchange*

The important ions held in this way are certain anions, especially phosphate, and some trace metals; the main surfaces involved are the iron and aluminium oxides. Unlike ion exchange, the surface charge plays no part in ligand exchange, which can take place onto positive, uncharged, or negative surfaces, as these surfaces have a variable charge, which depends on pH (Figure 7). The ion being chemisorbed replaces a water molecule or hydroxyl ion (OH^-) from the surface (Figure 14).

Phosphate, silicate, borate, arsenate, selenite, chromate, and fluoride are anions for which ligand exchange is important. Nitrate, chloride, bromide, and perchlorate are not held, while sulfate and selenate may be weakly held. As a consequence, leaching of nitrate and sulfate from soil in drainage water can be significant, but very little phosphate is lost in solution. Of the trace metals, Co, Cu, Ni, and Pb are strongly held on oxide surfaces by chemisorption, but the process is much less important for Cd and Zn.

Chemisorption is often described by use of the Langmuir equation:

$$x/x_m = Kc/(1 + Kc) \qquad (5.41)$$

where

c = concentration of chemisorbed ion in solution at equilibrium
x = amount of the ion chemisorbed on to the soil
x_m = maximum amount of ion chemisorbed onto the soil
K = a constant related to bonding.

The equation is often used in the linearized form

$$c/x = c/x_m + 1/Kx_m \qquad (5.42)$$

where c can be measured and, knowing the initial concentration and volume of the solution containing the chemisorbed ion, x can be calculated. By plotting c/x against c, a straight line will result (if the Langmuir equation is a suitable model for the process), which will have a slope of $1/x_m$ and an intercept of $1/Kx_m$, allowing the theoretical maximum chemisorption of the ion to be calculated. In practice, application of the Langmuir equation is often not strictly valid, as a straight line relationship is not found. But it is commonly used in order to obtain a value for x_m.

5.5.4 Complexation/Chelation

Humic compounds can react with metal cations to form organometallic complexes:

$$M^{x+} + \text{OM-LHy} \rightleftharpoons \text{OM-LM}^{(x-y)+} + yH^+ \qquad (5.43)$$

These are particularly stable if more than one bond is formed (chelation) and are important for heavy metals such as Cu^{2+} and Pb^{2+}, and also for Fe^{3+} and Al^{3+} for certain soil-forming processes. Complexation by low molecular weight organic compounds, such as those in root exudates, can act as a means of bringing metals into solution, making them more mobile and bioavailable, but reaction with high molecular weight humic material is an immobilizing process.

The stability of the metal complex can be expressed as the equilibrium constant (see also Chapter 4). For example, if Equation (5.43) is simplified to

$$M + L \rightleftharpoons ML \qquad (5.44)$$

(where M is the metal ion, L the organic ligand, and ML the metal complex)

$$K = [ML]/[M][L] \qquad (5.45)$$

While experimentally [M] can be measured and [ML] calculated, it is difficult to quantify [L], given the polymeric and chemically diverse nature of soil organic matter.

5.5.5 Precipitation/Dissolution

The concentration of some ions in soil solution is controlled mainly by the presence of poorly soluble solid phase components. Good examples of this are iron and aluminium, both of which are highly insoluble at the pH of most soils, and which are controlled by the solubility of the hydrous oxides. Control of solution concentration by dissolution of an oxide does not apply to calcium, as the solid phases are too soluble. When calcium carbonate is present, which is insoluble, it can control the Ca^{2+} concentration in soil solution.

Worked example 5.10 – control of ion concentrations in soil solution by solubility of a solid phase

(a) Amorphous iron oxide is commonly found in soil and its dissolution to release Fe^{3+} ions can be represented by the equation

$$Fe(OH)_{3(amorph)} + 3H^+ \rightleftharpoons Fe^{3+} + 3H_2O \quad K = 10^{3.54} \quad (5.46)$$

$$[Fe^{3+}]/[H^+]^3 = 10^{3.54}$$

$$\log Fe^{3+} = 3.54 - 3pH$$

$$At\, pH\, 6,\ \log Fe^{3+} = -14.46,$$

$$[Fe^{3+}] = 3.5 \times 10^{-15}\, mol\, L^{-1}$$

$$At\, pH\, 3,\ \log Fe^{3+} = -5.46,$$

$$[Fe^{3+}] = 3.5 \times 10^{-6}\, mol\, L^{-1}$$

$$At\, pH\, 1,\ \log Fe^{3+} = 0.54,$$

$$[Fe^{3+}] = 3.5\, mol\, L^{-1}$$

agreeing well with the observed behaviour of Fe^{3+}, which is soluble only under very acid conditions. This can be confirmed by using $K_{sp} = 10^{-36}$ for $Fe(OH)_3$ in Worked example 5.5.

(b) Calcium carbonate dissolves to release Ca^{2+} ions:

$$CaCO_3 + 2H^+ \rightleftharpoons Ca^{2+} + CO_2 + H_2O \quad K = 10^{9.74} \quad (5.47)$$

$$Log\, Ca^{2+} + 2pH = 9.74 - \log CO_2$$

For a soil atmosphere with $pCO_2 = 10^{-2.5}$ (see Worked example 5.5)

$$Log\, Ca^{2+} = 12.24 - 2pH$$

$$At\, pH\, 7, \log Ca^{2+} = -1.76,$$

$$[Ca^{2+}] = 0.02\, mol\, L^{-1}$$

$$At\, pH\, 5, \log Ca^{2+} = 2.24,$$

$$[Ca^{2+}] = 200\, mol\, L^{-1}$$

5.5.6 Soil Processes

A soil receives inputs of ions from the weathering of parent material, mineralization of organic matter, pollution, and specific additions, such as fertilizers. The processes described above control the mobility, or bioavailability, of these ions (Figure 15). Certain ions, such as sodium and nitrate, are poorly held in soil, and so readily leached out. On the other hand, phosphate and copper are strongly held. From the point of view of soil as a source of plant nutrients, it is beneficial for a soil to hold sufficient nutrients to meet plant requirements, and to hold the nutrients in a form that can be released to supply the plant over the growing season. Increasingly, we are using soils as a sink for pollutants, when the desirable result is that the pollutant is tightly held in order to prevent its dispersal into the wider environment.

QUESTIONS

(i) Explain the origin and characteristics of the charges on soil components that make up the CEC of a soil.

(ii) Using the information in Figure 8, how much N, P, and S are stored per hectare in the soil humus? How much N, P, and S are released annually per hectare by mineralization of the humus?

It is commonly found that about twice as much P and S is released from the soil than predicted, but that the amount of N agrees well with the prediction. What process could account for this?

A soil solution contains 3×10^{-6} mol PO_4-P L^{-1}, 5×10^{-4} mol SO_4-S L^{-1}, and 2×10^{-3} mol NO_3-N L^{-1}. If the annual drainage is 250 mm of water, how much P, S, and N are lost annually per hectare by leaching? ($P = 31$; $S = 32$; and $N = 14$).

(iii) A typical rate of addition of Pb in sewage sludge to agricultural soils in the UK is 1200 g Pb ha^{-1} $year^{-1}$. If this amount is added to a soil of bulk density 1.2 g cm^{-3} containing 5% humus with a binding capacity of 80 $cmol_c$ kg^{-1}, what percentage of the binding capacity to a depth of 20 cm is used annually to hold the lead? (Assume that all the added Pb is bound to the humus. $Pb = 207$.)

(iv) Using the values for solubility product below, and assuming that the solubility of the hydroxides controls the metal ion concentration in solution, calculate the theoretical concentrations of Fe^{2+}, Fe^{3+}, and Cr^{3+} at pH 2, 5, and 7.

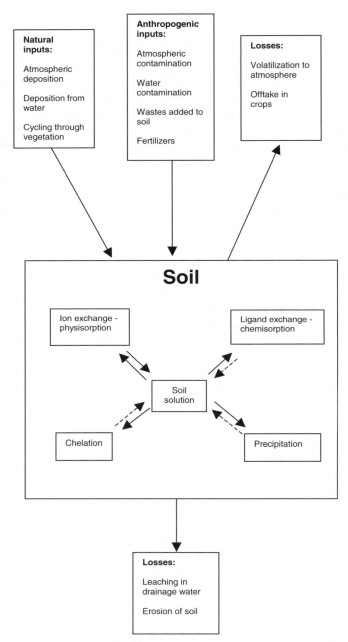

Figure 15   *An overview of soil processes*

$$K_{sp} \; Fe(OH)_2 = 10^{-14}; \; Fe(OH)_3 = 10^{-36}; \; Cr(OH)_3 = 10^{-33}$$

What does this tell you about the solubility of these ions? Does this agree with their observed behaviour in soil?

(v) A soil with a bulk density of $1.2 \; g \; cm^{-3}$ contains 1% FeS_2. How much $CaCO_3$ is required per hectare to neutralize the acid produced to a depth of 20 cm? (Assume that the FeS_2 oxidizes completely. Fe = 56; S = 32; Ca = 40; C = 12; and O = 16.). (Hint: look at Worked example 5.1 and combine Equations (5.16) and (5.17) to get an overall reaction for the oxidation of pyrite.)

(vi) Iron oxides in soil undergo reduction to produce Fe^{2+} ions-
$$Fe(OH)_3 + 3H^+ + e^- \rightleftharpoons Fe^{2+} + 3H_2O \quad E^0 = 1.057 \; V$$
What is the concentration of Fe^{2+} ions in solution in a soil of pH 7 at 100 and 0 mV?

(vii) 1.00 g samples of a soil were shaken with 50 cm^3 of different solutions of known zinc concentration. After 1 h, the soil suspensions were filtered and the concentration of zinc in the filtrate measured, with the following results:

Initial concentration (mg Zn L^{-1})	Final concentration (mg Zn L^{-1})
7.5	1
10	2
20	5
30	10
45	21
60	34

In each case, calculate the weight of zinc held by 1 g of soil. Plot the graph of weight of Zn held per gram (y-axis) against the final concentration of Zn.

Fit the data to the Langmuir equation and plot the graph. Obtain values for x_m and K.

(viii) The equilibrium between Fe^{3+} ions in soil solution and iron oxides in soil has been described by the equation

$$Fe(OH)_{3 \; soil} + 3H^+ \rightleftharpoons Fe^{3+} + 3H_2O \quad \log K = 2.70$$

(Note, this is a different form of iron oxide to that used in Worked example 5.10, hence a different value for K).

Derive the relationship between the concentration of Fe^{3+} ions in solution and pH.

Use this equation to plot the concentration of Fe^{3+} over the pH range 1–9.

How well does this information agree with the Fe^{3+} concentrations calculated in Question 4?

(ix) What properties of a soil would you investigate in order to learn about its ability to hold and supply P and K to crops?

(x) What properties of soil would you investigate in order to learn about the ability of a soil to act as a sink for pollutants?

REFERENCES

1. S.A. Barber, *Soil Nutrient Availability*, Wiley, New York, 1995.
2. J.B. Dixon and D.G. Schulze (eds), *Soil Mineralogy with Environmental Applications*, Soil Science Society of America, Madison, WI, 2002.
3. D.J. Greenland and M.H.B. Hayes (eds), *The Chemistry of Soil Constituents*, Wiley, Chichester, 1978.
4. D.J. Greenland and M.H.B. Hayes (eds), *The Chemistry of Soil Processes*, Wiley, Chichester, 1981.
5. W.L. Lindsay, *Chemical Equilibria in Soils*, Wiley, New York, 1979.
6. M.B. McBride, *Environmental Chemistry of Soils*, Oxford University Press, New York, 1994.
7. R.G. McLaren and K.C. Cameron, *Soil Science*, 2nd edn, Oxford University Press, Auckland, 1997.
8. J.J. Mortvedt, F.R. Cox, L.M. Shuman and R.M. Welch, *Micronutrients in Agriculture*, 2nd edn, Soil Science Society of America, Madison, WI, 1991.
9. D.L. Rowell, *Soil Science: Methods and Applications*, Longman Scientific & Technical, Harlow, 1994.
10. G. Sposito, *The Chemistry of Soils*, Oxford University Press, New York, 1989.
11. F.J. Stevenson, *Humus Chemistry*, Wiley, New York, 1982.
12. R.E. White, *Principles and Practice of Soil Science*, 3rd edn, Blackwell Science Ltd., Oxford, 1997.
13. A. Wild (ed), *Russell's Soil Conditions and Plant Growth*, 11th edn, Longman Scientific & Technical, Harlow, 1988.

CHAPTER 6

Environmental Organic Chemistry

CRISPIN J. HALSALL

Environmental Science Department, Lancaster University, Lancaster LA1 4YQ, UK

6.1 INTRODUCTION

Environmental Organic Chemistry (EOC) is an exciting branch of chemistry that has developed over the last few decades, and is probably best known for the study into the fate and behaviour of anthropogenic (man-made) pollutants. EOC, however, cannot be restricted to organic pollutants and their behaviour, but also encompasses the myriad of organic molecules produced by nature and their associated biogeochemistry. This area of chemistry is coupled to the recent advances in analytical techniques, particularly improvements in chromatography and the rise in 'hyphenated' techniques, such as GC-MS and LC-MS. The sensitivity of contemporary bench-top instruments allows detection of analytes in environmental samples, down to as low as a few picogrammes (pg) $(10^{-12}$ g) on the chromatography column. Recently, the improvements in LC-MS technology, has resulted in affordable, yet sensitive instrumentation that allows separation and detection of a mixture of polar, water-soluble analytes. The use of LC avoids derivatisation steps of highly polar analytes required as a pre-requisite to GC analysis. The different sources now available to LC-MS, including atmospheric pressure photoionisation and chemical ionisation, allows a high degree of selectivity for a wide number of analytes of varying molecular weights, polarity and aqueous solubility. This enables analysis of pollutant degradates and metabolites in aqueous systems and has led to the screening of pharmaceuticals, veterinary drugs and antibiotics in both surface and ground water.[1-3]

The traditional organic chemist often works in a highly controlled environment, where variables may be reduced to several reactants under carefully controlled temperatures and pressures; the goal to produce a product of the highest possible yield. The EOC, on the other hand, has to understand the chemistry and reactions of an organic molecule but under earth's near-surface conditions, with varying temperature ($-50\ ^\circ$C to $+50\ ^\circ$C) and atmospheric pressures (950 to > 1000 hectoPa (hPa)). In addition, the laboratory analyst in a chemical company, has the job of qualifying the purity of products to some standard specified by the customer. The environmental chemist however, has the difficult task of extracting, purifying and quantifying the same chemicals, their un-wanted by-products (and possibly their degradates) from environmental media (*e.g.* soil, water, biota, *etc.*) to a high degree of precision and accuracy. Measuring specific chemicals in different environmental media and understanding how the chemistry in these media 'fits together' presents a considerable challenge, and often results in the environmental chemist returning to the laboratory to simulate a particular process.

This chapter will explore aspects of EOC in relation to the environ-mental behaviour and fate of anthropogenic pollutants and will explore the physical and chemical processes that result in their environmental partitioning and degradation; specifically photodegradation for the latter. In each case practical examples and measurement techniques will be presented and illustrated.

6.2 THE DIVERSITY OF ORGANIC COMPOUNDS

The wide array of organic compounds subject to research within EOC, encompass a broad range of physical–chemical properties, but can be categorised according to either molecular weight, volatility and/or reactivity. Figure 1 provides examples of a range of molecules that have vapour pressures spanning orders of magnitude, and are subject to research because of their interesting chemistry either in the atmosphere, aquatic systems and/or soil and sediments.

Many of the more volatile compounds (C_1-C_6) emitted into the atmos-phere have an important impact on atmospheric photochemical processes. Oxidation of short-chain alkanes largely through reaction with the OH radical, form short-lived peroxy ($HO_2^{\cdot}$) and alkoxy radicals ($RO_2.$) that are important for converting NO to NO_2 in the polluted atmosphere and hence allowing the build up of ground-level ozone (O_3) during warm sunny weather (see Chapter 2). For higher molecular weight compounds ($\geq C_5$) the isomerisation of alkoxy radical intermediates gives rise to hydroxyl carbonyl products containing $-OH$, $-COOH$ and $-C{=}O$ groups. These products comprise the major fractions of organic aerosol, and a growing body of research literature is dedicated to understanding the oxidation products and reaction mechanisms of vapour hydrocarbons to less volatile,

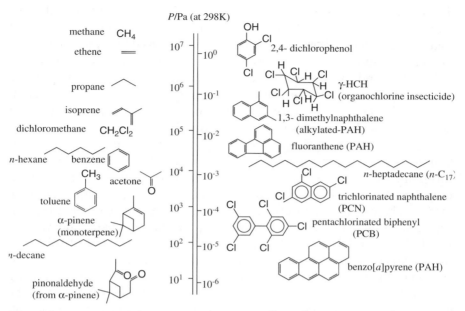

Figure 1 *Approximate vapour pressures (P) (10^7–10^{-6} Pa) for a diverse range of organic compounds, many of which are mentioned in this chapter. Note the use of common/trivial chemical names rather than IUPAC nomenclature*

more polar condensates involved in aerosol formation and growth.[4] Monoterpenes (C_{10}) released from different types of vegetation, contain unsaturated double bonds and are therefore susceptible to reaction with O_3. The reaction of α-pinene, for example, gives rise to a number of carbonyl and mono- and dicarboxylic acids, including pinonaldehyde and pinic acid which have been detected in forest aerosols.[5]

Semi-volatile organic compounds arise from a multitude of sources, both anthropogenic and natural, but include the persistent organic pollutants, many of which are halogenated aromatic compounds found at trace levels in the environment, but are dispersed widely across the globe. For aliphatic compounds such as petroleum hydrocarbons, the semi-volatile constituents include longer chain alkanes ($>C_{15}$), both normal (*n*-alkanes), branched (*iso*-alkanes) and cyclic alkanes (*cyclo*-alkanes or naphthenes) as well as a range of polycyclic compounds or terpanes including tricyclic carpanes, tetracyclic hopanes and pentacyclic steranes. Many of the hopanes and steranes are largely involatile, and in the atmosphere are predominantly associated with particles.[6] They are important constituents of crude oils, arising principally from the decay of bacterial/biotic matter and serve as useful biomarkers for different fuel types, and are also widespread in sediments and soils from both biogenic and petrogenic sources. The semi-volatile polycyclic aromatic hydrocarbons (PAHs) and their alkylated derivatives are generally less abundant than the alkanes and terpanes of similar carbon number, but are present in oil and liquid fuels and are also produced during combustion processes (pyrolysis), and are present in soil and sediments from both petrogenic and pyrolytic sources. The five-ringed benzo[*a*]pyrene is a notable example, and is classed as a probable human carcinogen, present in urban environments chiefly from the combustion of fossil fuels.

The involatile, complex polymeric organic matter present in soils and sediments is commonly referred to as 'humic substances', and probably forms through a series of oxidation and condensation reactions between polyphenols, polysaccharides and polyamino acids of plant and microbial origin.[7] There is no commonly accepted structure of humic substances, with molecular weights and molecular formulae open to debate, and varying widely between different soil and sediment types. Extraction of soils with water and/or organic solvents generally yield only a small fraction of soil-carbon present as humic substances, but the accepted practice of extracting soils with aqueous alkali (0.5% NaOH), followed by centrifugation of the soil/sediment suspension with subsequent acidification of the supernatant, yields the acid-insoluble humic fraction known as humic acid. Any organic material left in solution is referred to as fulvic acid and the remaining insoluble material is referred to as humin.[8] These operationally defined fractions prove very useful for conducting sorption and fate experiments for organic pollutants present in soil/sediment systems, where humic substances play an important role in the soil chemistry of organic compounds. Humic and fulvic acids give rise to dissolved organic material in water, and are complex organic macromolecules with a wide variety of functional groups, primarily carboxylic and phenolic groups, and to a lesser extent amino (*e.g.* -NH$_2$) and thiol (-SH) groups. These macromolecules have a wide range of molecular weights ($\approx <500$ to $>10,000$ g mol^{-1}) and various humic acid fractions have been separated by gel-permeation chromatography (or size-exclusion chromatography) with their structural and chemical characteristics investigated using ^{1}H- and ^{13}C-NMR, and pyrolysis-GC-MS.[9] The large surface areas of these molecules provide ideal surfaces for sorption by other organic molecules, including semi-volatile organic contaminants such as PAHs and pesticides, providing active sites for surface-based, heterogeneous chemical reactions.

6.2.1 Identifying Sources of Hydrocarbons

Distinguishing between different sources of hydrocarbons including alkanes and PAHs for example, is of fundamental importance for source apportionment of pollution, as well as in petroleum geochemistry to indicate the origin and thermodynamic maturity of organic matter in sediments. A commonly used index to broadly distinguish between anthropogenic and biogenic sources of hydrocarbons, present in any environmental compartment, is the Carbon Preference Index (CPI). CPI is usually applied to the homologous series of semi-volatile n-alkanes with carbon chain lengths ranging from $\sim C_{15}$ to C_{35}, where the CPI is simply the concentration ratio of the summed 'odd-numbered' chain lengths over the 'even-numbered' chain lengths over a specified chain-length range. The premise here is that recent biogenically derived n-alkanes (*e.g.* plant waxes, biota exudates) tend to favour odd-carbon numbered chain lengths (typically within the range C_{25} to C_{35}) whereas aged organic material, such as petrogenic hydrocarbons that have undergone ageing over geological timescales generally do not exhibit this odd-carbon number preference. Therefore, environmental samples with a CPI > 1, suggest that the aliphatic hydrocarbons, have arisen through biogenic input, whereas a CPI of < 1, suggests petrogenic (*e.g.* fossil fuel) sources. As an example, suspended particle matter (SPM) filtered from seawater samples collected in the Black Sea, showed high CPI values of 3.81–13.1 in offshore locations, where the SPM was considered to be of algal origin (where $CPI_{marine} = \Sigma C_{15-25}/\Sigma C_{14-24}$).[10] Much lower CPI values were observed closer to the Danube estuary, and were indicative of petrogenic derived hydrocarbons, presumably due to river discharges of petroleum products. There are other useful aliphatic indices or markers that are used in conjunction with the CPI to provide evidence of oil 'weathering' or to apportion carbon sources as either petrogenic or of recent biogenic origin. The ubiquitous and rather persistent branched alkanes, pristane (C_{19}) and phytane (C_{20}), are often ratioed to n-heptadecane (C_{17}) and n-octadecane (C_{18}) respectively, as supporting evidence of either biogenically derived or petrogenic hydrocarbons to sediments and aqueous systems (both pristine and phytane closely elute with these two n-alkanes on a low-polarity GC-column). For instance, a high n-C_{17}/pristane ratio (> 10), would indicate biogenically derived hydrocarbons, whereas a low ratio (< 3) would indicate a carbon source depleted in n-C_{17}, perhaps indicative of petrogenic carbon sources.

PAHs and their alkylated derivatives are also used to distinguish between pyrolytic and petrogenic inputs to sediments and soils and are often used in conjunction with aliphatic indices to discriminate hydrocarbon sources. The lower molecular weight 2- to 4-ringed PAHs and their alkylated versions comprise a small, but significant, component of crude oils and liquid fuels. Ratios of parent PAHs (non-alkylated) as well as parent to alkylated versions are frequently used to distinguish petrogenic sources vs. pyrolytic sources (both natural and anthropogenic) for aromatic hydrocarbons in soil/sediment systems.[11]

6.3 THE FATE OF ORGANIC CONTAMINANTS

There are thousands of organic chemicals entered into the international databases as part of legislative programmes aimed at assessing the risk they pose to human health and the environment. For example, in the European Union, the European Inventory for Existing Chemical Substances and the European List of Notified Chemical Substances include over 100,000 commercial substances.[12] Similarly in the US, the Office for Pollution Prevention and Toxics (OPPT) as part of the US-Environmental Protection Agency (EPA) maintains lists of high-production volume chemicals as well as persistent bioaccumlative toxins (PBTs). The increasing demand for risk assessment to examine substances with respect to their environmental persistence, ability to bioaccumulate and potential for toxicity, increases the demand for reliable physical–chemical property data, upon which to conduct environmental fate and degradation studies. In this respect, a risk assessment is conducted to ascertain the likelihood or probability of some harm arising to either humans or the environment due to exposure to that chemical substance. In order to assess exposure and understand chemical fate, then intimate knowledge of both the chemical use (quantity released) and behaviour (environmental transport, partitioning and transformation/degradation) are required. As Schwarzenbach *et al.*[13] have so elegantly described in their comprehensive text on EOC; organic chemicals once present in the environment are subject to essentially two types of process. The first involves their environmental transport within, and transfer between, environmental phases and the second involves their chemical alteration (transformation/degradation) driven by chemical and/or biological processes. In reality, within any given environmental system, these processes may occur simultaneously and different processes may strongly influence one another.

6.4 CHEMICAL PARTITIONING

The transfer of chemicals between two or more environmental compartments or phases can be described by equilibrium partitioning, and knowledge of this partitioning is essential for understanding and describing chemical fate in the environment. Chemical partitioning takes place between adjacent phases such as between a solid and a liquid (dissolution), a liquid and a gas (volatilisation), a solution and a solid surface (adsorption) or a solution and an immiscible liquid (solvent partitioning). The net transport of an organic compound from one phase to another (*e.g.* air to water) is limited by equilibrium constraints and can be quantified according to a partitioning constant or coefficient (K). This is simply defined as the ratio of the concentration (C) of the chemical present in each phase (*e.g.* 1 and 2) at equilibrium; the point at which the chemical potential in each phase is equal and there is no longer net transfer in one direction. K can be represented as

$$K_{1,2} = \frac{C_1}{C_2} \tag{6.1}$$

In strict terms, K is expressed as the ratio of the mole fractions (x) of a chemical within each phase (for example, this could be the mole fraction of a chemical present in a gaseous or liquid mixture). However, in environmental chemistry, the common way of expressing chemical concentration is not by mole fractions but as a molar concentration (mol L^{-1} or molarity, M), although it is worth pointing out that the volume (L or dm^3) may vary according to changes in temperature and pressure and should be normalised to standard temperature (298 K) and pressure (101,325 Pa (or 1 atm)). Furthermore, when dealing with one phase that is solid, it is often more appropriate to deal with concentrations as amount per mass to avoid having to estimate phase densities. A mole fraction can be converted to molar concentration (M) according to

$$M = \frac{x(mol/mol)}{V(L/mol)} \tag{6.2}$$

where V is the molar volume of the mixture or solution. For aqueous solutions relevant for environmental scenarios, it is possible to ignore the contribution of the organic chemical to the molar volume of the mixture due to the typically low concentrations of a chemical pollutant present in a body of water. In this case, V is simply set as the molar volume of water ($V_w = 0.018$ L mol^{-1} at 25 °C).[13]

The term partitioning constant or coefficient refers to one chemical species in each phase (for example, ionisable chemicals present in water may exist in both neutral and dissociated forms and therefore each would have a separate partitioning coefficient). For well-defined phases, such as pure water or the pure liquid state of the chemical, then partitioning with another equally well-defined phase such as air results in the use of the term partitioning *constant*. The Henry's Law constant describing chemical partitioning between pure air and water is one example. Distribution ratios on the other hand, such as the soil-water distribution ratio (K_d), describe partitioning between a heterogeneous solid phase (soil) and soil-water (containing a variety of dissolved solids and organic material). Furthermore, a distribution ratio accounts for all the speciated forms of a particular chemical, as well as the various different sorption mechanisms and correctly describes the *distribution* of a chemical rather than a particular partitioning process. An example of a useful partitioning coefficient applied to soils and sediments is the organic-carbon to water partitioning coefficient (K_{oc}), which specifically describes chemical partitioning between the organic-carbon fraction of a soil or sediment and water and is frequently used to describe pesticide fate in agricultural soils.

6.4.1 Important Partitioning Coefficients

There are several important partitioning coefficients, besides the fundamental properties of vapour pressure and aqueous solubility, which are essential for understanding chemical transfer in the environment and these are illustrated in Figure 2 and include the air/water (K_{aw}), *n*-octanol/water (K_{ow}) and *n*-octanol/air (K_{oa}) partitioning coefficients. Indeed, chemical vapour pressure (the 'solubility' of a chemical in air) and aqueous solubility explain the 'partitioning' of a chemical from the pure-liquid or solid-state into air and water, respectively. Transfer of either vapour or dissolved chemical to other environmental compartments, media or phases can be explained by specific partition coefficients.

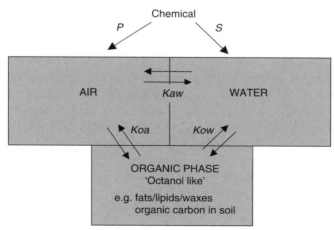

Figure 2 *Key physical–chemical properties that control the environmental partitioning of an organic chemical. The fundamental properties of vapour pressure (P) and aqueous solubility (S) determine the partitioning from the pure liquid or pure solid-state of the chemical into air and water, respectively*

6.4.1.1 Octanol-Water Partitioning Coefficient (K_{ow}). Octanol serves as a very useful surrogate of fats, lipids and organic phases present in the environment. This is largely attributable to octanol having a similar carbon to oxygen ratio (8:1) as fatty acids and lipids. As a result K_{ow} is used to represent chemical partitioning between lipids and water (where lipids and fats can be viewed as 'octanol-like solvents') and is an important chemical property used in the fields of biology and pharmacy, where K_{ow} is represented by the term '*P*' for *p*artitioning coefficient. In essence, K_{ow} is the ratio of the equilibrium concentration of a chemical between octanol (mol m^{-3}) and water (mol m^{-3}) and is a measure of a chemical's 'water hating' tendency or hydrophobicity. As Mackay[14] points out, most organic chemicals have a fairly constant solubility in octanol (200–2000 mol m^{-3}) but very different solubility in water and it may be misleading to describe K_{ow} as a chemical's 'love of fats' or lipophilicity. As the range of K_{ow} for environmental contaminants is large (over orders of magnitude), log K_{ow} values are often reported.

Perhaps the most common application of K_{ow} is for predicting the bioconcentration of low-polarity hydrophobic contaminants from water into aquatic biota. A bioconcentration factor (BCF) can be represented as

$$BCF = \frac{C_{\text{org}}}{C_{\text{w}}} \qquad (6.3)$$

where C_{org} is the actual concentration determined in an organism (*e.g.* fish), or some 'compartment' of the organism (*e.g.* lipids), and C_{w} is the truly dissolved concentration of the chemical in water. From empirical studies that have examined a range of different organisms and chemicals, BCF may be predicted from simple linear relationships with K_{ow}. For example, Mackay[15] described the BCF in fish, by examining hundreds of organic chemicals, as

$$BCF = 0.048 \, K_{ow} \qquad (6.4)$$

The constant 0.048 representing the $\approx 5\%$ lipid content of fish, or as Mackay describes, about 5% octanol by volume. However, relationships between BCF and K_{ow} appear to break down for chemicals with log $K_{ow} > 7$, as these 'superhydrophobic' substances may have difficulty crossing cellular membranes (*i.e.* they have a low bioavailability) or are rapidly egested from the organism. There is, in fact, little consensus on the reasons for the breakdown in the linear relationship between BCF and chemicals with log $K_{ow} > 7$, but both Borgå *et al.*[16] and Boethling *et al.*[17] point out that the experimental artefacts associated with measuring the freely dissolved chemical are likely to provide large uncertainties in the BCFs. Furthermore, bioconcentration is unlikely to be a significant exposure pathway for these 'superhydrophobic' chemicals relative to dietary intake.

K_{ow} is also used to describe partitioning to general organic phases within the environment and K_{oc} (introduced above) can be directly estimated from K_{ow} according to the empirically derived relationship derived initially by Karickhoff[18] and refined by Seth *et al.*[19]

where

$$K_{oc} = 0.35\, K_{ow} \tag{6.5}$$

this relationship is useful for calculating the sorption of chemicals to soils and sediments from water; specifically to the organic carbon fraction. It is worth noting that K_{oc} represents a solid/liquid partition coefficient and as such has units of L kg^{-1}, whereby K_{oc} can be represented as

$$K_{oc} = \frac{C_{oc}(moles/kg)}{C_{w}(moles/L)} \tag{6.6}$$

where C_{oc} is the chemical concentration sorbed to the natural organic carbon fraction in soil (~ 0.02 or 2%) and C_{w} is the dissolved water concentration. As K_{ow} is dimensionless, the constant in Equation 6.5 has units of L kg^{-1}. Experimentally, K_{oc} is determined by introducing (or 'spiking') a chemical into a water/soil mixture (slurry) that is stirred at a constant temperature, followed by filtration and analysis of the water at different time intervals. The resulting K_{d} value is then divided by the organic carbon fraction (f_{oc}) of the test soil, whereby

$$K_{oc} = \frac{K_{d}}{f_{oc}} \tag{6.7}$$

Multiplying K_{d} (which also has units of L kg^{-1}) by soil density (~ 2.4 kg L^{-1}) allows the dimensionless form of K_{oc} to be used.

6.4.1.2 Henry's Law Constant and Air-Water Partitioning (K_{aw}). Air-water partitioning is described by the Henry's Law constant (H), which is defined as the ratio of the partial pressure (p) of a chemical in the air to its mole fraction (or molar concentration) dissolved in water (C_{w}) at equilibrium, according to

$$H = \frac{p(Pa)}{C_{w}(mol/m^{3})} \tag{6.8}$$

H therefore has the units of Pa m^{-3} mol^{-1} and can be estimated by ratioing a chemical's liquid-phase saturation vapour pressure (usually the sub-cooled liquid vapour pressure) over its aqueous solubility, at some reference temperature. Note that the vapour pressure (P) can be readily converted into 'solubility' with units of molar concentration by applying the ideal gas law

$$PV = nRT \tag{6.9}$$

Rearranging this equation to n/V (mol m^{-3}) $= P/RT$, where RT is the gas constant – temperature product (8.314 Pa m^3 mol^{-1} K $\times$ K), allows the following expression

$$\frac{P}{RT \cdot C_{\mathrm{w}}} = \frac{C_{\mathrm{a}}}{C_{\mathrm{w}}} = H' \text{ or } K_{\mathrm{aw}} \tag{6.10}$$

where P/RT is the air concentration (C_{a}) with the same units as C_{w}, and H' is the 'dimensionless' H or K_{aw} and can also be obtained by simply dividing H by RT (note that the units cancel), where T is the relevant environmental temperature in K.

Like K_{ow}, values of H vary tremendously for the range of organic compounds of interest in the environment. For example, short-chain alkanes with their high vapour pressures and low aqueous solubilities have values of H that are orders of magnitude higher than alcohols, which are highly soluble in water and have lower vapour pressures. For certain pesticides and other semi-volatile organic compounds, which may persist in the environment, their lower vapour pressures generally result in relatively low values of H, but often their very low solubility can result in higher than expected values of H. As a consequence, these chemicals (particularly sparingly soluble organochlorine pesticides) appreciably partition to air from bodies of water in which they may be residing (*i.e.* run-off from agricultural regions). In addition, they will not be efficiently removed from the atmosphere by rainfall and as a result may be transported long-distances by the prevailing wind. Table 1 provides examples of select chemicals and their calculated H values using both saturated vapour pressure and aqueous solubility data. Clearly, *n*-hexane has the highest value of H out of this group of chemicals, but interestingly H for benzene is 557, $\sim$200-fold greater than DDT and $\sim$4000-fold greater than chlorpyrifos, yet both pesticides are relatively involatile, with vapour pressures some 600 million and 85 million times less than benzene, respectively. The low solubility of these pesticides ensure that H is higher than may be expected and both pesticides partition appreciably out of water.

Table 1   *Select physical–chemical property data for four relatively volatile chemicals and two semi-volatile insecticides: p,p'-DDT and chlorpyrifos (a currently used pesticide). Note that the chemicals are arranged in order of volatility (Data sourced from Mackay[14])*

Chemical	Molecular mass (g mol^{-1})	Vapour pressure (P) (Pa)	Aq. solubility (S) (mol m^{-3})	Henry's Law constant (H) (Pa m^{-3} mol^{-1})
Chloroform	119.4	23,080	68.7	336
n-hexane	86.2	20,200	0.11	18,364
Benzene	78.1	12,700	22.8	557
Phenol	94.1	70.6	871	0.081
Chlorpyrifos	350.6	0.00015	0.0011	0.14
p,p'-DDT	354.5	0.00002	0.0000087	2.3

Temperature is a key parameter for controlling H or K_{aw}, thereby strongly influencing air-water partitioning of chemicals in the environment. A good deal of effort is expended in deducing temperature-dependent H for contaminants of concern, which is not an easy task for low-polarity hydrophobic organic chemicals (HOCs) with poor solubility.

Figure 3 shows a schematic diagram of the typical apparatus used to measure H for these types of chemical. In essence the major component is the reactor vessel, which contains a large volume of deionised water (~ 0.5–10 L) spiked with a known concentration of the HOC in question. The chemical is introduced by first dissolving the chemical in an organic solvent which is miscible with water (*e.g.* acetone), and an aliquot added to the water in the reactor vessel. Care must be taken that the concentration does not exceed the aqueous solubility of the chemical and usually the concentration in the vessel is only a few percent of the aqueous solubility, resulting in a very low concentration, ~ 1–2 µg L^{-1}. Clean, water-saturated air (or N$_2$) is then bubbled or sparged through the reactor vessel and the exiting air passed through a vapour-trap to collect the chemical present in the air stream. The chemical quantity on the vapour trap can be turned into a time-averaged air concentration (C_a), by calculating the volume of air that has passed through the trap. H is calculated by ratioing C_a to the average water concentration in the reactor vessel (C_w) (see Equation 6.10) determined from analysis of water samples taken at the start and end of the experiment. Each gas-stripping run will be replicated to assess the precision, and new experiments conducted at different temperatures to provide the regression parameters to derive temperature-dependent values of H. Artefacts are always present, not least keeping the reactor vessel at a uniform temperature, but also preventing water from condensing in the vapour trap (particularly at

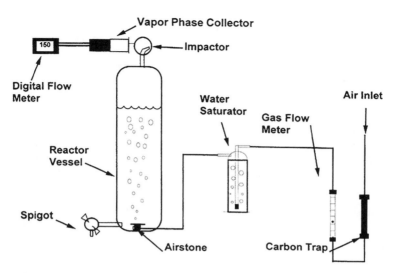

Figure 3 *Schematic diagram of the gas-stripping apparatus used to measure temperature-dependent Henry's Law constants (H) for semi-volatile organic compounds (SVOCs)*
From Bamford et al.[20] courtesy of SETAC press

higher temperatures), and accounting for the loss of water through evaporation. Furthermore, the change in water volume in the reactor vessel when extracting aliquots for chemical analysis should also be taken into account[21] and efforts directed at shielding the apparatus from laboratory light when determining H for light-sensitive chemicals.

6.4.1.3 Octanol-Air Partitioning Coefficient (K_{oa}). Partitioning of a chemical between some organic phase and air is often described using K_{oa}. Examples of terrestrial organic phases include organic carbon in soil, the waxy cuticle and lipid portion of vegetation and the organic film that coats atmospheric particle matter and partitioning between these phases and air for a number of chemical groups have been successfully described using K_{oa}.[22–24] Laboratory-based measurements of temperature-dependent K_{oa} values have been derived for a number of semi-volatile chemical groups, including PCBs[25,26] PAHs,[27] polychlorinated naphthalenes (PCNs)[27] and organochlorine pesticides.[28] K_{oa} can be described as the ratio of the chemical (solute) concentration in octanol to the concentration in air at equilibrium, represented as

$$K_{oa} = \frac{C_o}{C_a} \qquad (6.11)$$

where C_o and C_a represent the equilibrium concentrations in octanol and air respectively (mol m^{-3}) and hence K_{oa} is dimensionless. K_{oa} may be estimated as the ratio of K_{ow} over K_{aw} i.e. $K_{oa}=K_{ow}/K_{aw}$. As $K_{aw}=H/RT$ then K_{oa} can be expressed as

$$K_{oa} = \frac{K_{ow}RT}{H} \qquad (6.12)$$

As there are very few direct measurements of K_{oa}, application of Equation 6.12 is useful, although this approach may have limited utility, as reliable values of H and K_{ow} for a particular chemical may be scare. Furthermore, the validity of Equation 6.12 is brought into question when comparing calculated K_{oa} values with actual measurements[26] and extensive additional measurements of K_{oa} are required for a wide number of chemicals that occur in the environment. As with K_{ow} and K_{aw}, values of K_{oa} may vary over orders of magnitude within any particular chemical class, and therefore K_{oa} is often reported as a log value.

6.4.2 Temperature Dependence

Before proceeding, it is worth turning briefly to thermodynamics to provide some fundamental aspects to partitioning coefficients and, importantly, their relationship to temperature. Wide variations in temperature experienced in the environment can greatly alter the partitioning behaviour of a chemical, and this will be demonstrated later when comparing chemical behaviour between different latitudes with notable temperature differences.

As with any chemical reaction, movement of a chemical from one phase to another (i.e. liquid to gas) results in a change in the energy status of the system. A partitioning coefficient is related to the (Gibbs) free energy of transfer of the chemical between the two phases (ΔG, J mol^{-1}). Where ΔG relates both the enthalpic and entropic effects that result from the changes occurring in either phase (i.e. intermolecular interactions), through both the removal and addition of molecules of the chemical. The relationship between ΔG and the change in enthalpy (or heat energy) of transfer (ΔH, J mol^{-1}) and entropy (ΔS, J mol^{-1} K^{-1}) is given by the Gibbs–Helmholz equation: $\Delta G = \Delta H - T\Delta S$. Knowledge of the change in enthalpy for a certain partitioning process is particularly useful and is sought by environmental chemists to allow derivation of temperature-dependent partitioning. The relationship between K and ΔG is given by

$$K = constant.e^{-\Delta G/RT} \text{ or } \ln K = -\frac{\Delta G}{RT} + \ln(constant) \qquad (6.13)$$

K is therefore a function of temperature through the term RT, where R is the gas constant, 8.314 J mol^{-1} K and T is the temperature expressed in Kelvin (K). The value of the *constant* given in Equation 6.13 depends on how the abundance of the solute or chemical is expressed *i.e.* partial pressure, mole fraction or molar concentration. For mole fraction the constant has a value of 1, whereby ln 1 = 0 and can therefore be ignored. Rearranging Equation 6.13 to solve for ΔG, and then combining with the Gibbs–Helmholz equation results in

$$-RT \ln K = \Delta H - T\Delta S \tag{6.14}$$

$$\text{Hence, } \ln K = -\frac{\Delta H}{RT} + \frac{\Delta S}{R} \tag{6.15}$$

It therefore follows that an experimentally determined plot of ln K against $1/T$ (1/K) will yield a slope of $-\Delta H/R$, and an intercept of $\Delta S/R$ (ΔH and ΔS are considered to be constant over a small temperature range). In laboratories measuring partition coefficients for environmental fate, then -10 to 45 °C is a typical temperature range. Equation 6.15 is usually expressed in the form:

$$\ln K = -\frac{m}{T} + b \tag{6.16}$$

where m is the slope of the line and hence $Rm = \Delta H$. Once ΔH is established then K can be adjusted to any environmentally relevant temperature according to the integrated form of the van't Hoff equation

$$\ln K(T_2) = \ln K(T_1) - \frac{\Delta H}{R}\left(\frac{1}{T_2} - \frac{1}{T_1}\right) \tag{6.17}$$

where T_1 is the reference temperature (298 K) at which $K(T_1)$ and ΔH have been measured and T_2 is the temperature of interest. For many organic chemicals that persist in the environment, careful laboratory measurements of a particular partitioning coefficient over a range of temperatures may not exist, therefore ΔH^0(the enthalpy for the partitioning process at standard state, 101 kPa and 298 K) is often used in Equation 6.17.

Case study and worked example – Calculating water concentrations of Lindane
To illustrate the effect that temperature has on a partition coefficient and use some of the equations above, it is worth examining the

air-water partitioning of a pesticide. Lindane is an organochlorine insecticide that comprises of almost 100% γ-hexachlorocyclohexane (γ-HCH). The chemical has been recently nominated as candidate POP chemical and it is widely distributed in the global environment. Indeed this chemical is routinely monitored in arctic air[29,30] and is present in the polar environment largely through long-range atmospheric transport. Let's assume that γ-HCH has an air concentration of 100 pg m^{-3} within an agricultural region during the summer, where both air and surface water (lakes, rivers coastal seas) have a temperature of 25 °C. Using K_{aw} we can estimate the water concentration from the air concentration assuming equilibrium has been achieved. In the Arctic, the air concentration of γ-HCH is ~ 10 pg m^{-3} (10-fold lower), but the air and surface water temperatures in the high Arctic during the summer are ~ 1 °C. Again, the water concentration of γ-HCH can be estimated using K_{aw}, but this time adjusted to 1 °C. The surface water concentration in the temperate agricultural area is calculated from the air concentrations as follows:

H for γ-HCH has been measured as 0.24 Pa m^3 mol^{-1} at 298 K.[21]
Whereby H' or $K_{aw} = H/RT$ (T is 25 °C or ~ 298 K)
Therefore, $K_{aw} = 0.24/8.314 \times 298 = 9.69 \times 10^{-5}$ or $\ln K_{aw} = -9.24$
As $K_{aw} = C_a/C_w$, then $C_w = C_a/K_{aw} = 100/9.69 \times 10^{-5}$
C_w in the temperate agricultural region is 1.03×10^6 pg m^{-3} or ~ 1000 ng m^{-3} or ~ 1 ng L^{-1}

The water concentration in the Arctic is calculated in the same way but using the C_a for the Arctic. First, K_{aw} has to be adjusted for 1 °C (~ 274 K) using Equation 6.17, where:

$$\ln K_{aw}(274 K) = \ln K_{aw}(298\ K) - \frac{\Delta H}{R}\left(\frac{1}{274} - \frac{1}{298}\right)$$

ΔH is 61400 J mol^{-1} determined from the slope, m, in Equation 6.16 by Sahsuvar et al.[21] Note that in this study, ΔH actually represented the enthalpy of water to air volatilisation, rather than the enthalpy of air to water dissolution, and we consider 61,400 as approximate only. Therefore:

$$\ln K_{aw}(274\ K) = \ln K_{aw}(298\ K) - \frac{\Delta H}{R}\left(\frac{1}{274} - \frac{1}{298}\right)$$

The inverse of $\ln K_{aw} -11.4 = 1.12 \times 10^{-5}$. Therefore, the surface water concentration in the Arctic, attained through equilibrium partitioning with the air is

$C_w = C_a/K_{aw} = 10$ pg m$^{-3}/1.12 \times 10^{-5} = 892{,}857$ pg m^{-3} or ~ 890 ng m^{-3} or ≈ 1 ng L^{-1}

The two water concentrations (temperate and Arctic) are similar, even though the air concentration in the Arctic is an order of magnitude less than temperate air. This temperature driven phenomenon that favours chemical partitioning to surfaces is commonly known as the 'cold condensation effect' and has resulted in the worrying rise in persistent organic pollutants in the polar environments.

6.4.3 Partition Maps

The rise in legislation and the concern over the use and release of chemicals that possess physical–chemical properties that may be analogous to DDT and other persistent organic pollutants such as polychlorinated biphenyls (PCBs), has prompted demand for ways to rapidly screen or assess chemicals with respect to their environmental distribution and fate. The environmental transport, lifetime, and accumulation of organic chemicals in different compartments, such as biota, can be quantified through the use of multi-compartment (media) chemical fate models. It is simply not feasible to measure all chemicals of concern in a large number of compartments over an extended period of time. Identifying chemicals of concern, and the designation of the term 'POP' for a chemical substance, occurs if a chemical meets the criteria specified in Table 2. Importantly, the designation of 'persistence' is based on the

Table 2 *Criteria used to designate a chemical substance as a persistent organic pollutant or 'POP'. Information from the Stockholm Convention on POPs[31]*

Persistence	1.	$t_{1/2}{}^a$ in water > 2 months
	2.	$t_{1/2}$ in soil/sediments > 2 months
	3.	Other evidence that the chemical is sufficiently persistent to justify consideration
Bioaccumulation	1.	BCF or Bioaccumulation Factor (BAF) in aquatic species > 5000
	2.	Log $K_{ow} > 5$
	3.	Evidence of higher bioaccumulation in other species
Potential for long range transport	1.	Measured levels of the chemical in locations distant from sources of release that are of potential concern
	2.	Environmental fate properties and/or model results that demonstrate that the chemical has a potential for long-range environmental transport
	3.	$t_{1/2}$ in air > 2 days

$t_{1/2}$=half-life.

exceedence of specific environmental half-lives *i.e.* ≥ 6 months in soils/sediments, ≥ 2 months in water, ≥ 20 days in air. The latter is particularly relevant for those chemicals, which may be released to the air and are resistant to atmospheric degradation and/or deposition, and hence able to undergo long-range atmospheric transport, with the potential to contaminate remote areas. This multiple half-life approach is one adopted by several countries including Canada,[24] whereby a chemical that exceeds the criteria for any of these compartments may be declared persistent. However, if the chemical is unlikely to reside in that compartment then this may unfairly penalise a substance, because it fails the half-life criterion for a medium into which it does not appreciably partition.

To examine this issue, Gouin *et al.*[32] proposed a screening level method to allow the quick assessment of a large number of chemicals, based on their media-specific half-lives and their partitioning characteristics. An evaluative environmental fate model, the Equilibrium Criteria Model (EQM), was used for this study. EQM comprised of four hypothetical compartments air, water, soil and sediment, where octanol represented the organic matter found in soils and sediments. Each compartment was given a volume, in this case to represent the regional or country scale. For example, the air volume (V_a) was 10^{14} m^3, that is a land surface area of 100,000 m^2, with an atmospheric height of 1000 m. 10% of this area was considered to be covered with water to a depth of 20 m, resulting in a water volume (V_w) of 2×10^{11} m^3. Soil was considered to have a depth of 0.1 m with the soil volume (V_{soil}) calculated as 9×10^9 m^3, converted to an equivalent volume of octanol (V_o) according to

$$V_o = V_{soil}\Phi \times 0.35\rho \tag{6.18}$$

where Φ is the fraction of organic carbon in the soil (0.02), ρ is the density of soil (2.4 kg L^{-1} or 2400 kg m^{-3}), where 0.35 represents the constant relating the K_{oc} to K_{ow} (see Equation 6.5). Similarly for sediment, the equivalent V_o was 3.4×10^6 m^3 calculated using Equation 6.18, but with V_{sed} as 10^8 m^3, Φ as 0.04 and ρ the same as soil. For this scenario the ratios of the air/water/octanol volume was $\sim 650{,}000{:}1300{:}1$. The EQM model is a steady-state, equilibrium model into which a chemical was constantly discharged and allowed to attain equilibrium between the various media with no advective losses. Steady state was reached when total reactive losses equalled the discharge rate and the model was run for 233 chemicals using respective physical–chemical property data including K_{aw}, K_{ow} and K_{oa} as well as reactivity data in the various media (*i.e.* water, air, soil/sediment). At equilibrium the total mass of a chemical (M, g) can be given by

$$M = V_w C_w + V_a C_a + V_o C_o = C_w(V_w + K_{aw}V_a + K_{ow}V_o) \tag{6.19}$$

where C_o is the chemical concentration in octanol (*i.e.* 'soils/sediment') (g m^{-3}). Note that the volume-concentration product ($VxCx$) gives the mass of a chemical for a particular environmental medium. The right-hand side of Equation 6.19 illustrates how using the partition coefficients, the concentration of a chemical can be deduced for one compartment with knowledge of the chemical concentration in an adjacent compartment (*e.g.* $K_{aw}C_w = C_a$). The mass fractions of a chemical in each medium (F) are therefore

$$\text{Air (a)} : F_a = K_{aw}V_a/(V_w + K_{aw}V_a + K_{ow}V_o) \qquad (6.20)$$

$$\text{Water (w)} : F_w = V_w/(V_w + K_{aw}V_a + K_{ow}V_o) \qquad (6.21)$$

$$\text{Octanol (o)} : F_o = V_oK_{ow}/(V_w + K_{aw}V_a + K_{ow}V_o) \qquad (6.22)$$

Figure 4 displays the plot of log K_{aw} vs. log K_{ow} including points that represent all 233 chemicals used in the study. The chemicals were selected to represent environmental contaminants of concern and include a broad range of different chemical classes with varying properties. Figure 4 is effectively a partitioning map that illustrates the proportion of a chemical within each of the compartments. The 45° diagonals are lines of constant log K_{oa}, as $K_{oa} = K_{ow}/K_{aw}$ (or log K_{oa} = log K_{ow}−log K_{aw}). Lines of constant F_a, F_w and F_o were drawn using the volume ratios outlined above for this particular study. The 1% and 99% lines divide the K_{aw}/K_{ow} space into regions into which chemical partitioning is predominantly into one medium. For example, the region to the upper left of the figure is where more than 99% of a chemical is in air, to the lower right (beyond the 1% air line) is where more than 99% of a chemical is in the octanol phase and to the lower left (beyond the 1% octanol line) is where more than 99% of a chemical is in water. For chemicals that fall into these 'areas' then their half-lives in the respective medium will likely control their overall persistence and the half-lives in the other media are largely irrelevant. For example, a chemical present in the lower right of the figure, within the octanol 'area', will be strongly sorbed and only degradation data for soils/sediments is likely to be needed. Importantly, half-lives in air are generally shorter than those in water or organic phases due to rapid reactivity with the hydroxyl radical (OH·), so that even 0.5% partitioning to air may result in appreciable degradation with respect to overall loss. It is therefore useful to include a line that represents the 0.1% air. The lines corresponding to one third (33%) in each compartment, converge where log K_{ow} is 3.1 and log K_{aw} is −2.74; chemicals that fall within this area ('A & W & O') are truly multi-media and their half-lives

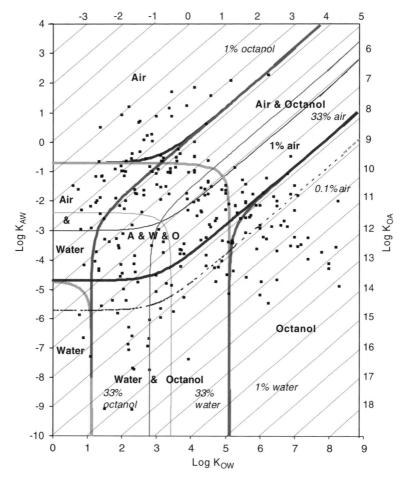

Figure 4   *$K_{aw} - K_{ow}$ partitioning map with the distribution of 233 organic chemicals and lines of equal F_a, F_w and F_o; that is the chemical mass fractions in air (F_a), water (F_w) and octanol (F_o). The central region denoted 'A&W&O' is occupied by chemicals which have mass fractions (>1%) in all compartments of Air, Water and Octanol and are therefore multi-media chemicals. Octanol represents organic-phases in the environment e.g. the organic carbon within soils and sediments*
From Gouin et al.[32], courtesy of the American Chemical Society

for each compartment or medium would be required to determine their overall persistence. Obviously the convergence point would change if a different scenario was modelled, whereby the water compartment volume was increased for example. However, out of the 233 chemicals included in the figure, ~34% had fractions >1% in each medium. If the chemicals between the 1% and 0.1% of air are included, then ∼40% of the chemicals become multi-media.

This approach of mapping chemicals into groups according to their partitioning tendencies provides a rapid method of identifying relevant environmental compartments where a chemical may reside. Key degradation data relevant for that specific compartment can then be further examined without the need for irrelevant data for other compartments.

6.5 CHEMICAL TRANSFORMATION AND DEGRADATION

Chemical degradation may proceed by a wide number of processes and the reactivity of a chemical is governed by its stereochemistry (3-dimensional structure), bond strengths and the presence of functional groups. Both abiotic and biotic processes will transform a chemical pollutant and the multitude of abiotic processes can be broadly grouped under the following types of reaction: oxidation, reduction, hydrolysis and photolysis. Indeed these generic reactions are not mutually exclusive, as the generation of oxidants present in the atmosphere for example, requires the action of sunlight on precursor molecules (*e.g.* the formation of the OH radical from ozone). Several of the chlorinated aromatic chemical groups illustrated in Figure 1 (and mentioned in Section 6.4) are largely resistant to these transformation processes, whereby their rates of reaction are very slow. However, even these chemicals do degrade eventually, although their degradates may have a similar longevity in the environment! Perhaps the fastest route of degradation for these semi-volatile compounds is through oxidation by the OH radical in the atmosphere, with half-lives typically on the order of hours to days, compared to months or years in soils and sediments. However, at any one time, the atmosphere may contain only a very small fraction of the overall environmental 'inventory' so its relevance as an important 'sink' for these chemicals is greatly reduced compared to volatile organic compounds (VOCs). Nonetheless, oxidation of PAHs present in the atmosphere can give rise to toxic analogues. The reaction of higher volatility PAHs (generally the 2- to 4-ring PAHs present predominantly in the vapour phase) with OH radicals during daylight hours, or with nitrate (NO_3) radicals during nighttime gives rise to more polar hydroxy- and nitro-PAHs. The atmospheric formation of nitro-PAHs may also occur *via* a daytime OH radical-initiated pathway in the presence of NO_x, although the formation yields of nitro-PAHs (or nitroarenes) *via* this pathway are low compared to the yield of hydroxy-PAHs.[33] Figure 5 illustrates a simplified reaction pathway of fluoranthene with the OH radical in the presence of NO_x, resulting in the

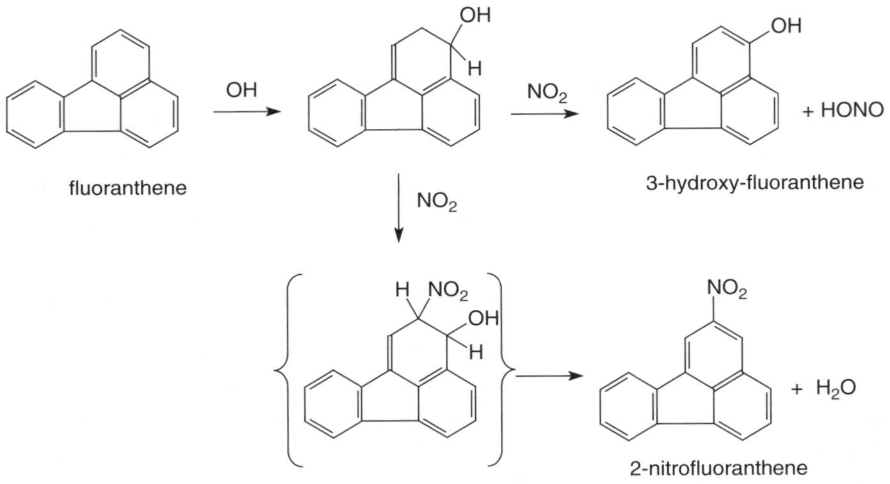

Figure 5 *The gas phase OH radical-initiated reaction of fluoranthene (a 3-ring PAH) in the presence of NOx. The reaction results in the formation of both hydroxy- and nitro-fluoranthene, although nitro-fluoranthene can also form from the nighttime reaction with the NO_3 radical*
Reaction scheme adapted from Arey et al.[33]

formation of both 3-hydroxyfluoranthene and 2-nitrofluoranthene (2-NF). Indeed, some of the nitroarenes are strongly mutagenic and present a risk to human health and have been reported in several recent air sampling campaigns.[34,35] Interestingly, 2-NF is one of the major particle-associated nitro-PAHs observed in a study conducted in southern California,[35] presumably due to its lower volatility and higher polarity than the parent fluoranthene, resulting in its condensation/sorption to atmospheric particles.

An important transformation process for halogenated organic compounds is reductive dehalogenation, a process which largely occurs in anoxic (oxygen free) environments, such as sediments, sewage sludges, water-logged soils and ground-water and catalysed by the presence of metal containing minerals, where the metal is in a low state of oxidation *e.g.* zero-valent or divalent iron. The reductive potential of a halogenated molecule (*i.e.* 'the willingness to accept electrons from the metal donor') is based on many factors, not least the number and type of halogen atoms around the molecule. In one common type of reaction the breaking of the carbon-halogen bond (through the halogen atom accepting electrons) is followed by replacement with a H-atom (*hydrogenolysis*), resulting in the gradual 'de-halogenation' of the molecule. For chlorinated solvents and chlorinated aliphatics in general, then this process may be driven solely by abiotic processes,[36] but for the more stable chlorinated aromatics, reductive dechlorination is likely to be assisted by microbial activity (or biotic processes) and is currently the focus of much research into the biologically driven remediation of contaminated soils and sediments.

6.6 CHEMICAL TRANSFORMATION THROUGH PHOTOCHEMISTRY

Organic pollutants are susceptible to photochemical transformation and degradation in both the atmosphere and in surface waters. In fact, photochemical degradation can occur in any environmental medium that is subject to sunlight (or more appropriately solar actinic radiation), and includes surface reactions whereby chemicals sorbed to environmental surfaces such as plant material, airborne particles and even snow and ice are affected by sunlight. Both direct and indirect photochemistry (the latter brought about by chemical reaction with short-lived oxidants such as the OH radical), is of interest to the organic chemist, and can lead to the destruction of the pollutant and/or its phototransformation. In some cases this may lead to more persistent, or toxic, photoproducts such as the photochemical transformation of chlorobenzenes to polychlorinated biphenyls in ice[37] or the soil-sorbed photooxidation of the organophosphorus pesticide parathion to the active toxin paraoxon.[38] The assessment of photochemistry on the fate of organic chemicals is therefore a necessary step in risk assessment procedures (required for pesticide registration, for example), with protocols for conducting laboratory photochemical fate studies now provided by organisations like the US-EPA, the Organisation for Economic Cooperation and Development (OECD) and the European Chemical Bureau; the aim here is to provide standard operating procedures to enable reproducibility in test results and intercomparsion between different studies. These protocols concern direct and indirect photolysis in aqueous systems, as both fresh and marine waters are important sinks for organic pollutants. For example, agrochemicals residing on soil and plant surfaces often end up in surface waters, and the multitude of organic contaminants released from waste water treatment works (WWTW) (*e.g.* non-ionic surfactants, pharmaceuticals, personal care products) and their impact on receiving waters is of growing concern.[39] However, atmospheric photochemistry relating to both vapour and particle-bound organic contaminants is perhaps less well studied, primarily due to the problem of maintaining a consistent, reproducible vapour in order to conduct gas-phase photochemical experiments. Nevertheless, there is a growing need to assess both direct photolysis and photooxidant chemistry to determine reaction rates to establish atmospheric half-lives and hence atmospheric transport potential – an important feature with respect to the contamination of remote environments by persistent chemicals (see Table 2). For the atmospheric chemist, much more emphasis is placed on photochemistry of VOCs, which play a major role in influencing air quality, by driving the formation of ground-level O_3[40] as well as initiating particle formation and growth.[41] In the following sections focus will be on semivolatile organic contaminants and their aqueous photochemistry.

6.6.1 Light Absorption and the Beer–Lambert Law

For the environmental photodegradation of organic chemicals, direct photolysis is an important removal process for those pollutants that absorb light at wavelengths above 290 nm; the cut-off of solar irradiation at the earth's surface. Visible light includes those wavelengths that cover the spectral range of 400–760 nm, whereas shorter wavelengths *i.e.* 290–400 nm are in the ultraviolet or UV region, and posses greater energy than light at longer wavelengths. Hence light in the UV-region is of most concern with respect to sunburn and skin cancers, and can be further divided into different wavelength regions, including UV-B (290–320 nm) and UV-A (320–400 nm).

The degree of absorption of UV and visible wavelengths (UV/vis) by organic compounds is related to the molecular structure of the chemical, in particular, by the presence of *chromophores*; structural moieties that exhibit a characteristic UV/vis absorption spectrum. Aromatic rings and conjugated double bonds are good examples of chromophores, but others include double bonds containing, for example, nitrogen and oxygen and other heteroatoms present within the carbon skeleton of the molecule. Light absorption by organic compounds at the wavelengths of environmental relevance is usually associated with the delocalised π-electron system of double bonds (*i.e.* those electrons involved in the multiple bonding of C=C and C=O groups, for example), rather than electrons in sigma orbitals as part of single bonds, as the absorption maxima of these electrons generally lie at much shorter wavelengths than the solar wavelength minimum of 290 nm. Nearly all environmental photochemistry involves the excitation of π or non-bonding electrons (*n*) to higher-energy, antibonding π* orbitals, thereby destabilising the molecule. Examples of *n* electrons are the non-bonding electron pairs associated with oxygen atoms of carbonyl groups (C=O) or nitrogen atoms (C=N). Chromophores consisting of a series of conjugated carbon double bonds (C=C–C=C) are readily found in nature, and a good example are the porphyrins (of which chlorophyll is a member). These provide useful 'light sensitive' systems that can harvest light energy and in the case of chlorophyll perform the vital role of photosynthesis.

Absorption of light at a given wavelength (A_λ) by a solution of a chromophoric molecule can be treated quantitatively according to the Beer–Lambert law. The ratio of the light intensity transmitted through the chemical solution (I_λ) to the incident light intensity ($I_{o,\lambda}$) can be related to the chemical concentration (C) according to,

$$A_\lambda = \log \frac{I_\lambda}{I_{o,\lambda}} = \varepsilon \cdot C \cdot l \qquad (6.23)$$

where C is expressed as molar concentration (M), ε the decadic (log) molar absorption coefficient (M^{-1} cm^{-1}) and l the light path length in the chemical solution (cm). I is often expressed in units of Einstein cm^{-2} s^{-1}, (where an Einstein is 6.02×10^{23} photons or a 'mole' of photons) with absolute light irradiance (expressed as Joules s^{-1} m^{-2} or Watts m^{-2}) measured using a chemical actinometer for a specified wavelength range.

6.6.2 Photolysis in Aqueous Systems

Light absorbance studies can be conducted using a UV/vis spectrophotometer programmed to irradiate a solution at individual wavelengths over part, or all, of the solar spectrum. Quartz-glass cuvettes ($l = 1$ cm) are usually employed which hold the chemical solution and a reference solution (*i.e.* typically the solvent but without the chemical to assess the absorbance artefacts of the holding solvent). Water is the solvent used for environmental fate studies, but apolar or monopolar organic compounds (including many pesticides) are only sparingly soluble in water so an organic solvent – water solution is often used, where the organic solvent is strongly polar and water-miscible such as methanol or acetonitrile. Figure 6 shows the spectral absorbance of chlorpyrifos (a currently used organophosphorus insecticide) in pure methanol between the wavelengths 200–400 nm. Chlorpyrifos displays several absorbance maxima, but within the UV-B/UV-A range of the solar spectrum the maximum absorbance occurs at 288 nm and $\varepsilon 288$ nm can be calculated by rearranging Equation 6.23 to:

$$\varepsilon_{288\,nm} = \frac{A_{288\,nm}}{C \cdot l} \qquad (6.24)$$

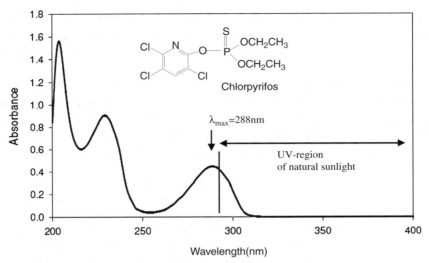

Figure 6 *Spectral absorbance of the insecticide chlorpyrifos (in methanol) over UV wavelengths. For the environmentally relevant range of the UV spectrum (UV-A/B), λ$_{max}$ occurs at 288 nm, with significant absorbance at higher wavelengths (to 310 nm), suggesting natural sunlight is capable of inducing phototransformation (thanks to J. Weber, Lancaster University, for generating the absorbance spectrum)*

From Figure 6, $A_{288\,nm} = 0.49$ and was generated with C as ~ 0.1 mM (or 1×10^{-4} M) and $l = 1$ cm, the value of $\varepsilon_{288\,nm}$ is therefore:

$$\varepsilon_{288\,nm} = \frac{0.49}{10^{-4} \times 1} = 4900\,\mathrm{M^{-1}\,cm^{-1}} \tag{6.25}$$

Values of λ_{max} and ε_{max} are useful to assess whether a compound might absorb ambient light and therefore be susceptible to environmental photodegradation. Values of ε however need to be calculated for each wavelength if full quantification of photochemical processes is required. Indeed for chlorpyrifos, significant absorbance occurs at wavelengths >290 nm suggesting that photodegradation is worth considering with respect to its environmental fate. It is worth noting that the solvent in which the chemical resides may have a strong influence on the absorption spectrum, and is important when trying to replicate environmental conditions for a chemical with a low aqueous solubility. It is therefore relatively common practice to replicate photochemical experiments using a number of solvent-water mixtures to assess the solvent effect on the spectral absorbance. Indeed, OECD guidelines with respect to aqueous photodegradation studies on compounds with low solubility, recommend that co-solvents should not comprise more than 1% by volume of the test solution. Furthermore, the co-solvent must not solvolyse the test chemical, must not absorb in the 290–800 nm region, and must not be a photosensitiser (that is promote chemical change through indirect photolysis due to exchange of molecular energy with the test chemical). To this end, aceto-nitrile and methanol (MeOH) are the co-solvents of choice (*e.g.* aqueous solutions comprising 99.5% H_2O/0.5% CH_3CN), but not acetone because of its strong photosensitizing effect.

Natural waters contain dissolved organic carbon (DOC) and also vary in pH (typically between pH 5 and 8). Both DOC and pH can have a marked effect on the photochemical degradation of organic pollutants and for realistic photochemical studies to be undertaken both factors need to be taken into account, particularly for ionisable chemicals that may dissociate in water depending on their acid dissociation constant (pKa) and the pH of the water. DOC comprises the principal light absorbing component in surface waters (comprising humic and fulvic acids) that provide the natural colouration of waters arising in organic-rich environments. Direct photochemical reactions can be slowed by the shielding effect of DOC, but additionally sensitised reactions may occur by indirect light-induced reactions of the dissolved material. One example of this type of light-induced reaction with DOC, is the possible coupling reactions of triclosan with humic matter, to form insoluble particles.[42] Triclosan is a relatively common anti-microbial agent added to hand-washes, tooth-paste and other personal care products and is frequently detected in surface waters throughout Europe and North America. Direct photo-transformation is responsible for the majority of triclosan loss in surface waters[43] and photobyproducts include 2,4-dichlorophenol and 2,8-dichlorodibenzo-*p*-dioxin.[42] Only the anion of triclosan is susceptible to significant phototransformation and as its pKa is 7.9 (weak acid), then the anion will predominate in water with a pH of >8. The photochemical reaction pathway is illustrated in Figure 7 and provides a useful example of photochemistry that is pH dependent and where DOC appears to influence the reaction by-products.

6.6.3 Photochemistry of Brominated Flame Retardants (BFRs)

Brominated flame retardants (BFRs) are an interesting and important group of chemicals that are widely incorporated into plastics, foams, electrical items and furnishings to reduce the risk of fire. Indeed, it is likely that the polyurethane foam used in an upholstered chair in which you maybe sitting, will have been treated with a BFR in order to conform to fire-safety standards. However, their widespread use, release

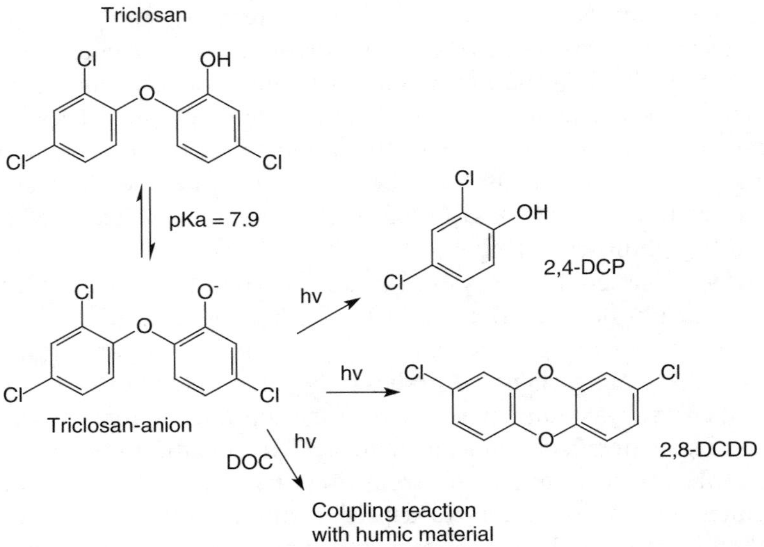

Figure 7 *Direct photochemical reaction pathway of Triclosan (5-chloro-2-(2,4-di-chlorophenoxy)phenol) in natural water (3–100 μM of Triclosan in pH-buffered solutions), irradiated with a Hg-vapour lamp. Only the triclosan phenolate anion (present in water at pH >8) is susceptible to significant photochemical transformation and will also react through indirect photolysis (not shown) with singlet oxygen ($^1O_2$) and OH radicals to form 2,4-dichloro-phenol (2,4-DCP). The photoproduct 2,8-dichlorodibenzo-p-dioxin (2,8-DCDD) belongs to a notoriously persistent group of chemicals commonly referred to as 'dioxins'*
From Latch et al.[42] courtesy of SETAC press

into the environment and occurrence in wildlife and human blood has resulted in growing concern about these groups of chemical.[44,45]

BFRs function by decomposing under heat to give rise to Br• radicals that react with hydrocarbons (flammable gases), given off by the burning substance, to form HBr molecules. These, in turn, 'mop-up' high energy OH• and H• radicals formed as part of the combustion chain reaction, thereby effectively retarding or slowing the combustion process. The effectiveness of a BFR is therefore dependent on the quantity of Br• radicals that can be released and the control of their release. Ideally, for the latter, the BFR should decompose at a lower temperature than that of the polymer in which it is incorporated and this will determine, among other things, the type of BFR used in a particular application.[45]

Two important BFRs are the polybrominated diphenyl ethers (PBDEs) and tetrabromobisphenol A (TBBPA) and their respective structures are presented in Figure 8. PBDEs are utilised as additives for treating plastics and polymers used within electronic components and

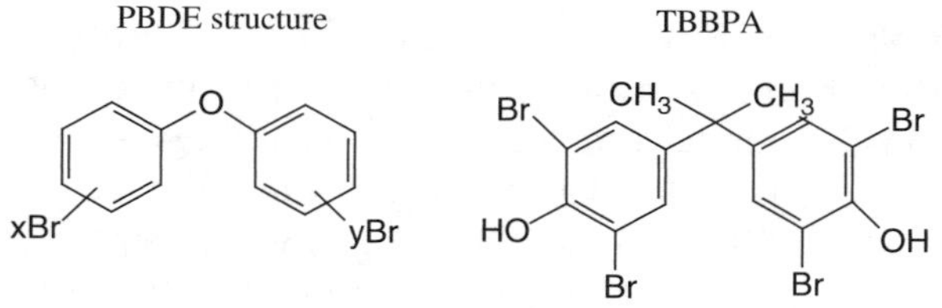

PBDE structure   TBBPA

Figure 8 *Generic structure of the polybrominated diphenyl ether (PBDE) and the structure of tetrabrominated bisphenol-A (TBBPA). Note the x and y on the PBDE structure represent the number of bromine atoms around each phenyl ring (where x = 1–5 and y = 1–5). The maximum number of 5 bromine atoms on each ring results in the fully brominated, decabrominated diphenyl ether*

electrical equipment in general, and are also used extensively as additives in textiles and flexible foams, whereby they are dissolved in the polymer before it is formed into shape or blown into its foam structure. As a result, PBDEs, will, over time, volatilise from the equipment or furnishing into which they have been added. TBBPA, on the other hand, is primarily used as a reactive flame retardant *i.e.* it is covalently bound into mainly epoxy and polycarbonate resins used in the manufacture of printed circuit boards within electronic equipment.

6.6.3.1 Demand for BFRs. PBDEs are grouped into homologues according to the number of bromine atoms around the phenyl rings and are available commercially as three different formulations; pentaBDE, octaBDE and decaBDE, whereby the penta- and octaBDE contain between 10 and 20 individual congeners and the decaBDE formulation largely comprising of the single congener, decabrominated diphenyl ether. Worldwide demand of PBDEs is high, and is estimated to be 7500 t for pentaBDE, 3760 t for octaPBDE and 56100 t for decaBDE for 2001,[46] although quantities of the pentaBDE and octaPDE are likely to change following a ban by the European Union and a voluntary phase-out by one of the main industrial producers in the US. Concern arises due to PBDE occurrence in human blood and breastmilk, with evidence of increasing PBDE levels in wildlife.[44] The physical–chemical properties of the PBDEs are analogous to 'legacy' persistent organic pollutants such as polychlorinated biphenyls (PCBs), previously used in the electrical industry. For example, PBDEs are semi-volatile (although the decabrominated congener is considered non-volatile at ambient temperatures) with log K_{ow} values of 5.9–6.2 for tetra-BDE, 6.5–7.0 for pentaBDE, 8.4–8.9 for octaBDE and 10 for decaBDE;[46] evidence of their propensity for movement in the environment and ability to strongly partition to fats/lipids.

It is estimated that worldwide demand for TBBPA in 2001 was some 120,000 tonnes[47] and concern over this chemical (and related phenols) has arisen due to its occurrence in sewage sludge, aquatic sediments and fish.

6.6.3.2   PBDE and TBBPA Photochemistry. For aryl-halides, the bromine-carbon bond is not as strong as the chorine-carbon bond, with a typical bond-dissociation energy of the bromine-carbon bond of 301 kJ mol^{-1} compared to 360 kJ mol^{-1} for the chlorine-carbon bond.[48] Therefore, unlike PCBs, many PBDEs are more susceptible to loss of the halogen atoms and photochemical decomposition has been recognised as an important route for both debromination and transformation for various PBDE congeners present in the wider environment. Several studies have investigated the photochemical degradation of PBDEs[46,49] to determine rates of degradation and identify photodegradation by-products in aqueous solutions under simulated light conditions. Figure 9 shows the absorption spectra (ε) of seven PBDE congeners dissolved in tetrahydrofuran (several of the higher brominated congeners are only sparingly soluble in MeOH/H$_2$O) and exposed to a UV-light source in the study of Eriksson *et al.*[46] Clearly the higher brominated congeners absorb light at longer wavelengths, possessing higher λ_{max} values than less brominated congeners. As such, the rate of degradation of PBDEs by UV light in the sunlight region was found to be dependent on the

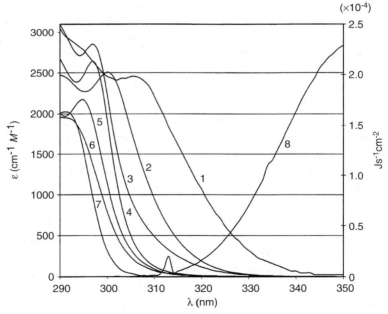

Figure 9   *Absorbance spectra (expressed as ε, the molar absorbance coefficient) of seven PBDE congeners. The line numbers ar: (1) BDE-209; (2) BDE-206; (3) BCDE-203; (4) BDE-183; (5) BDE-155; (6) BDE-85; (7) BDE-77; line 8 represents the UV light irradiance in units of J s^{-1} cm^{-2} (or W cm^{-2})*
From Eriksson et al.[46] courtesy of the American Chemical Society

degree of bromination, whereby the lower brominated congeners generally degraded at a much slower rate than that of the higher brominated congeners. For example, the observed first order rate constant (k) for the photodecomposition of BDE-47 (a tetrabrominated diphenyl ether) in MeOH/H$_2$O solution was 0.07×10^5 s^{-1} compared to 40×10^5 s^{-1} for BDE-209 (the decabrominated diphenyl ether). The relatively slow rate of reaction of BDE-47 (approximately 600-fold less than BDE-209) can be largely attributed to its lower absorbance of light at longer wavelengths, which in turn may help to explain the prevalence of PBD-47 in environmental samples. Interestingly, Eriksson *et al.*[46] observed that the decaBDE decomposed to lower brominated diphenyl ethers and also tentatively identified additional photodegradation by-products including methoxylated polybrominated dibenzofurans through interpretation of mass spectra following analysis of the photolysis solution by GC-MS.

A similar aqueous photochemistry study on TBBPA[47] has found much faster rates of transformation for this chemical relative to the PBDEs. However, the rate was dependent on the pH of the solution, with phototransformation favouring the anion of TBBPA, present at a significant fraction when the pH of the test solution was >8 (the pKa of TBBA is ~ 7.5–8.5). At pH 10, λ_{max} and ε_{max} were 310 nm and 9170 cm^{-1} M^{-1} respectively, compared to 290 nm and ~ 2000 cm^{-1} M^{-1} at pH 5.5, illustrating that at the higher pH the TBBA-anion is absorbing strongly at a higher wavelength compared to TBBPA and therefore more likely to be photodegraded under natural sunlight. The value of k for the photodecomposition of TBBPA in an aqueous solution at pH 10 was measured as 0.7×10^3 s^{-1}, compared to 0.033×10^{-3} s^{-1} at pH 5.5, and the photodegradates included a number of bromophenols and alkylated bromophenols.

6.7 CONCLUSIONS

Knowledge of the partitioning properties of organic contaminants is fundamental to understanding their environmental distribution and hence fate. For many organic chemicals that pose a threat to the environment, there is a lack of robust temperature-dependent partition coefficients, and in many instances these have to be estimated either from other partitioning descriptors or modelled from the molecular properties of the chemical. Partition 'maps' provide a useful way to visualise the environmental distribution of a wide range of chemical classes, and to identify those relevant media where a chemical may reside. Increasingly the results from multi-media chemical fate models are displayed in this way and partition maps will likely provide a useful tier for chemical risk assessment in the future.

Unlike the older 'legacy' POPs, such as DDT and PCBs, many current chemicals used in agriculture, industry and for domestic purposes, are generally more susceptible to degradation processes once present in the environment (although, ironically, the persistent nature of DDT can be attributed to its biological degradate *p,p'*-DDE). In some instances these degradates may be more persistent and/or harmful to the environment than the parent chemicals. The photolytic degradation of Triclosan and certain BFRs are a case in point. Interestingly, a recent review on pesticide-degradates highlighted their higher frequency of detection over the parent compounds in surface water sampled across the US.[50] These pesticide-degradates typically had longer half-lives and lower log K_{oc} values than the parent pesticides, making them relatively long-lived and more mobile in the environment. Future research efforts will be required to understand the fate of these chemicals and the threat they pose to both humans and the environment.

6.8   QUESTIONS

(i) A pesticide has an experimentally determined K_d value of 5.6, calculate its K_{oc} value assuming the soil has a 2% organic carbon content. What type of polymeric material comprises a large proportion of the organic carbon fraction in soil? Describe how you could operationally separate its base-soluble fraction?

(ii) The Henry's Law constant for a pesticide at 25 °C has been experimentally determined as 0.008 Pa m^3 mol^{-1}. Calculate K_{aw} (dimensionless Henry's Law constant)? If this pesticide has a log K_{ow} 2.9, which environmental compartment is the chemical likely to reside in (use the partition map of Figure 3 to help with your answer)?

(iii) The pesticide outlined in Question (i) and (ii) does not absorb sunlight in aqueous solution, however upon dissociation its anion strongly absorbs wavelengths >300 nm. If this pesticide has a pKa of 6 and is present in surface waters with a pH ~ 8, describe the probable fate of this chemical.

REFERENCES

1. D. Ashton, M. Hilton and K.V. Thomas, *Sci. Tot. Environ.*, 2004, **333**, 167–184.

2. A. Bruchet, C. Hochereau, C. Picard, V. Deottignies, J.M. Rodrigues and M.L. Janex-Habibi, *Water Sci. Technol.*, 2005, **52**, 53–61.
3. P.A. Blackwell, H.C. Lutzhoft, H.P. Ma, B. Halling-Sorensen, A.B.A. Boxall and P. Kay, *J. Chromatog. A*, 2004, **1045**, 111–117.
4. J. Yu, R.C. Flagan and J.H. Seinfeld, *Environ. Sci. Technol.*, 1998, **32**, 2357–2370.
5. I.G. Kavouras, N. Mihalopoulos and E.G. Stephanou, *Environ. Sci. Technol.*, 1999, **33**, 1028–1037.
6. M.P. Fraser, G.R. Cass, B.R.T. Simoneit and R.A. Rasmussen, *Environ. Sci. Technol.*, 1997, **31**, 2356–2367.
7. R.A. Larson and E.J. Weber, *Reaction Mechanisms in Environmental Organic Chemistry*, Lewis Publishers, CRC Press LLC, Florida,1994.
8. C.T. Chiou, D.E. Kile, D.W. Rutherford, G. Sheng and S.A. Boyd, *Environ. Sci. Technol.*, 2000, **34**, 1254–1258.
9. L. Li, Z.Y. Zhao, W.L. Huang, P. Peng, G.Y. Sheng and J.M. Fu, *Org. Geochem.*, 2004, **35**, 1025–1037.
10. C. Maldonado, J.M. Bayona and L. Bodineau, *Environ. Sci. Technol.*, 1999, **33**, 2693–2702.
11. M.B. Yunker and R.W. Macdonald, *Org. Geochem.*, 2003, **34**, 1525–154.
12. European Chemicals Bureau (http://ecb.jrc.it/).
13. R.P. Schwarzenbach, P.M. Gschwend and D.M. Imboden, *Environmental Organic Chemistry*, 2nd edn, Wiley Interscience, Wiley, NJ, 2003.
14. D. Mackay, *Multimedia Environmental Models – The Fugacity Approach*, Lewis Publishers, Michigan, USA, 1991.
15. D. Mackay, *Environ. Sci. Technol.*, 1982, **16**, 274–278.
16. K. Borgå, A.T. Fisk, P.F. Hoekstra and D.C.G. Muir, *Environ. Toxicol. Chem.*, 2004, **23**, 2367–2385.
17. R.S. Boethling, P.H. Howard and W.M. Meylan, *Environ. Toxicol. Chem.*, 2004, **23**, 2290–2308.
18. S.W. Karickhoff, *Chemosphere*, 1981, **10**, 833–849.
19. R. Seth, D. Mackay and J. Muncke, *Environ. Sci. Technol.*, 1999, **33**, 2390–2396.
20. H.A. Bamford, D.L. Poster and J.E. Baker, *Environ. Toxicol. Chem.*, 1999, **18**, 1905–1912.
21. L. Sahsuvar, P.A. Helm, L.M. Jantunen and T.F. Bidleman, *Atmos. Environ.*, 2003, **37**, 983–992.
22. J.L. Barber, G.O. Thomas, R. Bailey, G. Kerstiens and K.C. Jones, *Environ. Sci. Technol.*, 2004, **38**, 3892–3900.
23. S.N. Meijer, M. Shoeib, K.C. Jones and T. Harner, *Environ. Sci. Technol.*, 2003, **37**, 1300–1305.

24. A. Finizio, D. Mackay, T. Bidleman and T. Harner, *Atmos. Environ.*, 1997, **31**, 2289–2296.
25. T. Harner and D. Mackay, *Environ. Sci. Technol.*, 1995, **29**, 1599–1606.
26. P. Kömp and M.S. McLachlan, *Environ. Toxicol. Chem.*, 1997, **16**, 2433–2437.
27. T. Harner and T.F. Bidleman, *J. Chem. Eng. Data*, 1998, **43**, 40–46.
28. M. Shoeib and T. Harner, *Environ. Toxicol. Chem.*, 2002, **21**, 984–990.
29. H.H. Hung, C.J. Halsall, P. Blanchard, H.H. Li, P. Fellin, G. Stern and B. Rosenberg, *Environ. Sci. Technol.*, 2002, **36**, 862–868.
30. C.J. Halsall, *Environ. Poll.*, 2004, **128**, 163–175.
31. Stockholm Convention on Persistent Organic Pollutants (http://www.pops.int/).
32. T. Gouin, D. Mackay, E. Webster and F. Wania, *Environ. Sci. Technol.*, 2000, **34**, 881–884.
33. J. Arey, B. Zielinska, R. Atkinson, A.M. Winer, T. Ramdahl and J.N. Pitts, *Atmos. Env.*, 1986, **20**, 2339–2345.
34. H. Soderstrom, J. Hajslova, V. Kocourek, B. Siegmund, A. Kocan, W. Obiedzinski, M. Tysklind and P.A. Bergqvist, *Atmos. Env.*, 2005, **39**, 1627–1640.
35. F. Reisen and J. Arey, *Environ. Sci. Technol.*, 2005, **39**, 64–73.
36. M.M. McGuire, D.L. Carlson, P.J. Vikesland, T. Kohn, A.C. Grenier, L.A. Langley, A.L. Roberts and D.H. Fairbrother, *Anal. Chim. Acta*, 2003, **496**, 301–303.
37. P. Klán and I. Holoubek, *Chemosphere*, 2002, **46**, 1201–1210.
38. W.F. Spencer, J.D. Adams, R.E. Hess, T.D. Shoup and R.C. Spear, *Agric. Food Chem.*, 1980, **28**, 366–371.
39. T.A. Ternes, A. Joss and H. Siegrist, *Environ. Sci. Technol.*, 2004, **38**, 393A–399A.
40. A. Hakami, M.S. Bergin and A.G. Russell, *Environ. Sci. Technol.*, 2004, **38**, 6748–6759.
41. J.H. Seinfeld and J.F. Pankow, *Ann. Rev. Phys. Chem.*, 2003, **54**, 121–140.
42. D.E. Latch, J.L. Packer, B.L. Stender, J. Vanoverbeke, W.A. Arnold and K. McNeill, *Environ. Toxicol. Chem.*, 2005, **24**, 517–525.
43. H. Singer, S. Muller, C. Tixier and L. Pillonel, *Environ. Sci. Technol.*, 2002, **36**, 4998–5004.
44. R.A. Hites, *Environ. Sci. Technol.*, 2004, **38**, 945–956.
45. F. Rahman, K.H. Langford, M.D. Scrimshaw and J.N. Lester, *Sci. Tot. Environ.*, 2001, **275**, 1–17.

46. J. Eriksson, N. Green, G. Marsh and A. Bergman, *Environ. Sci. Technol.*, 2004, **38**, 3119–3125.
47. J. Eriksson, S. Rahm, N. Green, A. Bergman and E. Jakobsson, *Chemosphere*, 2004, **54**, 117–1126.
48. R.T. Morrison and R.N. Boyd, *Organic Chemsitry*, 3rd edn, Allyn and Bacon Inc., USA, 1973.
49. I. Hua, N. Kang, C.T. Jafvert and J.R. Fabrega-Duque, *Environ. Toxicol. Chem.*, 2003, **22**, 798–804.
50. A.B.A. Boxall, C.J. Sinclair, K. Fenner, D. Kolpin and S.J. Maund, *Environ. Sci. Technol.*, 2004, **38**, 369A–375A.

CHAPTER 7

Biogeochemical Cycling of Chemicals

ROY M. HARRISON

Division of Environmental Health & Risk Management, School of Geography, Earth & Environmental Sciences, University of Birmingham, Edgbaston, Birmingham B15 2TT, UK

7.1 INTRODUCTION: BIOGEOCHEMICAL CYCLING

The earlier chapters of this book have followed the traditional sub-division of the environment into compartments (*e.g.* atmosphere, oceans, *etc.*). While the sub-divisions accord with human perceptions and have certain scientific logic, they encourage the idea that each compartment is an entirely separate entity and that no exchanges occur between them. This, of course, is far from the truth. Important exchanges of mass and energy occur at the boundaries of the compartments and many processes of great scientific interest and environmental importance occur at these interfaces. A physical example is that of transfer of heat between the ocean surfaces and the atmosphere, which has a major impact upon climate and a great influence upon the general circulation of the atmosphere. A chemically based example is the oceanic release of dimethylsulfide to the atmosphere, which may, through its decomposition products, act as a climate regulator (see Chapters 2 and 4).

Pollutants emitted into one environmental compartment will, unless carefully controlled, enter others. Figure 1 illustrates the processes affecting a pollutant discharged into the atmosphere.[1] As mixing processes dilute it, it may undergo chemical and physical transformations before depositing in rain or snow (wet deposition) or as dry gas or particles (dry deposition). The deposition processes cause pollution of land, freshwater, or the seas, according to where they occur. Similarly, pollutants discharged into a river will, unless degraded, enter the seas. Solid wastes are often disposed into a landfill. Nowadays these are

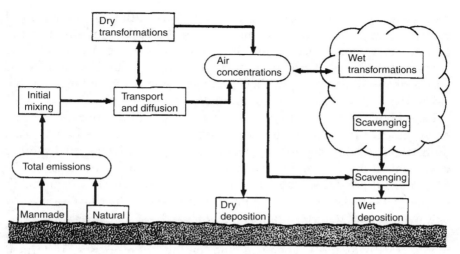

Figure 1 *Schematic diagram of the atmospheric cycle of a pollutant[1]*
(Reprinted from Environmental Science and Technology by permission of the American Chemical Society.)

carefully designed to avoid leaching by rain and dissemination of pollutants into groundwaters, which might subsequently be used for potable supply. In the past, however, instances have come to light where insufficient attention was paid to the potential for groundwater contamination, and serious pollution has arisen as a result.

Another important consideration regarding pollutant cycling is that of degradability, be it chemical or biological. Chemical elements (other than radioisotopic forms) are, of course, non-degradable and hence once dispersed in the environment will always be there, although they may move between compartments. Thus, lead, for example, after emission from industry or motor vehicles, has a rather short lifetime in the atmosphere, but upon deposition causes pollution of vegetation, soils, and waters.[2] On a very long timescale, lead in these compartments will leach out from soils and transfer to oceans, where it will concentrate in bottom sediments.

Some chemical elements undergo chemical changes during environmental cycling, which completely alter their properties. For example, nitrate added to soil as fertilizer can be converted to gaseous nitrous oxide by biological denitrification processes. Nitrous oxide is an unreactive gas with a long atmosphere lifetime, which is destroyed only by breakdown in the stratosphere. As will be seen later, nitrogen in the environment may be present in a wide range of valence states, each conferring different properties.

Some chemical compounds are degradable in the environment. For example, methane (an important greenhouse gas) is oxidized *via* carbon monoxide to carbon dioxide and water. Thus, although the chemical elements are conserved, methane itself is destroyed and were it not continuously replenished would disappear from the environment. The breakdown of methane is an important source of water vapour in the stratosphere, illustrating another, perhaps less obvious, connection between the cycles of different compounds.

Degradable chemicals that cease to be used will disappear from the environment. Polychlorinated biphenyls (PCBs) are no longer used industrially to any significant degree, having been replaced by more environmentally acceptable alternatives. Their concentrations in the environment are decreasing, although because of their slow degradability (*i.e.* persistence), it will take many years before their levels decrease below analytical detection limits.

The transfer of an element between different environmental compartments, involving both chemical and biological processes, is termed biogeochemical cycling. The biogeochemical cycles of the elements lead and nitrogen will be discussed later in this chapter.

7.1.1 Environmental Reservoirs

To understand pollutant behaviour and biogeochemical cycling on a global scale, it is important to appreciate the size and mixing times of the different reservoirs. These are given in Table 1. The mixing times are a very approximate indication of the timescale of vertical mixing of the reservoir.[3] Global mixing can take very much longer as this involves some very slow processes. These mixing times should be treated with considerable caution as they oversimplify a complex system. Thus, for example, a pollutant gas emitted at ground level mixes in the boundary layer (approximately 0–1 km altitude) on a timescale typically of hours. Mixing into the free troposphere (1–10 km) takes days, while mixing into the stratosphere (10–50 km) is on the timescale of several years. Thus, no one timescale describes atmospheric vertical mixing, and the same applies to other reservoirs. Such concepts are useful, however,

Table 1 *Size and vertical mixing of various reservoirs (from Brimblecombe[3])*

	Mass (kg)	Mixing time (years)
Biosphere	4.2×10^{15}	60
Atmosphere	5.2×10^{18}	0.2
Hydrosphere	1.4×10^{21}	1600
Crust	2.4×10^{22}	$>3 \times 10^7$
Mantle	4.0×10^{24}	$> 10^8$
Core	1.9×10^{24}	

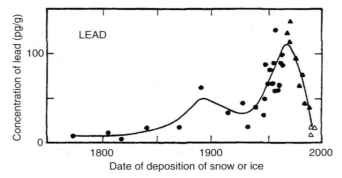

Figure 2 *Changes in lead concentrations in snow/ice deposited in central Greenland from 1773 to 1992*
(Adapted from Candelone *et al.*[4])

when considering the behaviour of trace components. For example, a highly reactive hydrocarbon emitted at ground level will probably be decomposed in the boundary layer. Sulfur dioxide, with an atmospheric lifetime of days, may enter the free troposphere but it is unlikely to enter the stratosphere. Methane, with a lifetime of several years, extends through all of the three regions.

It should be noted from Table 1 that the atmosphere is a much smaller reservoir in terms of mass than the others. The implication is that a given pollutant mass injected into the atmosphere will represent a much larger proportion of total mass than in other reservoirs. Because of this, and the rather rapid mixing of the atmosphere, *global* pollution problems have become serious in relation to the atmosphere before doing so in other environmental media. The converse also tends to be true, that once emissions into the atmosphere cease, or diminish, the beneficial impact is seen on a relatively short timescale. This has been seen in relation to lead, for instance, where lead in Antarctic ice (derived from snow) has shown a major decrease resulting from diminishing emissions from industry and use of leaded petrol[4] (Figure 2). Improved air quality in relation to CFCs will take longer to achieve because of the much longer atmospheric lifetimes (> 100 years) of some species (see Chapter 1).

7.1.2 Lifetimes

A very useful concept in the context of pollutant cycling is that of the lifetime or residence time of a substance in a given reservoir. We can think in terms of substances having sources, magnitude S, and sinks, magnitude R. At equilibrium

$$R = S$$

An analogy is with a bath; the inflow from a tap (S) is equal to the outflow (R) when the bath is full. An increase in S is balanced by an increase in R. If the total amount of substance A in the reservoir (analogy = mass of water in the bath) is A, then the lifetime, τ is defined by

$$\tau = \frac{A \ (\text{kg})}{S \ (\text{kg s}^{-1})} \tag{7.1}$$

In practical terms, the lifetime is equal to the time taken for the concentration to fall to $1/e$ (where e is the base of natural logarithms) of its initial concentration, if the source is turned off. If the removal mechanism is a chemical reaction, its rate may be described as follows

$$R' = \frac{d[A]}{dt} = k[A] \tag{7.2}$$

(In this case $d[A]/dt$ describes the rate of loss of A if the source is switched off; obviously with the source on, at equilibrium $d[A]/dt=0$.) The latter part of Equation (7.2) assumes first order decay kinetics, *i.e.* the rate of decay is equal to the concentration of A, termed $[A]$, multiplied by a rate constant, k. As discussed later this is often a reasonable approximation.

Taking Equation (7.1) and dividing both numerator and denominator by the volume of the reservoir, allows it to be rewritten in terms of concentration. Thus

$$\tau = \frac{[A] \ (\text{kg m}^{-3})}{S' \ (\text{kg m}^{-3} \ \text{s}^{-1})} \tag{7.3}$$

since $S' = R'$

$$\tau = \frac{[A]}{k[A]} = k^{-1} \tag{7.4}$$

Thus the lifetime of a constituent with a first order removal process is equal to the inverse of the first order rate constant for its removal. Taking an example from atmospheric chemistry, the major removal mechanism for many trace gases is reaction with hydroxyl radical, OH. Considering two substances with very different rate constants[5] for this reaction, methane and nitrogen dioxide

$$CH_4 + OH \rightarrow CH_3 + H_2O \tag{7.5}$$

$$\frac{-d}{dt}[CH_4] = k_2 \, [CH_4] \, [OH] \quad k_2 = 6.2 \times 10^{-15} \, \text{cm}^3 \, \text{molec s}^{-1} \tag{7.6}$$

$$NO_2 + OH \rightarrow HNO_3 \quad k_2 = 1.4 \times 10^{-11}\,cm^3\,molec\,s^{-1} \qquad (7.7)$$

Making the crude assumption of a constant concentration of OH radical[6] (more justifiable for the long-lived methane, for which fluctuations in OH will average out, than for short-lived nitrogen dioxide)

$$\frac{-d}{dt}[CH_4] = k_2[CH_4]\,[OH]$$
$$= k_1'[CH_4]$$

where $k_1' = k_2[OH]$

Worked example

What are the atmospheric lifetimes of CH_4 and NO_2 if the diurnally averaged concentration of OH radical is $1 \times 10^6\,molec\,cm^{-3}$?

$$k_1' = 6.2 \times 10^{-15} \times 1 \times 10^6$$
$$= 6.2 \times 10^{-9}\,s^{-1}$$

Then from Equation (7.4)

$$\tau = k^{-1}$$
$$= (6.2 \times 10^{-9})^{-1}\,s$$
$$= 5.1\ \text{years for } CH_4$$

By analogy, for nitrogen dioxide, the lifetime
$$\tau = 20\ h$$

This general approach to atmospheric chemical cycling has proved useful in many instances. For example, measurements of atmospheric concentration, $[A]$, for a globally mixed component may be used to estimate source strength, since

$$S' = R' = \frac{-d[A]}{dt} = k_2[A][OH]$$

and

$$S = S' \times V$$

where V is the volume of atmosphere in which the component is mixed. Source strengths estimated in this way, for example, for the compound methyl chloroform, CH_3CCl_3, known to destroy stratospheric ozone, may be compared with known industrial emissions to deduce whether natural sources contribute to the atmospheric burden.

7.1.2.1   Influence of Lifetime on Environmental Behaviour.   Some knowledge of environmental lifetimes of chemicals is very valuable in predicting their environmental behaviour. In relation to the atmosphere, there is an interesting relationship between the spatial variability in the concentrations of an atmospheric trace species and its atmospheric lifetime.[3] Compounds such as methane and carbon dioxide with a long lifetime with respect to removal from the atmosphere by chemical reactions or dry and wet deposition (see Section 7.2 of this chapter) show little spatial variability around the globe, as their atmospheric lifetime (several years) exceeds the timescale of mixing of the entire troposphere (of the order of a year). On the other hand, for a short-lived species such as nitrogen dioxide, removal by chemical means or dry or wet deposition occurs much more quickly than atmospheric mixing and hence there is very large spatial variability, with concentrations sometimes exceeding 100 ppb in urban areas, while remote atmosphere concentrations can be a the level of a few parts per trillion. By analogy, short-lived species also show a much greater hour-to-hour and day-to-day variation at a given measuring point than long-lived species for which local sources impact only to a modest degree on the existing background concentration.

This illustration using the atmosphere can be taken somewhat further in relation to other environmental media. Lifetimes of highly soluble species such as sodium and chloride in the oceans are long compared to the mixing times and therefore variations in salinity across the world's oceans are relatively small (see Chapter 4). In contrast, where soils are concerned, mixing times will generally far exceed lifetimes and extreme local hot spot concentrations can be found where soils have become polluted.

Lifetime also influences the way in which we study the environmental cycles of pollutants. In the case of reactive atmospheric pollutants, it is the reaction rate, or rate of dry or wet deposition, which determines the lifetime. We are therefore concerned mainly with the rates of these processes in determining the atmospheric cycle. In the case of longer-lived species, such as persistent organic compounds like PCBs and dioxins, chemical reaction rates are rather slow and these compounds can approach equilibrium between different environmental media such as the atmosphere and surface ocean or the atmosphere and surface soil, with evaporation exceeding deposition during warmer periods and wet and dry depositions replacing the contaminant in the soils or oceans in cooler weather conditions. Both the kinetic approach dealing with reaction rates and the thermodynamically based approach considering partition between environmental media will be introduced in this chapter. In general the kinetic or reaction rate approach will be most appropriate to the study of short-lived reactive substances, while the equilibrium approach will be more applicable to long-lived substances.

7.2 RATES OF TRANSFER BETWEEN ENVIRONMENTAL COMPARTMENTS

7.2.1 Air–Land Exchange

The land surface is an efficient sink for many trace gases. These are absorbed or decomposed on contact with plants or soil surfaces. Plants can be particularly active because of their large surface area and ability to absorb water-soluble gases. The deposition process is crudely described by the deposition velocity, v_d,

$$v_d (\text{cm s}^{-1}) = \frac{\text{Flux} \ (\mu g \, m^{-2} \, s^{-1})}{\text{Atmospheric concentration} \ (\mu g \, m^{-3})}$$

The term *flux* is analogous to a flow of material, in this case expressed as micrograms of substance depositing per square metre of ground surface per unit time. In the case of rough surfaces the square metre of area refers to the area of hypothetical horizontal flat surface beneath the true surface rather than the sum of the area of all the rough elements such as plant leaves, which make up the true surface.

Since the deposition process itself causes a gradient in atmospheric concentration, v_d is defined in relation to a reference height, usually 1 m, at which the atmospheric concentration is measured. For reasons described later, v_d is not constant for a given substance, but varies according to atmospheric and surface conditions. However, some typical values are given in Table 2, which exemplify the massive variability.

Table 2 *Some typical values of deposition velocity*

Pollutant	Surfaces	Deposition velocity (cm s⁻¹)
SO_2	Grass	1.0
SO_2	Ocean	0.5
SO_2	Soil	0.7
SO_2	Forest	2.0
O_3	Dry grass	0.5
O_3	Wet grass	0.1
O_3	Snow	0.1
HNO_3	Grass	2.0
CO	Soil	0.05
Aerosol ($>2.5 \ \mu m$)	Grass	0.1

For some trace gases, for example, nitric acid vapour, dry deposition represents a major sink mechanism. In this case the process may have a major impact upon atmospheric lifetimes.

Worked example
Dry deposition is frequently the main sink for ozone in the rural atmospheric boundary layer. What is the lifetime of ozone with respect to this process?
Assuming a typical dry deposition velocity of 1 cm s^{-1} and a boundary layer height of 1000 m, (H),

$$\frac{-d}{dt}[O_3] \ (\mu g \ m^{-3} \ s^{-1}) = \frac{Flux(\mu g \ m^{-2} \ s^{-1})}{Mixing \ depth \ (m)}$$

$$= \frac{\nu_d \times [O_3]}{H}$$

$$= k[O_3]$$

where $k = \nu_d/H$
By analogy with Equation (7.4),

$$\tau = \frac{H}{\nu_d}$$

$$= \frac{1000}{0.01 \ s}$$

$$= 28 \ h$$

Thus taking the boundary layer as a discrete compartment, the lifetime of ozone with respect to dry deposition is around 1 day. The lifetime in the free troposphere (the section of the atmosphere above the boundary layer) is longer, being controlled by transfer processes in and out, and chemical reactions. The stratosphere lifetime of ozone is controlled by photochemical and chemical reaction processes.

Dry deposition processes are best understood by considering a resistance analogue. In direct analogy with electrical resistance theory, the major resistances to deposition are represented by three resistors in series. Considering the resistances in sequence, starting well above the ground, these are as follows:

(i) r_a, the aerodynamic resistance describes the resistance to transfer downwards towards the surface through normally turbulent air;

(ii) r_b, the boundary layer resistance describes the transfer through a laminar boundary layer (approximately 1 mm thickness) at the surface;

(iii) r_s, the surface (or canopy) resistance is the resistance to uptake by the surface itself. This can vary enormously, from essentially zero for very sticky gases such as HNO_3 vapour, which attach irreversibly to surfaces, to very high values for gases of low water solubility, which are not utilised by plants (*e.g.* CFCs).

Since these resistances operate essentially in series, the total resistance, R, which is the inverse of the deposition velocity, is equal to the sum of the individual resistances.

$$R = \frac{1}{\nu_d} = r_a + r_b + r_s \qquad (7.8)$$

Some trace gases have a net source at the ground surface and diffuse upwards; an example is nitrous oxide.

Whether the flux is downward or upward, it is driven by a concentration gradient in the vertical, dc/dz. The relationship between flux, F, and concentration gradient is

$$F = K_z \frac{dc}{dz}$$

where K_z is the eddy diffusivity in the vertical (a measure of the atmospheric conductance). Fluxes, and thus deposition velocities, can be estimated by measurement of a concentration gradient simultaneously with the eddy diffusivity.[7] It is usually assumed that trace gases transfer in the same manner as sensible heat (*i.e.* convective heat transfer, not radiative or latent heat) or momentum. Thus, the eddy diffusivity for either of these parameters is measured usually from simple meteorological variables (gradients in temperature and wind speed).

A few substances are capable of showing both upward and downward fluxes. An example is ammonia. Ammonium in the soil, NH_4^+, is in equilibrium with ammonia gas, $NH_{3(g)}$

$$NH_4^+ + H_2O \rightleftharpoons NH_{3(g)} + H_3O^+ \qquad (7.9)$$

when atmospheric concentrations of ammonia exceed equilibrium concentrations at the soil surface (known as the compensation point), the net flux of ammonia is downwards. When atmospheric concentrations are below the equilibrium value, ammonia is released into the air.[8]

7.2.2   Air–Sea Exchange

The oceans cover some two-thirds of the Earth's surface and consequently provide a massive area for exchange of energy (climatologically important) and matter (an important component of geochemical cycles).

The seas are a source of aerosol (*i.e.* small particles), which transfer to the atmosphere. These will subsequently deposit, possibly after chemical modification, either back in the sea (the major part) or on land (the minor part). Marine aerosol comprises largely unfractionated seawater, but may also contain some abnormally enriched components. One example of abnormal enrichment occurs on the eastern coast of the Irish Sea. Liquid effluents from the Sellafield nuclear fuel reprocessing plant in west Cumbria are discharged into the Irish Sea by pipeline. At one time, permitted discharges were appreciable and as a result radioisotopes such as ^{137}Cs and several isotopes of plutonium have accumulated in the waters and sediments of the Irish Sea. A small fraction of these radioisotopes were carried back inland in marine aerosol and deposited predominantly in the coastal zone.[9] While the abundance of ^{137}Cs in marine aerosol was reflective only of its abundance in seawater (an enrichment factor – see Chapter 4 – of close to unity), plutonium was abnormally enriched due to selective incorporation of small suspended sediment particles in the aerosol. This has manifested itself in enrichment of plutonium in soils on the west Cumbrian coast,[10] shown as contours of $^{239+240}$Pu deposition (pCi cm^{-2}) to soil in Figure 3.

The seas may also act as a receptor for depositing aerosol. Deposition velocities of particles to the sea are a function of particle size, density, and shape, as well as the state of the sea. Experimental determination of aerosol deposition velocities to the sea is almost impossible and has to rely upon data derived from wind tunnel studies and theoretical models. The results from two such models appear in Figure 4, in which particle size is expressed as aerodynamic diameter, or the diameter of an aerodynamically equivalent sphere of unit specific gravity.[11,12] If the airborne concentration in size fraction of diameter d_i is c_i, then

$$\text{Total Flux} = \sum^{i} \nu_d(d_i)c_i$$

where $\nu_d(d_i)$ is the mean value of deposition velocity appropriate to the size fraction d_i. Measurements show that while most of the lead, for example, is associated with small, sub-micrometre particles, the larger particles compose the major part of the flux.

Airborne concentrations of particulate pollutants are not uniform over the sea. The spatial distribution of zinc over the North Sea[13]

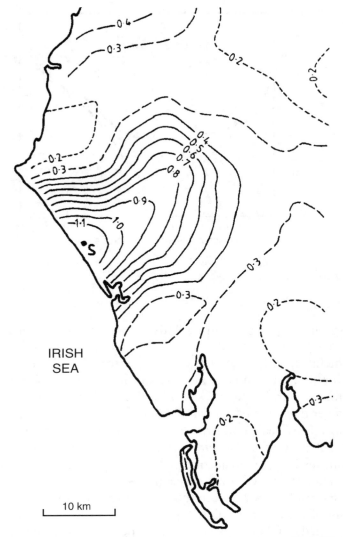

Figure 3 *Concentrations of plutonium in soils of West Cumbria ($^{239\ +\ 240}$Pu to 15 cm depth; pCi cm^{-2}). The point marked S indicates the position of the Sellafield reprocessing works*
(From Cawse[10])

averaged over a number of measurement cruises appears in Figure 5. Spatial patterns of other metals and many artificial pollutants are similar, reflecting the impact of land-based source regions, with concentrations falling towards the north and centre of the sea.

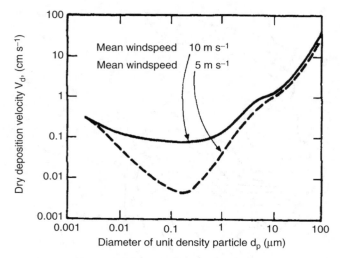

Figure 4 *Calculated values of deposition velocity to water surfaces as a function of particle size and wind speed*

Because of its position and relatively high pollution loading, the North Sea is a focus of considerable interest. An inventory of inputs of trace metals (*e.g.* Pb, Cd, Zn, Cu, *etc.*) accords similar importance to riverine inputs and atmospheric deposition.[14] Controls have been applied to many source categories and total inputs of the metals indicated have in general declined appreciably. One particular example is lead, for which most European countries introduced severe controls on use in gasoline (petrol) during the 1980s and a total ban in 2000; atmospheric concentrations have fallen accordingly. Although the data are less clear, it might be anticipated that concentrations in river water will also decline as a result of reduced inputs from direct atmospheric deposition and in runoff waters from highways and land surfaces.

As explained in Chapter 4, the sea may be both a source and a sink of trace gases. The direction of flux is dependent upon the relative concentration in air and seawater.[15] If the concentration in air is C_a, the equilibrium concentration in seawater, $C_{w\,(equ)}$ is given by

$$C_{w(equ)} = C_a H^{-1} \qquad (7.10)$$

where H is the Henry's Law Constant. The Henry's Law Constant can be expressed as follows:

$$H = \frac{p_s}{S_{aq}}$$

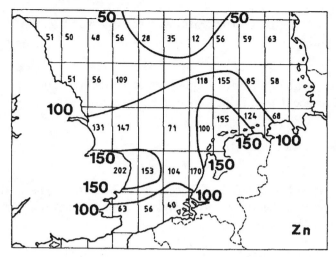

Figure 5 *Spatial distribution of zinc concentrations (in ng m^{-3}) in air over the North Sea during 1989*
(From Ottley and Harrison.[13])

where p_s = saturation vapour pressure and S_{aq} = equilibrium solubility in water.

Worked example
For benzene
$$p_s = 12.7 \text{ kPa at } 25°C$$

$$S_{aq} = 1.78 \text{ g L}^{-1} = \frac{1.78 \times 1000}{78} = \text{mol m}^{-3}$$

Calculate H for benzene at 25°C.
For benzene
$$H = \frac{p_s}{S_{aq}}$$

$$H = \frac{12.7 \times 10^3}{(1.78 \times 10^3)/78}$$
$$= 556 \text{ Pa m}^3 \text{ mol}^{-1}$$

If C_w is the actual concentration of the dissolved gas in the surface seawater and

$$C_w = C_{(equ)}$$

the system is at equilibrium and no net transfer occurs. If, however, there is a concentration difference, ΔC, where

$$\Delta C = C_a H^{-1} - C_w$$

there will be a net flux. If

$$C_a H^{-1} > C_w$$

the water is sub-saturated with regard to the trace gas and transfer occurs from air to water. Conversely, gas transfers from supersaturated water to the atmosphere if

$$C_a H^{-1} < C_w$$

The rate at which gas transfers occurs is expressed by

$$F = K_{(T)w} \Delta C$$

where $K_{(T)w}$ is termed the total transfer velocity. This can be broken down into component parts as follows:

$$\frac{1}{K_{(T)w}} = \frac{1}{\alpha k_w} + \frac{1}{H k_a} = r_w + r_a$$

where k_a and k_w are the individual transfer velocities for chemically unreactive gases in air and water phases, respectively and α ($=k_{reactive}/k_{inert}$) is a factor that quantifies any enhancement of gas transfer in the water due to chemical reaction. The terms r_w and r_a are the resistances to transfer in the water and air phases, respectively, and are directly analogous to the resistance terms in Equation (7.8). For chemically reactive gases, usually $r_a \gg r_w$ and atmospheric transfer limits the overall flux. For less reactive gases the inverse is true and $K_{(T)w} \equiv k_w$; the resistance in the water is the dominant term.

Much research has gone into evaluating k_w and $K_{(T)w}$, both in theoretical models, and in wind tunnel and field studies. The results are highly wind speed dependent due to the influence of wind upon the surface state of the sea. The results of some theoretical predictions and experimental studies[16] for CO_2 (a gas for which k_w is dominant) are shown in Figure 6.

In addition to dry deposition, trace gases and particles are also removed from the atmosphere by rainfall and other forms of precipitation (snow, hail, *etc.*), entering land and seas as a consequence. Wet deposition may be simply described in two ways. First,

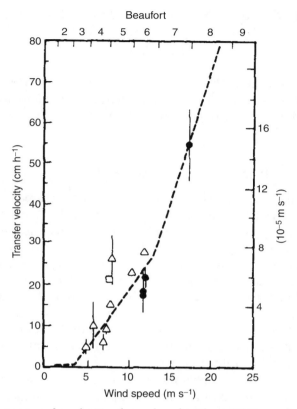

Figure 6 *Air–sea transfer velocities for carbon dioxide at 20°C as a function of wind speed at 10 m (m s⁻¹ or Beaufort Scale). The graph combines experimental data (points) and a theoretical line*
(From Watson *et al.*[16]) (Reprinted by permission from *Nature* (London), **349**, 145; Copyright© 1991 Macmillan Magazines Ltd.)

$$\frac{1}{K_{(T)w}} = \frac{1}{\alpha k_w} + \frac{1}{H k_a} = r_w + r_a$$

Typical values of scavenging ratio[17] lie within the range 300–2000. Scavenging ratios are rather variable, dependent upon the chemical nature of the trace substance (particle or gas, soluble or insoluble, *etc.*) and the type of atmospheric precipitation. Incorporation of gases and particles into rain can occur both by in-cloud scavenging (also termed rainout) and below-cloud scavenging (termed washout).

Numerical modellers often find it convenient to describe wet deposition by a scavenging coefficient, actually a first order rate constant for removal from the atmosphere. Thus, for trace substance A,

$$\frac{d[A]}{dt} = -\Lambda[A]$$

where Λ is the washout coefficient, with units of s^{-1}. A typical value of Λ for a soluble substance is $10^{-4}\ s^{-1}$ although actual values are difficult to measure and are highly dependent upon factors such as rainfall intensity.

7.3   TRANSFER IN AQUATIC SYSTEMS

When rain falls over land some drain off the surface directly into surface water courses in surface runoff. A further part of the incoming rainwater percolates into the soil and passes more slowly into either surface waters or underground reservoirs. Water held in rock below the surface is termed groundwater, and a rock formation that stores and transmits water in useful quantities is termed an aquifer. Water that passes through soil or rock on its way to a river is chemically modified during transit, generally by addition of soluble and colloidal substances washed out of the ground. Some substances are removed from the water; for example, river water often contains less lead than rainwater; one mechanism of removal is uptake by soil.

   River waters carry both dissolved and suspended substances to the sea. The concentrations and absolute fluxes vary tremendously. The suspended solids load is largely a function of the flow in the river, which influences the degree of turbulence and thus the extent to which solids are held in suspension and resuspended from the bed, once deposited. Table 3 shows a comparison of 'average' riverine suspended particulate

Table 3   *A comparison of the concentration of major elements in 'average' riverine particulate material and surficial rocks*

Element	Concentrations ($g\ kg^{-1}$)	
	Riverine particulate material	*Surface rocks*
Al	94.0	69.3
Ca	21.5	45.0
Fe	48.0	35.9
K	20.0	24.4
Mg	11.8	16.4
Mn	1.1	0.7
Na	7.1	14.2
P	1.2	0.6
Si	285.0	275.0
Ti	5.6	3.8

matter and surficial rock composition[18] for the major elements. Elements resistant to chemical weathering or biological activity (*e.g.* aluminium, titanium, iron, phosphorus) show some enrichment in the riverine solids, while more soluble elements are subject to weathering and are depleted in the solids, being transported largely in solution (sodium, calcium). Some pollutant elements such as the metals lead, cadmium, and zinc tend to be highly enriched in the solids relative to surficial rocks or soils due to inputs from human activities.

The dissolved components of river water typically exhibit significantly higher concentrations than in rainwater[19] (Table 4), due to leaching from rocks and soils. Some insight into the processes governing river water composition may be gained from Figure 7. Starting from the point of lowest dissolved salts concentrations, the ratio of $Na/(Na + Ca)$ approaches one. This is similar to rainwater, and is termed the precipitation dominance regime. It is typified by rivers in humid tropical areas of the world with very high rainwater inputs and little evaporation. As the dissolved solids concentration increases the ratio $Na/(Na + Ca)$ declines, indicating an increasing importance for calcium in the rock dominance regime. Here, increased weathering of rock provides the major source of dissolved solids. As dissolved solids increase further, the abundance of calcium decreases relative to sodium as the water becomes saturated with respect to $CaCO_3$, and this compound precipitates. Waters in the evaporation/precipitation regime are typified by rivers in very arid parts of the world (*e.g.* River Jordan) and the major seas and oceans of the world.[20,21] In total, the rivers of the world carry around 4.2×10^{12} kg per year of dissolved solids to the oceans and 18.3×10^{12} kg per year of suspended solids.

Table 4 *Average concentrations of the major constituents dissolved in rain and river water*

Constituent	Concentrations $(mg\ L^{-1})$	
	Rain water	*River water*
Na^+	1.98	6.3
K^+	0.30	2.3
Mg^{2+}	0.27	4.1
Ca^{2+}	0.09	15
Fe		0.67
Al		0.01
Cl^-	3.79	7.8
SO_4^{2-}	0.58	11.2
HCO_3^-	0.12	58.4
SiO_2		13.1
pH	5.7	

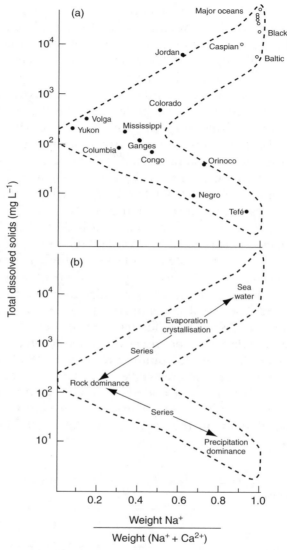

Figure 7   *The chemistry of the Earth's surface waters: (a) typical values of the ratio $Na^+/$*
$(Na^+ + Ca^{2+})$ as a function of dissolved solids concentrations from various
major rivers and oceans and (b) the processes leading to the observed ratios
(From Gibbs[20]) copyright© 1970, American Association for the Advance-
ment of Science

In slow-moving water bodies such as lakes and ocean basins, suspended
solids falling to the bottom produce a well-stratified layer of bottom
sediment. This is stratified in terms of age with the oldest sediment at the
bottom (where when suitably pressurised in can form rock) and the

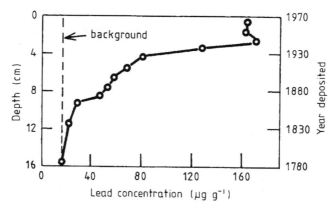

Figure 8 *Lead profile in a lake sediment in relation to depth and the year of incorporation* (From Davies and Galloway.[22])

newest at the top, in contact with the water. If burrowing organisms do not provide too much disturbance (termed bioturbation), the sediment can preserve a record of depositional inputs to the water body. An example is provided by Figure 8 in which lead is analysed in sediment core dated from its radioisotope content.[22] The concentration rises from a background around the year 1800, corresponding to the onset of industrialisation. Considerably increased deposition is seen after 1930 due to the introduction of leaded petrol. While some of the lead input is *via* surface waters, the majority probably arises from atmospheric deposition.

7.4 BIOGEOCHEMICAL CYCLES

A general model of a biogeochemical cycle appears in Figure 9. Although biota are not explicitly included, their role is a very important one in mediating transfers between the idealised compartments of the model. For example, the role of marine phytoplankton in transferring sulfur from the ocean to the atmosphere in the form of dimethyl sulfide has been highlighted in Chapter 4. Biota play a major role in determining atmospheric composition. Photosynthesis removes carbon dioxide from the atmosphere and replenishes oxygen. In a world without biota, lightning would progressively convert atmospheric oxygen into nitrogen oxides and thence to nitrate, which would reside in oceans. Biota also exert more subtle influences. In aquatic sediments, micro-organisms often deplete oxygen more quickly than it can be replenished from the overlying water, producing anoxic conditions. This leads to chemical

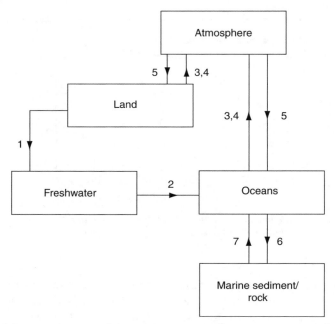

Figure 9 *Schematic diagram of the major fluxes and compartments in a biogeochemical cycle: (1) runoff; (2) streamflow; (3) degassing; (4) particle suspension; (5) wet and dry deposition; (6) sedimentation and (7) remobilisation*

reduction of elements such as iron and manganese, which has implications for their mobility and bioavailability.

Biological reduction processes in sediments may be viewed as the oxidation of carbohydrate (in its simplest form CH_2O) with accompanying reduction of an oxygen carrier. In the first instance, dissolved molecular oxygen is used. The reaction is thermodynamically favoured, as reflected by the strongly negative ΔG.

$$CH_2O + O_2 \rightarrow CO_2 + H_2O \quad \Delta G = -125.5 \text{ kJ mol}^{-1} \text{ e}^-$$

When all of the dissolved oxygen is consumed, anaerobic organisms take over. Initially, nitrate-reducing bacteria are favoured

$$2CH_2O + NO_3^- + 2H^+ \rightarrow 2CO_2 + H_2O + NH_4^+$$
$$\Delta G = -82.2 \text{ kJ mol}^{-1} \text{ e}^-$$

Once the nitrate is utilised, sulfate reduction takes over

$$SO_4^{2-} + H^+ + 2CH_2O \rightarrow HS^- + 2H_2O + 2CO_2$$
$$\Delta G = -25.6 \text{ kJ mol}^{-1} \text{ e}^-$$

Finally, methane-producing organisms dominate in a sediment depleted in oxygen, nitrate, and sulfate.

$$2CH_2O \rightarrow CH_4 + CO_2 \quad \Delta G = -23.5 \text{ kJ mol}^{-1} \text{ e}^-$$

Thus highly anoxic waters are commonly sources of hydrogen sulfide, H_2S, from sulfate reduction and of methane (marsh gas). The formation of sulfide in sediments has led to precipitation of metal sulfides over geological time, causing accumulations of sulfide minerals of many elements, *e.g.* PbS, ZnS, HgS, *etc.*

7.4.1 Case Study 1: The Biogeochemical Cycle Of Nitrogen

Nitrogen has many valence states available and can exist in the environment in a number of forms, depending upon the oxidising ability of the environment. Figure 10 indicates the most important oxidation states and the relative stability (in terms of free energy formation).[23] The oxides of nitrogen represent the most oxidised and least thermodynamically stable forms. These exist only in the atmosphere. Ammonia can exist in gaseous form in the atmosphere but rather rapidly returns to the soil and waters as ammonium, NH_4^+. Fixation of atmosphere N_2 by leguminous plants leads to ammonia, NH_3. In aerobic soils and aquatic systems, NH_3 and NH_4^+ are progressively oxidised by micro-organisms *via* nitrite to nitrate. The latter is taken up by some biota and used as a nitrogen source in synthesising amino acids and proteins; the most thermodynamically stable form of nitrogen. After the death of the organism, microbiological processes will convert organic nitrogen to ammonium (ammonification), which is then available for oxidation or use by plants. Conversion of ammonia to nitrate is termed nitrification, while denitrification involves conversion of nitrate to N_2.

Figure 11 shows an idealised nitrogen cycle. The numbers in boxes represent quantities of nitrogen in the various reservoirs, while the arrows show fluxes.[23] It is interesting to note that substances involving relatively small fluxes and burdens can have a major impact upon people. Thus nitrogen oxides, NO, NO_2, and N_2O are very minor constituents relative to N_2 but play major roles in photochemical air pollution (NO_2), acid rain (HNO_3 from NO_2), and stratospheric ozone depletion (N_2O).

Nitrate from fertilisers represents a very small flux but has major implications in terms of eutrophication of surface waters.

7.4.2 Case Study 2: Aspects of Biogeochemical Cycle of Lead

Lead is a simpler case to study than nitrogen due to the small number of available valence states. The major use of lead until recently was a

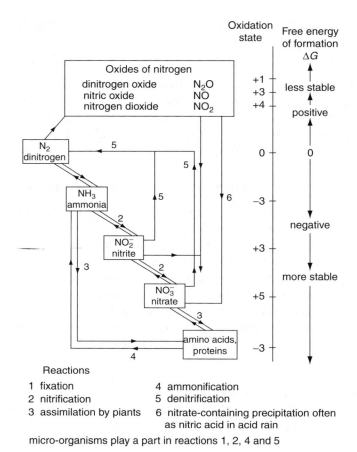

Figure 10 *Chemical forms and cycle of nitrogen*
(From O'Neill.[23])

tetraalkyl lead gasoline additives in which lead is present as Pb^{IV}. The predominant compounds used were tetramethyl lead, $Pb(CH_3)_4$, and tetraethyl lead, $Pb(C_2H_5)_4$. These are lost to the atmosphere as vapour from fuel evaporation and exhaust emissions from cold vehicles, but comprised only 1–4% of lead in polluted air.[2] Leaded gasoline also contains the scavengers 1,2-dibromoethane, CH_2BrCH_2Br, and 1,2-dichloroethane, CH_2ClCH_2Cl, which convert lead within the engine to lead halides, predominantly lead bromochloride, $PbBrCl$, in which lead is in the Pb^{II} valence state, its usual form in environmental media. About 75% of lead alkyl burned in the engine is emitted as fine particles of inorganic lead halides. Atmospheric emissions of lead arise also from industry; both these and vehicle-emitted lead have declined massively. Figure 12 shows trends in United Kingdom emissions of lead to

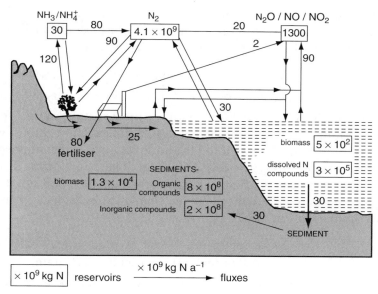

Figure 11 *Schematic representation of the biogeochemical cycle of nitrogen, indicating the approximate magnitude of fluxes and reservoirs (After O'Neill.[23])*

atmosphere from leaded petrol upto 1996, since when emissions from leaded petrol have declined almost to zero.[24] Lead emitted to the atmosphere has a lifetime of around 7–30 days and hence may be subject to long-range transport. Concentrations of trace elements in polar ice provide a historical record of atmospheric deposition. Measurements (Figure 2) have shown a marked enhancement in lead accompanying the increase in leaded gasoline usage, and a major decline in recent years attributable to reduced emissions to atmosphere.

Atmospheric lead is deposited in wet and dry deposition. Lead is relatively immobile in soil, and agricultural surface soils in the UK exhibit concentrations approximately double those of background soil, which contain *ca.* 15–20 mg kg^{-1} derived from soil parent materials, other than in areas of lead mineralisation where far greater concentrations can be found. Local perturbations to the cycle of lead can be important. For instance, the lead content of garden soils correlates strongly with the age of the house. This is probably due to the deterioration of leaded paintwork on older houses and the former practices of disposing of household refuse and fire ashes in the garden. Lead is also of low mobility in aquatic sediments and hence the sediment may provide a record of historical lead deposition (see Figure 8).

Plants can take up lead from soil, thus providing a route of human exposure. Careful research in recent years has established transfer factors, terms the Concentration Factor (CF), where

$$CF = \frac{\Delta \text{ Concentration of lead in plant (mg kg}^{-1} \text{ dry wt.)}}{\Delta \text{ Concentration of lead in soil (mg kg}^{-1} \text{ dry wt.)}}$$

The value of CF for lead is lower than for most metals and is typically within the range $10^{-3} - 10^{-2}$. Much higher values had been estimated from earlier studies, which ignored the importance of direct atmospheric deposition as a pathway for contamination. The direct input from the air to leaves of plants is often as great, or greater than soil uptake.[24,25] This pathway may be described by another transfer factor, termed the Air Accumulation Factor (AAF), where

$$AAF(m^3 \text{ g}^{-1}) = \frac{\Delta \text{ Concentration of lead in plant (}\mu\text{g g}^{-1} \text{ dry wt.)}}{\Delta \text{ Concentration of lead in air (}\mu\text{g m}^{-3})}$$

Values of AAF are plant dependent, due to differences in surface characteristics, but values of 5–40 are typical.[25,26] Thus, a plant grown on an agricultural soil with 50 mg kg^{-1} lead will derive 0.25 mg kg^{-1} dry weight lead from the soil (CF $= 5 \times 10^{-3}$), while airborne lead of 0.1 μg m^{-3} will contribute 2.0 μg g^{-1} ($\equiv$ mg kg^{-1}) of lead (AAF $= 20$ m^3 g^{-1}). Thus, in this instance airborne lead deposition is dominant. The air lead concentration of 0.1 μg m^{-3} was typical of rural areas of the UK until 1985. Since that time, the drastic reduction of lead in gasoline has led to appreciably reduced lead-in-air concentrations in both urban and rural localities.

Human exposure to lead arises from four main sources:[2,27]

(i) Inhalation of airborne particles. The adult human respires approximately 20 m^3 of air per day. Thus for an urban lead concentration of 0.1 μg m^{-3}, *intake* is 2 μg per day. This is rather efficiently absorbed (*ca.* 70%) and therefore *uptake* is around 1.4 μg per day in this instance.

(ii) Indigestion of lead in foodstuffs. The concentrations of lead in food obviously vary between different foodstuffs and even between different batches of the same food. Typical fresh weight concentrations (much of the weight of some foods is water) are from 10 to 50 μg Pb kg^{-1}. Thus a food consumption of 1.5 kg per day represents an *intake* of around 50 μg per day and an *uptake* (10–15% efficient) of around 6 μg per day.

(iii) Drinking water and beverages. Concentrations of lead in drinking water vary greatly, related particularly to the presence or absence of lead in the household plumbing system. Most households in the UK conform to the EC standard of 50 µg L^{-1} and a concentration of 4 µg L^{-1} may be taken as representative. Gastrointestinal absorption of lead from water and other beverages is highly dependent upon food intake. After long fasting, absorptions of 60–70% have been recorded, 14–19% with a short period of fasting before and after the meal, and only 3–6% for drinks taken with a meal. If 15% is taken as typical, for a daily consumption of 1.5 L, *intake* is 7.5 µg and uptake 1.1 µg.

(iv) Cigarette smoking exposes the individual to additional lead.

While both individual exposure to lead and the uptake efficiencies of individuals are very variable, it is evident that exposure arises from a number of sources and control or human lead intake, if deemed to be desirable, requires attention to all of those sources. An additional pathway of exposure, not easily quantified, and not included above is ingestion of lead-rich surface dust by hand-to-mouth activity in young children.

The above calculations estimate that for a typical adult in a developed country, daily uptake of lead from air, diet, and drinking water is, respectively 1.4 µg, 6 µg, and 1.1 µg. Exposure to lead from all of these sources has fallen rapidly over the past 20–30 years. Figure 12 contrasts the temporal trends in use of lead in petrol (gasoline) and blood leads in the general population of the UK over the period when much of this decline took place. It is interesting to note that from 1971 to 1985 use of lead in petrol was relatively steady, but blood leads declined by a factor of more than two over this period mainly as a response to reductions in dietary exposure, particularly associated with the cessation of use of

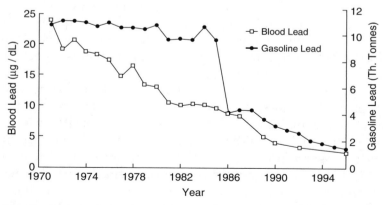

Figure 12 *Trends in lead use in petrol (gasoline) and of lead in the blood of the general population in the United Kingdom, 1970–1995*

leaded solder to seal food cans. A dramatic reduction in gasoline lead usage occurred at the end of 1985 when the maximum permissible lead content of petrol was reduced from 0.4 g L^{-1} to 0.15 g L^{-1}, and there has been a steady reduction in lead use since, with the increases in market penetration of unleaded fuel, until it became mandatory in the year 2000. Despite the ability of a vehicle emitting lead to cause direct lead exposure through the atmosphere, as well as indirect exposure through contamination of food and water, the lack of any obvious step change in blood lead associated with the reduction of lead in petrol shows clearly that at that time leaded petrol was not a major source of direct exposure for the general population.

7.5 BEHAVIOUR OF LONG-LIVED ORGANIC CHEMICALS IN THE ENVIRONMENT

A very large number of organic chemicals are manufactured and have a wide variety of uses. In most cases of manufacture, distribution or use, some of the chemical becomes released and dispersed in the environment. In some instances, the chemical is highly reactive and degrades very rapidly; often the main mechanism is breakdown by bacteria or other organisms and there are standard tests to evaluate the biodegradability of synthetic chemicals. On the other hand, many such chemicals, in order to be useful in products, are relatively long-lived and hence may be persistent within the environment. This raises the question of which environmental compartment (air, water, soil, *etc.*) the chemical will partition into predominantly. Such knowledge is critical to predicting possible environmental effects and directing monitoring programmes. Over the past few years Mackay and colleagues[28] have developed methods for predicting the behaviour of chemicals released into the environment and at the simplest level, these are based solely upon an assumption of equilibrium partitioning between the various environmental compartments. The basic principles are described in Chapter 6, and this section is concerned with how these methods are applied in a practical way.

The approach depends upon assuming a model environment consisting of specified volumes of air, water, soil, and sediment (see Table 5). Within those compartments, there are other components such as particles in water, fish, or soil solids, and some of these are assumed to contain an organic fraction of defined abundance (see Table 5). The presence of this organic matter is important since most organic compounds will preferentially partition into an organic-rich medium.

Having defined a model environment, partitioning between the various compartments is predicted on the basis of equilibrium relationships,

Table 5   *Properties of a "model world" for modelling the behaviour of organic chemicals*[28]

	Area (m^2)	Depth (m)	Volume (m^3)	Advective residence Time (h)
Air	10^{11}	1000	10^{14}	100
Water	10^{10}	20	2×10^{11}	1000
Soil	9×10^{10}	0.2	1.8×10^{10}	
Sediment	10^{10}	0.05	5×10^8	50,000

	Volume fractions	Organic carbon or lipid mass fraction	Density, $kg\ m^{-3}$
Aerosol	2×10^{-11}		2000
Particles in water	5×10^{-6}	0.2	1500
Fish	10^{-6}	0.5	1000
Soil pore air	0.2		
Soil pore water	0.3		
Soil solids	0.5	0.02	2400
Sediment pore water	0.8		
Sediment solids	0.2	0.04	2400

which relate to the fundamental physico–chemical properties of the organic chemical under consideration. Important in this context is the Henry's Law constant, which describes the equilibrium partitioning between air and water that is defined as

$$H = K_{aw} = \frac{p}{C_W} = \frac{p_s}{S_{aq}}$$

where p is the vapour pressure in air

C_W, the concentration in water,
S_{aq}, the equilibrium aqueous solubility, and
p_s, the saturation vapour pressure for the pure substance.

Partition between compartments within the aqueous environment is typically described through the octanol–water partition coefficient, K_{ow}, or the organic carbon–water partitioning coefficient, K_{oc} (see Chapter 5). This approach assumes that partitioning from water into fish or suspended solids within the water is determined by the availability of organic matter whose properties as a solvent can be described by a parallel with octanol for which partitioning data are widely available

$$K_{ow} = \frac{\text{Concentration in octanol}}{\text{Concentration in water at equilibrium}}$$

A further partition coefficient is the octanol–air partition coefficient, which can be used to describe the partitioning between the organic matter in soil, for which octanol is taken as a surrogate, and the atmosphere. Thus,

$$K_{oa} = \frac{\text{concentration in octanol}}{\text{concentration in air at equilibrium}}$$

Values of these partition coefficients have been measured for a wide variety of substances, but for those which measurements have not been made, there are methods of calculation based upon fundamental properties of the molecule.

Frequently, the organic matter comprises only a fraction, f, of the medium. Therefore for partition between air and soil containing 20% organic matter content

$$K_{oa} = \frac{\text{concentration in organic matter}}{\text{concentration in air}}$$
$$= \frac{C_{om}}{p}$$
$$= \frac{C_{soil}}{fp}$$

and $C_{soil} = K_{oa}fp = 0.2\,K_{oa}.p$

Worked example

What concentration of naphthalene in a soil of 20% organic matter will be in equilibrium with a concentration of 10 ppb vapour in air (water solubility = 31 g m^{-3}; saturation vapour pressure = 10.4 Pa and log K_{ow} = 3.37).

In a saturated system,

$$K_{ow} = 2.35 \times 10^3 = \frac{C_{octanol}}{31}$$

and

$$C_{octanol} = 7.28 \times 10^4 \text{ g m}^{-3} \text{ at equilibrium}$$

Then,

$$K_{oa} = \frac{7.28 \times 10^4}{10.4}$$
$$= 7.00 \times 10^3 \text{ g Pa}^{-1} \text{ m}^{-3}$$

A concentration of 10 ppb in air $= 1.013 \times 10^5 \times 10^{-8}$ Pa $= 1.013 \times 10^{-3}$ Pa

$$
\begin{aligned}
C_{\text{octanol}} &= 1.013 \times 10^5 \times 10^{-8}\,\text{Pa} \\
&= 1.013 \times 10^{-3}\,\text{Pa} \\
&= 1.013 \times 10^{-3} \cdot K_{\text{oa}} \\
&= 1.013 \times 10^{-3} \cdot 7.00 \times 10^3 \\
&= 7.09\,\text{g m}^{-3}
\end{aligned}
$$

and

$$
\begin{aligned}
C_{\text{soil}} &= 0.2 \cdot C_{\text{octanol}} \\
&= 0.2 \cdot 7.09 \\
&= 1.42\,\text{g m}^{-3}
\end{aligned}
$$

In order to calculate the equilibrium partitioning of a substance within the model world, a mass balance exercise is conducted. Thus, for a system comprising simply air, water, and soil, the mass balance equation would be

$$
\begin{aligned}
M &= V_1 C_1 + V_2 C_2 + V_3 C_3 \\
&= V_1 C_1 + V_2 C_1 K_{21} + V_3 C_1 K_{31} \\
&= C_1 (V_1 + V_2 K_{21} + V_3 K_{31})
\end{aligned}
$$

where M is the total mass of substance and V_i and C_i represent, respectively the volumes and concentrations of each medium, i.

It may be seen that by rearranging this equation, the concentration C_1 in one of the media can be estimated from knowledge of the total mass of compound in the system, the volume of each of the three media and the equilibrium partition coefficients between them.

It may also be seen that if the various partition coefficients are known, then concentrations in all three compartments, C_1, C_2, and C_3 can be calculated and therefore the overall distribution of the chemical within the environment inferred. For partitioning, for example, into soil, it is assumed that the substance partitions solely into the organic carbon fraction of the soil solids, and since this is a relatively small percentage of the total soil, the mass fraction of organic matter has to be included in the calculation.

This describes one of the most basic kinds of calculations, which nonetheless can be quite powerful. More sophisticated methods, which take account of transport times within the environment and breakdown rates of chemicals are also available. Such models are widely used in predicting the environmental fate of organic chemicals.

QUESTIONS

(i) Discuss what is meant by a biogeochemical cycle, describing the major facets and illustrating the processes involved with examples.

(ii) Discuss the temporal trends in lead emissions and concentrations in the environment and how environmental media can be used to elucidate historical trends in environmental lead.

(iii) Explain what is meant by an environmental lifetime and derive an expression for environmental lifetime in terms of a chemical rate constant. Compare and contrast the typical atmospheric lifetimes of methane, nitrogen oxide, and the CFCs and explain how this relates to the atmospheric distribution and properties of these compounds.

(iv) Explain the processes by which trace substances can exchange between the atmosphere and the oceans and show how rates of exchange can be calculated. Give examples of substances whose exchange between these media is important.

(v) Explain why the waters in rivers in different parts of the world have differing composition and relate this to the climatology of the region. Explain carefully what is meant by dissolved and suspended solids and explain how both arise.

(vi) Explain the environmental pathways followed by lead emissions from road traffic after emission to the atmosphere and explain how this can lead to pollution of a range of environmental media. Indicate the quantitative ways in which such transfer can be expressed.

(vii) Estimate atmospheric lifetimes for the following:

(i) methane, if the globally and diurnally averaged concentration of hydroxyl radical is 5×10^5 cm^{-3}

(ii) nitrogen dioxide in the middle of a summer day when the concentration of hydroxyl radical is 8×10^6 cm^{-3}

(iii) nitrogen dioxide at nighttime if the sole mechanism of removal is dry deposition with a deposition velocity of 0.1 cm s^{-1}, and the mixing depth is 100 m.

(viii) If the atmospheric concentration of sulfur dioxide is 10 ppb, calculate the following:

    (i) the atmospheric concentration expressed in $\mu g\ m^{-3}$ at one atmospheric pressure and 25°C. (Use the fact that 1 mol of any gas (64 g in the case of sulfur dioxide) occupies 22.4 L at STP)

    (ii) the deposition flux to the surface if the deposition velocity is $1.0\ cm\ s^{-1}$

    (iii) the atmospheric lifetime with respect to dry deposition for a mixing depth of 800 m

    (iv) the atmospheric lifetime with respect to oxidation by hydroxyl radical if the diurnally averaged OH radical concentration is $8 \times 10^5\ cm^{-3}$ and the rate constant for the SO_2-OH reaction is $9 \times 10^{-13}\ cm^3\ molec^{-1}\ s^{-1}$

    (v) the lifetime with respect to wet deposition if the washout coefficient is $10^{-4}\ s^{-1}$.

(ix) Explain the thermodynamic controls on biological reduction processes in aquatic sediments and explain how these influence the chemical forms of nitrogen in the environment.

(x) If any organic chemical has a log K_{ow} of 4.3 and a river water concentration of 10 $\mu g\ L^{-1}$, what concentration is predicted to occur in fish if assumed to have a 50% organic matter (lipid) content (assume that the fish and the water are of equal density).

REFERENCES

1. W.H. Schroeder and D.A. Lane, *Environ. Sci. Technol.*, 1988, **22**, 240.
2. R.M. Harrison and D.P.H. Laxen, *Lead Pollution: Causes and Control*, Chapman & Hall, London, 1981.
3. P. Brimblecombe, *Air Composition and Chemistry*, 2nd edn, Cambridge University Press, Cambridge, 1996.
4. J.-P. Candelone, S. Hong, C. Pellone and C.F. Boutron, *J. Geophys. Res.*, 1995, **100**, 16605–16616.
5. B.J. Finlayson-Pitts and J.N. Pitts Jr., *Atmospheric Chemistry*, Wiley, Chichester, 1986.
6. C.N. Hewitt and R.M. Harrison, *Atmos. Environ.*, 1985, **19**, 545.
7. J.A. Garland, *Proc. R. Soc. London, Ser. A*, 1977, **354**, 245.
8. S. Yamulki, R.M. Harrison and K.W.T. Goulding, *Atmos. Environ.*, 1996, **30**, 109–118.
9. R.S. Cambray and J.D. Eakins, *Nature*, 1982, **300**, 46.
10. P.A. Cawse, UKAEA Report No. AERE-9851, 1980.

11. S.A. Slinn and W.G.N. Slinn, *Atmos. Environ.*, 1980, **14**, 1013.

12. R.M. Williams, *Atmos. Environ.*, 1982, **16**, 1933.

13. C.R. Ottley and R.M. Harrison, Eurotrac ASE Annual Report, Garmisch-Partenkirchen, 1990.

14. R.F. Critchley, *Proceedings of International Conference on Heavy Metals in the Environment*, Heidelberg, Germany; CEP Consultants, Edinburgh, 1983, 1109.

15. P.S. Liss and L. Merlivat, *The Role of Air-Sea Exchange in Geochemical Cycling*, P. Buat-Menard (ed), Reidel, Dordrecht, 1986, 113.

16. A.J. Watson, R.C. Upstill-Goddard and P.S. Liss, *Nature*, 1991, **349**, 145.

17. R.M. Harrison and A.G. Allen, *Atmos. Environ.*, 1991, **25A**, 1719.

18. J.M. Martin and M. Meybeck, *Mar. Chem.*, 1979, **7**, 177–206.

19. R.M. Garrels and F.T. MacKenzie, *Evolution of Sedimentary Rocks*, W.W. Norton (ed), New York, 1971.

20. R.J. Gibbs, *Science*, 1970, **170**, 1088.

21. R.M. Harrison and S.J. de Mora, *Introductory Chemistry for the Environmental Sciences*, 2nd edn, Cambridge University Press, Cambridge, 1996.

22. A.O. Davies and J.N. Galloway, *Atmospheric Pollutants in Natural Waters*, S.J. Eisenreich (ed), Ann Arbor, MI, 1981, 401.

23. P. O'Neill, *Environment Chemistry*, George, Allen, and Unwin, London, 1985.

24. Department of Environment, Transport and the Regions, 'Digest of Environmental Statistics', No. 19, The Stationary Office Ltd., Edinburgh, 1997.

25. R.M. Harrison and M.B. Chirgawi, *Sci. Total Environ.*, 1989, **83**, 13.

26. R.M. Harrison and M.B. Chirgawi, *Sci. Total Environ.*, 1989, **83**, 47.

27. Royal Commission on Environmental Pollution, 'Ninth Report: Lead in the Environment', HMSO, London, 1983.

28. D. Mackay, E. Webster and T. Gouin, Partitioning, persistence and long-range transport of chemicals in the environment, in: *Chemicals in the Environment: Assessing and Managing Risk*, R.E. Hester and R.M. Harrison (eds), Issues in Environmental Science and Technology, No. 22, Royal Society of Chemistry, Cambridge, pp 132–153, 2006.

Glossary

abiotic: physical or chemical processes in the environment as opposed to biologically driven processes (biotic).

adiabatic compression: compression of a fluid without extraction of heat resulting in increased temperature.

advection: horizontal or vertical transfer of material, heat, *etc.*, due to the mass movement of water in the ocean.

aeolian transport: wind-borne transfer of material.

aggregation: process of combining particulate material into larger particles.

algae: simple unicellular or multi-cellular photosynthetic plants, which thrive in a wet environment, *e.g.*, lakes.

alkalinity: a measure of the proton deficit in solution, which is operationally defined by titration with a strong acid to the carbonic acid end point.

alluvium(al): unconsolidated material, consisting of weathered or eroded particles of minerals and rocks, *i.e.*, transported by a flowing river and deposited at points along the flood plain of the river.

ammonification: the bacterial transformation of dissolved organic nitrogen (DON) into dissolved inorganic nitrogen (DIN) with the first product being NH_3.

amphoteric: ability of a substance to act as an acid (proton donor) or base (proton acceptor).

androgenic: general name for any substance with male sex hormone activity in vertebrates.

anoxic: waters devoid of oxygen because circulation is restricted vertically due to thermal or saline stratification and horizontally by topographic boundaries thereby producing stagnant conditions. Examples of this atypical marine environment are the Black Sea and Saanich Inlet (Canada).

apolar compound: a molecule that does not possess functional groups and undergoes molecular interactions that are dominated by Van-der-Waals interactions only (*e.g.*, *n*-alkanes).

aragonite: a solid phase of $CaCO_3$.

authigenic: formed, *e.g.*, minerals, in place during or after deposition of sediment.

authigenic precipitation: solid-phase production in seawater due to a substance exceeding its solubility product, such as the formation of ferromanganese nodules.

bacteria: cellular microorganisms incapable of photosynthesis.

benthic: living on the bottom of a water body, *e.g.*, lake.

bioaccumulation: the progressive increase in concentration of a substance in an organism because the rate of intake *via* the body surface (*e.g.*, from water) or in food is greater than the organism's ability to remove the substance from the body.

bioavailable: the portion of a chemical substance (nutrient, metal, toxin) that can be absorbed, transported, and utilized physiologically.

biogenic hydrocarbon: an organic compound originating from a recent biological source, *e.g.*, biogenic volatile organic compounds (VOCs) emitted from vegetation.

biogenous material: matter produced by the fixation of mineral phases by marine organisms; notable examples being calcite and opaline silica in the marine environment.

biomagnification: a food chain or food web phenomenon, whereby a substance or element increases in concentration at successive trophic levels.

biomethylation: biologically induced reactions by which methyl groups are added to metals, metalloids, and organometallic complexes, with consequent and varied influence on the volatility, bioavailability, and toxicity of the element concerned.

bioremediation: the use of living organisms, generally bacteria, to clean up oil spills or remove other pollutants from soil, water, and wastewater.

calcite: a solid phase of $CaCO_3$.

carbonate compensation depth (CCD): the depth at the sediment–water interface at which no calcareous material is preserved in the sediments.

chelate: a complex formed when a central metal atom shares more than one electron pair with a given ligand thereby forming a ring structure, which exhibits enhanced stability largely due to the entropy effect of releasing large numbers of molecules from the co-ordinated water envelopes.

chelation: the formation of a chelate.

chlorinity (Cl‰): the chloride concentration in seawater, expressed in g kg^{-1}, as measured by Ag^+ titration (*i.e.*, ignoring other halide contributions by assuming Cl^- to be the only reactant).

chromophore: a chemical group or moiety within a molecule that absorbs light.

cloud condensation nuclei (CCN): small particles in the atmosphere onto which water can condense.

conservative behaviour: the concentration of a constituent or absolute magnitude of a property varies only due to mixing processes.

conglomerates: a group of detrital sedimentary rocks consisting of rounded or sub-rounded fragments.

convergence: a region in which the streamlines (*i.e.*, currents) come together causing water to sink.

Coriolis force: the acceleration due to the earth's rotation deflecting moving fluids (*i.e.*, both air and water) to the right in the northern hemisphere and to the left in the southern hemisphere; the magnitude of the effect is a function of latitude, being nil at the equator and increasing towards the poles.

cyanobacteria: commonly known as blue–green algae, which are capable of reproducing rapidly to form algal blooms in nutrient-enriched freshwater and, after death, of releasing toxins.

denitrification: the reduction of nitrogen species to molecular nitrogen (N_2), which can occur under conditions of hypoxia or anoxia.

detritus: disintegrated material of both inorganic and biological origins.

diagenesis: the collection of processes that alter the sediments following deposition; they may be physical (compaction), chemical (cementation, mineral segregation, ion exchange reactions), or biological (respiration).

diatoms: a class of planktonic one-celled algae with skeletons of silica.

dielectric constant: the ratio of electric flux density to electric field, which expresses the degree of non-conductivity of different substances.

divergence: a zone in which the flow fields (*i.e.*, currents) separate.

endocrine disruptor: an exogenous substance or mixture that alters function(s) of the endocrine system and subsequently causes adverse health effects in an intact organism, or its progeny, or (sub-) populations.

enrichment factor: the ratio in a sample compared to a reference substance (*i.e.*, a possible source) of concentration ratios of a given element, X, to a reference element; typically Al is used as a reference element to denote a terrestrial source and Na for a marine source.

eutrophication: over-enrichment of a water body with nutrients, resulting in excessive growth of organisms (*i.e.*, algal blooms) and depletion of oxygen concentrations (hypoxia or anoxia depending upon the organic loading) in sub-surface waters and/or sediments.

flocculation: process of combining particulate material into larger particles facilitated by particle surfaces being coated with organic material.

halmyrolysis: low-temperature reactions in seawater reactions producing secondary material from components of continental or volcanic origin; essentially an extension of chemical weathering of lithogenous components.

halocline: a region in the water column that exhibits a sharp change in salinity.

hermaphroditic: having both male and female sexual organs.

humic substances: large, refractory molecules formed as a result of decomposition of organic matter.

hydrogenous material: matter produced abiotically within the water column.

hydrophobic: the low affinity for water or 'water-hating tendency' of a chemical as measured by its octanol/water partition coefficient.

hydrothermal waters: formed when seawater circulates into a fissured rock matrix and, under conditions of elevated temperature and pressure, results in compositional changes in the aqueous phase occur due to seawater–rock interactions.

hyperkeratosis: skin lesion condition, often in the form of nodules on the palms of the hands and soles of the feet, which can be brought about by exposure to inorganic arsenic in drinking water.

hyperpigmentation: in the case of chronic arsenic poisoning, a finely freckled, 'raindrop' pattern of pigmentation that is particularly pronounced on the trunk and extremities.

hypolimnion: the stratum of water below the thermocline.

hypoxia: waters that are deficient in oxygen.

imposex: the imposition of male sex organs on female marine gastropods, notably resulting from exposure to tributyltin compounds.

ion pair: the transient coupling of a cation and anion, formed by the collision of oppositely charged ions due to electrostatic attraction, during which each ion retains its own co-ordinated water envelope.

isopycnal: line (imaginary or on a chart) or an imaginary surface connecting points, which have the same density.

latent heat of evaporation: the heat required (or released) to convert a unit mass of a substance from a liquid to a gas state (or from gas to liquid) at the same temperature and pressure.

latent heat of fusion: the heat required (or released) to convert a unit mass of a substance from a solid to the liquid state (or from liquid to solid) at the same temperature and pressure.

ligand: a moiety acting as a Lewis base that shares a pair of electrons with a metal atom thereby forming a complex; ligands can be neutral (*e.g.*, H_2O) or anionic (*e.g.*, Cl^-, HCO_3^-) species.

lithogenous material: substances coming from the continents as a result of weathering processes; the most important components being quartz and the clay minerals (kaolinite, illite, montmorillonite, and chlorite).

lysocline: the depth at the sediment–water interface at which appreciable dissolution of $CaCO_{3(s)}$ starts to occur.

marine boundary layer: lowest portion of the atmosphere (from 0.25 to 3.5 km) in contact with the ocean surface.

metallothionein: A protective protein that binds heavy metals such as cadmium and lead thereby detoxifying such contaminants.

molal concentration: the number of moles of solute per kg of solution.

molar concentration: the number of moles of solute in a litre (dm^3) of solution (*i.e.*, molarity).

molecular viscosity: small-scale internal fluid friction that is due to the random motion of the molecules within a smooth-flowing fluid.

mutagen: a chemical substance that induces a change in the of DNA of cells.

nitrification: the stepwise oxidation of NH_4^+ to NO_2^- and eventually to NO_3^-.

nitrogen fixation: the direct uptake and assimilation of molecular nitrogen (N_2) by organisms.

ocean conveyor belt: see thermohaline circulation.

ochre: a natural hydrated form of ferric oxide, often precipitated from acid mine drainage.

oestrogenic: related to, behaving like, or causing effects similar to the natural female sex hormone in vertebrates.

oligotrophic: a term used of lakes that are poorly productive in terms of organic matter formed because of a low nutrient supply, *i.e.*, the opposite of eutrophic.

osmoregulatory: a term used of an organism that actively regulates the osmotic concentration of its internal fluids.

oxygen compensation depth: the horizon in the water column at which the rate of O_2 production by photosynthesis equals the rate of respiratory O_2 oxidation.

permeability: the capacity, *e.g.*, of a rock, to transmit water.

petrogenic hydrocarbon: an organic compound originating from unburnt fossil fuels.

photic zone: the upper surface of the ocean in which photosynthesis can occur, typically taken to be the depth at which sunlight radiation has declined to 1% of the magnitude at the surface.

photosensitiser: a substance that absorbs light and subsequently initiates a photochemical reaction or photophysical alteration in the chemical of interest.

photosynthetically active radiation (PAR): electromagnetic energy in the 400–700 nm wavelength range that can be absorbed by chlorophyll or other light-harvesting pigments in plants and is used for photosynthesis.

physico-chemical speciation: the various physical and chemical forms in which an element may exist in a specified system.

phytoplankton: microscopic plants living in the oceanic water column.

phytoremediation: the use of living plants to reduce the risk posed by contaminated soil, sludges, sediments, and groundwater by removing, degrading, immobilizing, or containing contaminants *in situ*.

planetary albedo: the reflectivity of the world with respect to incident sunlight: calm seawater ($\sim 2\%$), vegetated regions (10–25%), deserts ($\sim 35\%$), and snow covered surfaces ($\sim 90\%$).

podsolization: the leaching of soluble complexes of aluminum and ferrous iron from the A horizon and the subsequent deposition of these metals, together with organic matter, in the B horizon.

potential density, σ_θ: the density defined on the basis of potential temperature instead of *in situ* temperature.

potential temperature: the temperature that a water parcel would have if raised adiabatically to the ocean surface.

pycnocline: a region in the water column that exhibits a sharp change in density.

pyrolytic hydrocarbon: an organic compound originating from combustion sources.

quantum yield: number of molecules of reactant consumed per photon of light absorbed.

radiolaria: single-celled planktonic organisms that build a skeleton of silica.

Redfield ratio: the constant relative amount of nutrients taken up by phytoplankton during photosynthesis, being 106C:16N:1P.

refractometer: instrument for measuring the indices of refraction of various substances.

remineralisation: bacterial degradation of particulate organic nitrogen (POC) into dissolved organic nitrogen (DON).

residence time: the average lifetime of the component in a specified system, which is, in effect, a reciprocal rate constant.

respiration: biochemical and cellular processes within an organism by which carbon is combined with absorbed oxygen to produce carbon dioxide.

Revelle factor (R): a resistance to the dissolution of $CaCO_{3(s)}$ by carbonic acid indicating that the ocean is relatively well buffered against changes in ΣCO_2 in response to variations in atmospheric pCO_2.

salinity (S‰): a measure of the salt content of seawater; a typical value for oceanic waters being 35 g kg^{-1}.

scavenge: the process by which sinking particles remove dissolved substances (metals and organic material) by surface adsorption.

siderophore: a low molecular weight substance synthesized by a variety of micro-organisms to sequester Fe that is often in limited supply in the environment.

sigma-tee (σ_t): the density (actually the specific gravity and hence it is a dimensionless number) of water at atmospheric pressure based on temperature and salinity *in situ*.

sorption: either or both adsorption or absorption of a chemical substance to a surface.

specific heat: the amount of heat required to raise the temperature of 1 g of a substance by 1 °C.

super-saturation: the unstable state of a solution or vapour that contains more solute or gas than its solubility or vapour pressure allows.

surface tension: the work required to expand the surface of a liquid by unit area; the cohesive force exerted at the surface of a liquid that makes it tend to assume a spherical shape because molecules at the surface are not surrounded by molecules on all sides and accordingly they interact more strongly with those directly adjacent to them on and below the surface.

surfactants: surface-active or wetting agents that decrease the surface tension of a liquid.

thermocline: that zone between the warm surface water and colder deeper water, *e.g.*, of a lake, in which the temperature changes most rapidly.

thermohaline circulation: the global deepwater circulation driven by density differences, which in turn are caused by variations in salinity or temperature.

tidal prism: the difference in the volume of water in a water body between low and high tides.

troposphere: the lowest region of the atmosphere, extending to an altitude of 8–18 km.

T–S diagram: a plot of temperature against salinity, with applications in defining water masses of different origin.

turbidity currents: the muddy and turbulent flow of sediment-laden waters along the seafloor caused by sediment slumping that transport copious amounts of material, including coarse-grained sediments, to the deep sea.

upwelling: the transport of sub-surface water to the ocean surface.

zooplankton: animal component of the planktonic community.

Subject Index

A **Glossary** of terms can be found on pages 347–353.